AF411168

Refinements to Animal Models for Biomedical Research

Refinements to Animal Models for Biomedical Research

Editor

Gabrielle C. Musk

MDPI • Basel • Beijing • Wuhan • Barcelona • Belgrade • Manchester • Tokyo • Cluj • Tianjin

Editor
Gabrielle C. Musk
University of Western Australia
Australia

Editorial Office
MDPI
St. Alban-Anlage 66
4052 Basel, Switzerland

This is a reprint of articles from the Special Issue published online in the open access journal *Animals* (ISSN 2076-2615) (available at: https://www.mdpi.com/journal/animals/special_issues/ Refinements_Animal_Models_Biomedical_Research).

For citation purposes, cite each article independently as indicated on the article page online and as indicated below:

LastName, A.A.; LastName, B.B.; LastName, C.C. Article Title. *Journal Name* **Year**, *Volume Number*, Page Range.

ISBN 978-3-0365-0332-5 (Hbk)
ISBN 978-3-0365-0333-2 (PDF)

Cover image courtesy of Gabrielle C. Musk.

Contents

About the Editor

Gabrielle C. Musk graduated from Murdoch University in Western Australia with a Bachelor of Science (Veterinary Biology) and a Bachelor of Veterinary Medicine and Surgery in 1995. Since then, she has worked in clinical practice in rural South Australia and the United Kingdom. She undertook a Residency training program in Veterinary Anesthesia and Analgesia at the University of Glasgow from 2001 to 2004 and became a Diplomat of the European College of Veterinary Anesthesia and Analgesia in 2005. Returning to Western Australia in 2006, Gabrielle worked in veterinary academia at Murdoch University as a Veterinary Anesthetist and then embarked upon a Ph.D. at the University of Western Australia, investigating the optimal use of positive end-expiratory pressure during high-frequency jet ventilation in preterm lambs as a model for preterm babies. At the end of her Ph.D., she became involved in the care of animals used in research and is currently a Veterinary Officer and Veterinary Anesthetist for Animal Care Services at the University of Western Australia.

Editorial

Refinements to Animal Models for Biomedical Research

Gabrielle C. Musk

Animal Care Services, University of Western Australia, Perth 6107, Australia; gabrielle.musk@uwa.edu.au

Received: 26 November 2020; Accepted: 16 December 2020; Published: 18 December 2020

This collection includes the manuscripts published in the Special Issue of Animals: Refinements to Animal Models for Biomedical Research. Twelve peer-reviewed papers covering a range of approaches to the concept of refinement have been included in this contemporary resource. The principles of the 3Rs (reduction, refinement and replacement) form the basis of ethically acceptable research using animal models for human and/or animal benefits. If alternatives to the use of animals (replacement) are not available, the importance of reduction and refinement cannot be overstated.

Reduction and refinement are closely linked. One contributor to the number of animals maintained in a study is the frequency and severity of unexpected or adverse events, which may result in the animal being excluded from the study. Refinement should not only result in optimisation of well-being and welfare of research animals, but should also mitigate the risk of the unplanned occurring, which will in turn directly reduce the number of animals wasted in the process. Society, funding bodies and governments all agree that adverse effects such as pain, fear and distress should be appropriately mitigated and that animals should be housed in conditions that promote their health and well-being [1]. Opportunities for refinement should be continuously sought and explored to meet the expectations of these stakeholders, to benefit the animals involved and to reduce the number of animals used overall.

Nevertheless, what is refinement? The original definition of refinement proposed by Russell and Burch (1959) was "any decrease in the incidence or severity of inhumane procedures applied to those animals which still have to be used" [2]. These authors also stated that the "object of refinement is simply to reduce to an absolute minimum the amount of distress imposed on those animals that are still used" [2]. Despite the apparent inconsistency—is it "any decrease" or is it "to reduce to an absolute minimum"?—and along with various other contemporary definitions, the overriding aim of refinement should be that, whenever possible, the elimination of animal distress consistent with the conduct of sound science is both a scientific and ethical aim [3].

With this aim in mind, it is worth reinforcing that refinements should be evidence-based wherever possible. This collection of manuscripts contributes to the body of evidence to support the introduction of refinements in a range of species including mice [4], rats [5], rabbits [6], guinea pigs [7], pigs [8] and sheep [9]. The systematic review included herein by Whittaker and Barker investigated the question of the impact of blood sampling on the well-being of mice and the quality of the blood sample [4]. Unfortunately, the findings of the review were largely inconclusive as the nature of the primary research introduced a high risk of bias. These authors acknowledge that despite significant and repeated efforts to improve the standard of reporting of the use of animals for scientific purposes, there is still much to be achieved [4]. Adoption of the principles within the Animal Research: Reporting of In Vivo Experiments (ARRIVE) and Planning Research and Experimental Procedures on Animals: Recommendations for Excellence (PREPARE) guidelines, amongst other such recommendations, will help achieve more rigorous, transparent and transferable animal-based research [10,11]. Unfortunately, these principles are still not commonly applied and strong evidence to support proposed methods of animal care, especially for the purpose of refinement, is difficult to collate at the current time. Nevertheless, refinements can still be made and in the process the body of evidence will grow. In the future, the basis for refinements will hopefully shift from less of an empirical mode to more of an evidence-based approach. In turn, there may be stronger justification to standardize aspects of animal care between projects performed at different locations which will improve our capacity to compare results. For example, Kint et al. describe such an approach for the anaesthesia of rats undergoing functional magnetic resonance imaging in this Special Issue [5].

Further consideration of Russell and Burch (1959) and their opinion that refinement may be considered a distinct method of removing inhumanity, focusing on the actual conduct of research and how sentient animals are treated [3]: the indisputable value of anaesthesia and analgesia in refinements in animal experimentation has been acknowledged for some time [12]. Clutton provides a comprehensive Anglocentric history of anaesthetics and analgesics in the refinement of animal experiments in this Special Issue. The importance of anaesthesia as the greatest single advance in humane technique is described along with the role of anaesthetics and analgesics prescribed and administered by appropriately trained people being acknowledged [12]. This comprehensive commentary of Clutton's is thoroughly researched and is a useful resource to understand the scope of progress that has been made in the use of anaesthetics and analgesics in general, but specifically in the care of research animals.

Other publications in this Special Issue provide useful practical contributions to the often sparse evidence for the use of analgesia for animals, relating to the use of transdermal fentanyl patches in rabbits [6] and sheep [9]. The greatest challenge in providing analgesia for animals is pain assessment and, within this book, the utility of pain assessment strategies is also discussed. The use of nociceptive threshold testing in multiple species is described [13] along with the value of facial grimace scales [14,15]. In addition, the quest for an optimal approach to pain assessment in neonatal piglets, in the context of castration, is described [8]. Perhaps the ultimate aim of pain assessment for both individual animals and for groups of animals in analgesic efficacy studies is an objective measurement. Circulating compounds such as inflammatory cells may be useful and Kongara et al. describe quantification of gene expression of peripheral leucocyte inflammatory cytokines in calves undergoing disbudding [16]. This collection of papers describes the evolution of various approaches to pain assessment in animals with constructive commentary on how to exploit their use. Ongoing efforts to develop strategies for pain assessment in animals are necessary until we are able to reliably and consistently identify and interpret signs or markers of pain in animals.

In classic laboratory animal species, the health status of the animal is well documented and understood. Some species, including sheep (and pigs), are not usually bred specifically for research and are often sourced from commercial farming enterprises. Berset et al. present the results of a European survey, which elucidated the importance of health status documentation and monitoring as essential components to refinement in the use of sheep [17]. These authors highlight the frequency and negative impact of pre-existing animal health issues on the research objectives.

A single manuscript on refinements for euthanasia was also included in this Special Issue. The case study explores the use of an irreversible penetrating spring-loaded captive bolt method of euthanasia in guinea pigs with favourable results [7]. This manuscript is a good example of the translation of techniques from one species to another and the collation of preliminary evidence to establish safety and efficacy in a novel species.

The contents of this Special Issue are evidence for the enthusiasm for the concept of refinement across a broad range of species and contexts within biomedical and animal research.

Funding: This work received no external funding.

Conflicts of Interest: The author declares no conflict of interest.

References

1. Prescott, M.J.; Lidster, K. Improving quality of science through better animal welfare: The NC3Rs strategy. *Lab. Anim.* **2017**, *46*, 152–156. [CrossRef] [PubMed]
2. Russell, W.M.S.; Burch, R.L. *The Principles of Humane Experimental Technique*; Universities Federation for Animal Welfare: Wheathhampstead, UK, 1959.
3. Tannenbaum, J.; Bennett, B.T. Russell and Burch's 3Rs then and now: The need for clarity in definition and purpose. *J. Am. Assoc. Lab. Anim. Sci.* **2015**, *54*, 120–132. [PubMed]
4. Whittaker, A.L.; Barker, T.H. The Impact of Common Recovery Blood Sampling Methods, in Mice (Mus Musculus), on Well-Being and Sample Quality: A Systematic Review. *Animals* **2020**, *10*, 989. [CrossRef] [PubMed]
5. Kint, L.T.; Seewoo, B.J.; Hyndman, T.H.; Clarke, M.W.; Edwards, S.H.; Rodger, J.; Feindel, K.W.; Musk, G.C. The Pharmacokinetics of Medetomidine Administered Subcutaneously during Isoflurane Anaesthesia in Sprague-Dawley Rats. *Animals* **2020**, *10*, 1050. [CrossRef] [PubMed]
6. Mirschberger, V.; von Deimling, C.; Heider, A.; Spadavecchia, C.; Rohrbach, H.; Zeiter, S. Fentanyl Plasma Concentrations after Application of a Transdermal Patch in Three Different Locations to Refine Postoperative Pain Management in Rabbits. *Animals* **2020**, *10*, 1778. [CrossRef] [PubMed]
7. Cohen, S.; Kwok, M.; Huang, J. Humane Euthanasia of Guinea Pigs (Cavia porcellus) with a Penetrating Spring-Loaded Captive Bolt. *Animals* **2020**, *10*, 1356. [CrossRef] [PubMed]
8. Sheil, M.; Polkinghorne, A. Optimal Methods of Documenting Analgesic Efficacy in Neonatal Piglets Undergoing Castration. *Animals* **2020**, *10*, 1450. [CrossRef] [PubMed]
9. Buchholz, T.; Hildebrand, M.; Heider, A.; Stenger, V.; Arens, D.; Spadavecchia, C.; Zeiter, S. Transdermal Fentanyl Uptake at Two Different Patch Locations in Swiss White Alpine Sheep. *Animals* **2020**, *10*, 1675. [CrossRef] [PubMed]
10. Percie du Sert, N.; Hurst, V.; Ahluwalia, A.; Alam, S.; Avey, M.T.; Baker, M.; Browne, W.J.; Clark, A.; Cuthill, I.C.; Dirnagl, U.; et al. The ARRIVE guidelines 2019: Updated guidelines for reporting animal research. *bioRxiv* **2019**, 703181. [CrossRef]
11. Smith, A.J.; Clutton, R.E.; Lilley, E.; Hansen, K.E.A.; Brattelid, T. PREPARE: Guidelines for planning animal research and testing. *Lab. Anim.* **2018**, *52*, 135–141. [CrossRef] [PubMed]
12. Clutton, R.E. An Anglocentric History of Anaesthetics and Analgesics in the Refinement of Animal Experiments. *Animals* **2020**, *10*, 1933. [CrossRef] [PubMed]
13. Taylor, P. Remote Controlled Nociceptive Threshold Testing Systems in Large Animals. *Animals* **2020**, *10*, 1556. [CrossRef] [PubMed]
14. Cohen, S.; Beths, T. Grimace Scores: Tools to Support the Identification of Pain in Mammals Used in Research. *Animals* **2020**, *10*, 1726. [CrossRef] [PubMed]
15. Mota-Rojas, D.; Olmos-Hernández, A.; Verduzco-Mendoza, A.; Hernández, E.; Martínez-Burnes, J.; Whittaker, A.L. The Utility of Grimace Scales for Practical Pain Assessment in Laboratory Animals. *Animals* **2020**, *10*, 1838. [CrossRef] [PubMed]
16. Kongara, K.; Dukkipati, V.S.R.; Tai, H.M.; Heiser, A.; Murray, A.; Webster, J.; Johnson, C.B. Differential Transcription of Selected Cytokine and Neuroactive Ligand-receptor Genes in Peripheral Leukocytes from Calves in Response to Cautery Disbudding. *Animals* **2020**, *10*, 1187. [CrossRef] [PubMed]
17. Berset, C.M.; Lanker, U.; Zeiter, S. Survey on Sheep Usage in Biomedical Research. *Animals* **2020**, *10*, 1528. [CrossRef] [PubMed]

Publisher's Note: MDPI stays neutral with regard to jurisdictional claims in published maps and institutional affiliations.

Commentary

An Anglocentric History of Anaesthetics and Analgesics in the Refinement of Animal Experiments

R. Eddie Clutton

The Wellcome Trust Critical Care Laboratory for Large Animals, Roslin Institute, Easter Bush Veterinary Centre, Roslin, Midlothian EH25 9RG, UK; e.clutton@ed.ac.uk; Tel.: +44-131-650-6220 or +44-07749-887-342

Received: 14 September 2020; Accepted: 15 October 2020; Published: 21 October 2020

Simple Summary: In simultaneously describing the history of animal experimentation and the development of anaesthetics and analgesics from an Anglocentric perspective, this article reveals how the latter have considerably refined animal experiments and brought benefits to both science and the animals involved—particularly in the 19th and 20th centuries. The more recent development of training and educational programmes in laboratory animal anaesthesia and their role in maintaining desirable trends in experimental refinement are also described.

Abstract: Previous histories of animal experimentation, e.g., Franco (2013) have focused on ethics, the law and the personalities involved, but not on the involvement of anaesthetics or analgesics. Given that these were major subjects of (UK) Parliamentary debates on vivisection in the mid-19th century and viewed as "indisputable refinements in animal experimentation" (Russell and Burch 1959), it seemed that an analysis of their role was overdue. This commentary has, in interweaving the history of animal experimentation in the UK with the evolution of anaesthesia, attempted to: (1) clarify the evidence for Russell and Burch's view; and (2) evaluate anaesthesia's ongoing contribution to experimental refinement. The history that emerges reveals that the withholding or misuse of anaesthetics and, or analgesics from laboratory animals in the UK has had a profound effect on scientists and indirectly on the attitudes of the British public in general, becoming a major driver for the establishment of the anti-vivisection movement and subsequently, the Cruelty to Animals Act (1876)—the world's first legislation for the regulation of animal experimentation. In 1902, the mismanaged anaesthetic of a dog in the Department of Physiology, University College London resulted in numerous events of public disorder initiated by medical students against the police and a political coalition of anti-vivisectionists, trade unionists, socialists, Marxists, liberals and suffragettes. The importance of anaesthesia in animal experiments was sustained over the following 150 years as small mammalian species gradually replaced dogs and cats as the principle subjects for vivisection. In discussing experimental refinement in their 1959 report, "The Principles of Humane Experimental Technique" Russell and Burch described anaesthetics as " … the greatest single advance in humane technique, (which) has at the same time been virtually indispensable for the advance of experimental biology". Since then, the role of anaesthetics and in particular analgesics has become an unavoidable consideration whenever animal experiments are planned and conducted. This has been accompanied by a proliferation of training and educational programmes in laboratory animal anaesthesia.

Keywords: animal research; animal testing; biomedical research; anaesthesia; analgesia; history of science

---

1. Introduction

The progression of animal experimentation from classical times to the present has been directed by the rarely compatible views of "scientists", philosophers and legislating authorities and, more recently, vested interest and the public have become influential. Arguably, the greatest hostility between those for and against animal vivisection—based largely on the questions of animal sentience and the scientific value of noxious animal experiments—was seen in the years between the late 18th and early 19th centuries, when, co-incidentally, the first influential observations on the anaesthetic properties of gases were reported. The introduction of general anaesthetics into medical practice in the mid-19th century greatly facilitated surgery by alleviating the (human) patient's agonies. At the same time, and for the same reason, many felt that anaesthetics increased the ethical defensibility of animal experiments. The similarly opining would also claim that in eliminating the widespread and profound physiological effects of pain and suffering—an effect which had been recognized for some 2300 years—anaesthetics improved data value and increased the validity of animal experiments. Some scientists countered this (and continue to do so) by emphasising that anaesthetics and analgesics can produce similar if not greater degrees of physiological perturbation than the pain itself, and so should be excluded from experimental protocols. Fortunately (for animals) Russell and Burch's "The Principles of Humane Experimental Technique" (1959) [1] established anaesthesia and analgesia as indisputable refinements in animal experimentation and which, along with reduction and replacement, pervades all stages of contemporary animal experimentation in the UK, i.e., from funding application and ethical approval through to the publication of results. A focused historical examination of the contribution of anaesthesia and analgesia to the refinement of animal experiments has not previously been attempted. This article reports an Anglocentric view of the role of anaesthetics, analgesics and anaesthetists in reshaping the attitudes of scientists, philosophers, the authorities and the public to animal experimentation.

Galen, Vesalius and Paracelsus (330BC-1541)

Beginning with a description of surgical procedures conducted on animals *without* the use of anaesthetics, i.e., vivisection, reveals an early concern with its morality and an ongoing conviction that pain has a disruptive effect on experimental results. Thus, scholars of the Empiric School of Medicine (330 BCE–ca 400 AD) dismissed the study of physiology by vivisection on the grounds of its cruelty and clinical irrelevance [2]. Early descriptions also reveal an ambivalence by vivisectors towards animal suffering and a need to justify its practice. Galen preferred using pigs and goats over apes because, during brain dissections, the unpleasant expression of the ape discomfited him [3]. Vesalius recognized the cruelty of vivisection but justified it for what it revealed: his thesis *"De humani corporis fabrica libri septem"* ("On the Fabric of the Human Body in Seven Books") (1543) features a small capital letter "Q" historiating the dissection of a pregnant bitch. (An historiated initial is an enlarged letter at the beginning of a paragraph that contains a picture.) Whilst conceding that the animal "is cruciata", i.e., is crucified or tortured, "it allowed the demonstration of the unborn puppies struggle to breathe once the placental blood flow was ended" [4]. In the same publication, an historiated capital "Q" shows a conscious pig—immobilized with chains—undergoing tracheal surgery for the placement of a tube allowing periodic lung inflation, the first convincing account of artificial ventilation [5]. That restraining chains were required to immobilize the animal suggests that Vesalius was unaware that, at this time, Paraclesus was using ether (known as sweet oil of vitriol) to produce analgesia in chickens (although the first edition of *Opera Medico-chemica sive paradoxa* ("On the Field of Medicinal Chemistry or Paradoxes") which contains an account of this was not published until 1605). In this thesis, Paracelsus proposed that the sweet oil of vitriol:

> *"...quiets all suffering without any harm and relieves all pain, and quenches all fevers, and prevents complications in all disease."* [6]

It is unfortunate for both humans and animals that 300 years were to elapse before both innovations, i.e., positive pressure lung ventilation and inhalant anaesthetics, were re-introduced into clinical anaesthetic practice.

Harvey, Descartes and the Oxford Group (1578–1665)

Had they been contemporaries, Paracelsus' interest in pain relief might have benefitted William Harvey's early studies of "the heart's motion" in conscious animals whose chest walls had been surgically removed. Harvey complained that the heart's action was too fast, a complication which *later* prompted him to use "colder animals, such as toads, frogs and serpents . . . " or "dying dogs and hogs" [7]. That the excessive heart rate was the inescapable result of undergoing thoracic surgery whilst conscious was not overlooked: O'Meara (1665) argued that, " . . . the miserable torture of vivisection places the body in an unnatural state and that amid the terrible pains of vivisection all the juices are brought to flow together, thus denying the validity of animal experimentation" [3]. Such concerns may have puzzled French philosopher and vivisector Renee Descartes who believed that animals, "When burnt with a hot iron or cut with a knife their writhing and screaming are like the creaking of a hinge, no more" [8]. This belief emboldened some scientists "with less responsible and reflective minds" [3] to conduct even more gruesome experiments [1] whilst elsewhere, it had the opposite effect and promoted consternation—even amongst established vivisectionists. Johann Brunner, Professor of Medicine at Heidelburg, described the animals he dissected as "the martyrs of the anatomists" [3], whilst members of the Oxford Group, specifically Christopher Wren, Robert Boyle and Robert Hooke, not only recognized the cruelty of vivisection and its adverse effects on scientific data, but began preliminary experiments with intravenous anaesthesia. They also considered the use of opium to alleviate the suffering of experimental dogs [9]. Writing to Boyle on the 10th November 1664, Hooke, having that year completed a thoracic dissection and lung inflation study in a dog, complained:

"I shall hardly be induc'd to make any further trials of this kind because of the torture of the creature but certainly the inquiry would be very noble if we could any way find a way soe to stupify the creature as that it might not be sensible which I fear there is hardly any opiate will performe".

The reference to opiate use is important because both Wren and Boyle had been conducting observational studies on the use of alcoholic opiate concoctions in dogs for the previous 8 years [9]. In *The Usefulness of Experimental Philosophy* (1663), published one year before Hooke's regretted thoracotomy study, Boyle described an experiment performed by Wren, who injected a warm solution of opium in sack (a sweet white dessert-type wine) into the vein of a dog's hind leg:

"We had scarce untied the dog..., before the opium began to disclose its narcotick quality; and almost as soon as he was upon his feet, he began to nod with his head, and faulter and reel in his pace, and presently after appeared so stupified, that there were wagers offered his life could not be saved". (The dog survived).

Perhaps aware that "some narckotic" may have alleviated his animal's suffering, in 1665 a remorseful Hooke wrote:

"The microscope enables one to look at nature 'acting according to her usual course and way, undisturbed, whereas when we endeavour to pry into her secrets by breaking open the doors upon her, and dissecting and mangling creatures whilst there is life yet within them, we find her indeed at work, but put into such disorder by the violence offered that we cannot tell if the results are of any significance",

He thus revealed his prescient belief that the extreme physiological responses to painful vivisection were not representative of the normal state. That, he asserted, was only discernible by non-invasive means, e.g., microscopy [10].

Davy (1778–1829)

In April 1799, Humphry Davy reported his initial findings concerning the inhalation of nitrous oxide (N_2O) and then embarked upon a series of animal experiments culminating with the publication of, "Researches Chemical and Philosophical: Chiefly Concerning Nitrous Oxide and its Respiration", in 1800. In this, Davy describes guinea pig experiments and the administration of pure nitrous oxide to a "stout and healthy cat".

> *"after 5 min the pulse was hardly perceptible; he made no motions and appeared wholly senseless. After 5 min and a quarter he was taken out … in 8 or 9 min he was able to walk … in half an hour he was completely recovered".*

He frequently inhaled the gas himself, once whilst suffering severe pain from a wisdom tooth eruption after which he wrote:

> *"As nitrous oxide in its extensive operation appears capable of destroying physical pain, it may probably be used with advantage during surgical operations in which no great effusion of blood takes place."* [11]

Bentham (1748–1841) and Seturner (1783–1841)

That Davy's findings remained unadopted in medical practice until 1844 is attributed by Cartwright (1950) to the callousness and unprecedented brutality of the 18th-century Englishman [11]. Such callousness may have fuelled a growing philosophical interest in animal suffering, as exemplified by Bentham's proposal that animals be treated according to their capacity to suffer. In 1789, he asked:

> *"The question is not, Can they reason? nor, Can they talk? but, Can they suffer?"* [12]

Paradoxically, Bentham's utilitarianism necessitated his support of "justifiable" vivisection. In a letter to the editor of the Morning Chronicle, 4th March 1825, he wrote:

> *"Sir—I never have seen, nor ever can see, any objection to the putting of dogs and other inferior animals to pain, in the way of medical experiment, when that experiment has a determinate object, beneficial to mankind, accompanied with a fair prospect of the accomplishment of it. But I have a decided and insuperable objection to the putting of them to pain without any such view".*

While human beings had been capitalising on the analgesic effects of *Papaver somniferum* derivatives since prehistory, the use of opioids in animals remained largely unreported until 1805, following Sertürner's isolation of morphine [13]. He tested an aqueous alcoholic extraction of the salt on four dogs and a mouse "that he found wandering in the laboratory". He gave 6 grains to a dog, followed an hour later by another 6 grains. The dogs vomited, had convulsions, and were sleepy, but did not sleep. One "gentle little dog" died. He reported the results of his animal studies in the *Journalder Pharmacie fuer Aerzte und Apotheker* in 1806, a reading of which would probably not have convinced anyone of the new compound's analgesic potential.

Magendie (1783–1855)

Francoise Magendie was to become the most influential vivisectionist of all time although when he first began dissecting conscious animals is unknown. Elliott (1987) puts it within a few years of his first publication (1809) because the essay did not describe vivisection, only an intent to conduct it [14]. By the time of his death in 1855, Magendie had been dubbed the "father of experimental physiology" and had done more to foment anti-vivisectionist sentiment in the UK than any other "scientist" in Europe [15]. Descriptions of his vivisections, often performed publicly, abound and reveal absolute indifference to animal suffering and to those who judged his actions cruel. Admittedly, there was a lack of effective anaesthesia—at least for the first forty years of his career—but not the final nine (see Figure 1 for the approximate dates when anaesthetics and analgesics became available for use in animal experiments). France's first successful ether anaesthetic was administered on 22 December 1846, with reports on its general and obstetrical use being published in January and February 1847, respectively [16]. Magendie himself examined the effects of rectal ether and morphine in dogs [17].

Furthermore, as the first Professor of Physiology at the College de France and Vice-President of the Academie des Sciences, it seems improbable that Magendie could have remained ignorant of the efforts of French physiologists and pharmacologists, e.g., Gerdy, Longet, Flourens, Figuier and Soubeiran to investigate and improve anaesthetic methods [16]. To an astute scientist, the advantages of ether (and later chloroform) in at least some animal experiments should have been clear—and if not for the animals' benefit—for his own and those of his like-minded mentees, against whom the charges of disgraceful cruelty were being hurled with increasing ferocity from a growing European anti-vivisectionist movement. Magendies defence was anthropocentric: he experimented on animals because he did not wish to experiment on humans. In response to a Quaker who challenged him to:

"......desist from these experiments, because thou hast not the right to cause animals to die or make them suffer...."

Magendie rejoined that if his experiments did not serve humanity, they would indeed be cruel. However, to use animals in order to make discoveries useful to medicine did not merit such reproach [18]. This position applied to anaesthetics. As early as 1847, Magendie had observed that "intoxication caused by sulphuric ether was "little understood" and that only when it was thoroughly understood could one safely and with a clear conscience apply it to man" [19].

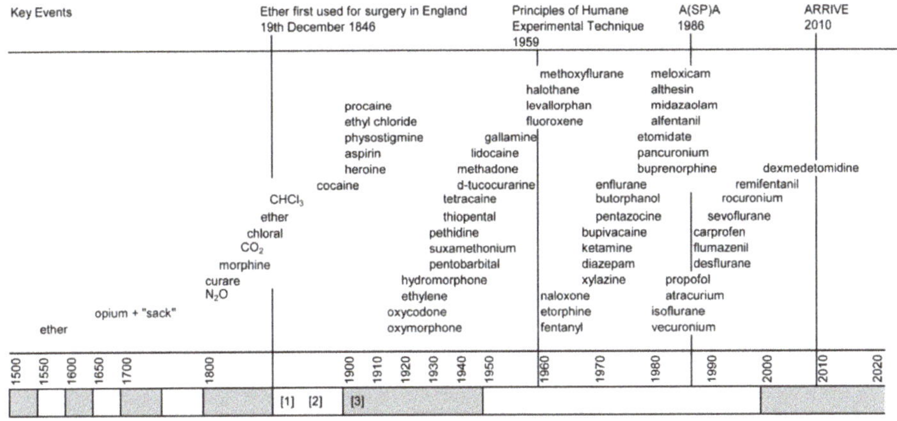

Figure 1. Dates of anaesthetic and analgesic drug discoveries in relation to key events in the refinement of animal experimentation. Alternatively shaded x axis blocks each represent a 50 year epoc. A(SP)A 1986: The Animals (Scientific Procedures) Act 1986. ARRIVE: Animals in Research: Reporting in Vivo Experiments; Reporting guidelines. Additional dates for reference: [1]: 1873 Sanderson publishes, "Handbook for the Physiological Laboratory"; [2]: the Magnan "affair"; [3]: the brown dog affair. CHCl3: chloroform.

Hickman (1800–1830)

Magendies' animals may have suffered less had Henry Hill Hickman been more persuasive when he visited France in April 1826 to present his ideas on surgical anaesthesia to King Charles Xth. Hickman chose CO_2 to examine his concept of "suspended animation" with the unique a priori study objective of inducing reversible unconsciousness in order to facilitate surgery [20]. Consequently, Hickman went beyond inducing hypercapnic coma (as early as the 1820s) by testing CO_2's ability to produce surgical anaesthesia in dogs and mice. In one of his first experiments, a one-month-old puppy was enclosed beneath a glass cover (bell jar) and

"in ten Minutes he showed great marks of uneasiness, in 12 respiration became difficult, and in 17 Minutes ceased altogether, at 18 Minutes I took off one of the ears, which was not followed by

haemorrhage, respiration soon returned and the animal did not appear to be the least sensible to pain, in three days the ear was perfectly healed" [21]

Hickman's subsequent pamphlet entitled: "A letter on Suspended Animation containing experiments showing that it may safely be employed during operations on animals with the view of ascertaining its probable utility in surgical operations on the Human Subject," was ignored by the Royal Society probably because Hickman's chosen sponsor lacked the necessary enthusiasm for its fair promotion. Furthermore, an 1826 article in The Lancet authored by "Antiquack" (purportedly a professionally envious Davy) entitled 'Surgical Humbug' ruthlessly criticized his work and prompted his defection to France, where, despite the support of Napoleon's field surgeon, he met a similar response. Since his premature death in 1830, this and previous failures to recognize the potential of his work have been the subject of anguished review [22] because it meant Hickman's prescient desire to develop surgical anaesthesia in humans took another 20 years for others to achieve.

Martin (1754–1834)

Richard Martin MP in conjunction with Edinburgh-born Lord Erskine succeeded in getting the "Cruel Treatment of Cattle Act 1822" or 'Martin's Act', passed into British law—the first legislation against animal cruelty introduced by means of a Parliamentary procedure anywhere in the world. It received Royal assent on the 22 July 1822. Martin's Act sought to prevent the cruel and improper treatment of horses, mares, geldings, mules, asses, cows, heifers, steers, oxen, sheep and other cattle. The introduction of this and subsequent legislation is shown in Figure 2 (legislation).

Two years later, on 16 June 1824, a meeting was called in Old Slaughters Coffee House, Martins Lane, London at which various clergymen and anti-slavery campaigners launched the Society for the Prevention of Cruelty to Animals (SPCA). Martin was also present and tasked to investigate the markets and streets of the Metropolis, slaughterhouses and the behaviour of coachmen. In 1840, Victoria I granted the "Royal" prefix, and so the RSPCA was born. Neither Act nor Royal Society were initially motivated to control animal vivisection although this was soon to change. The foundation of this and other organisations concerned with vivisection is shown in Figure 2.

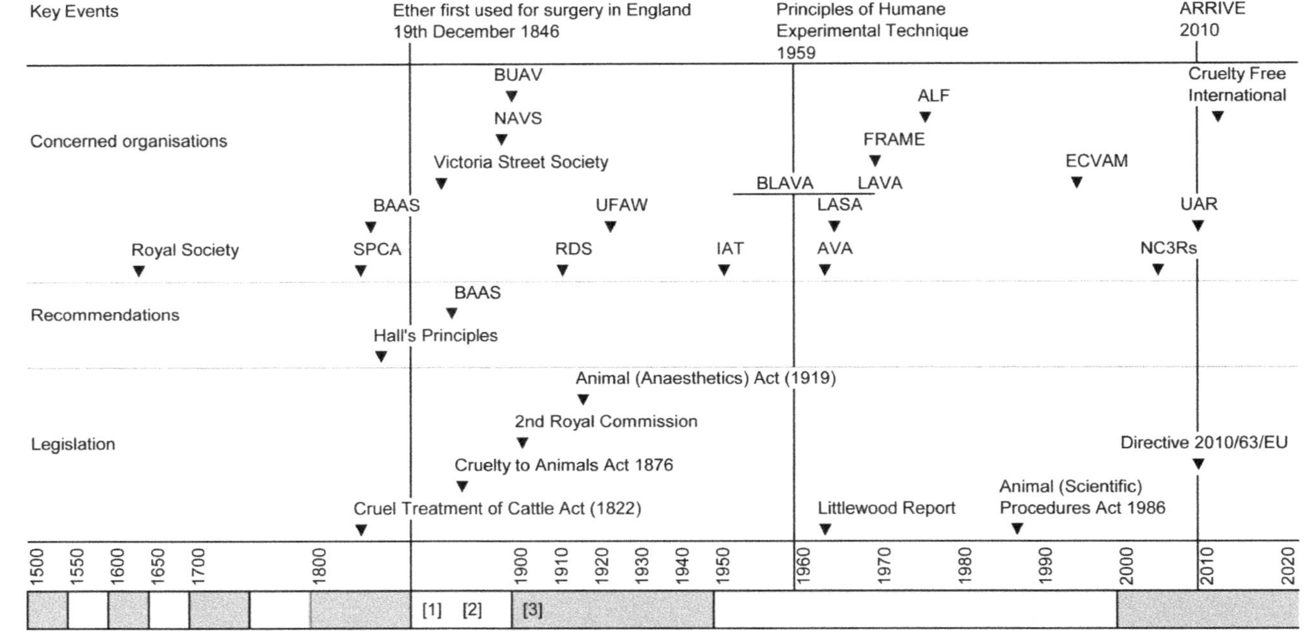

Figure 2. Key events in animal experimentation in the UK partitioned into the introduction of legislative and optional (recommended) constraints, and the formation of societies that have influenced the use of anaesthetics and analgesics in animal experiments. Alternatively shaded x axis blocks each represent a 50 year epoc. A(SP)A 1986: The Animals (Scientific Procedures) Act 1986. ARRIVE: Animals in Research: Reporting in Vivo Experiments; Reporting guidelines. Additional dates for reference: [1]: 1873 Sanderson publishes, "Handbook for the Physiological Laboratory"; [2]: the Magnan "affair"; [3]: the brown dog affair. ALF: Animal Liberation Front; AVA: Association of Veterinary Anaesthetists; BAAS: British Association for the Advancement of Science; BLAVA: British Laboratory Animal Veterinary Association; BUAV: British Union of Anti-vivisectionists; ECVAM: European Centre for the Validation of Alternative Methods; FRAME: Fund for the Replacement of Animals in Medical Experiments; IAT: Institute of Animal Technology; LASA: Laboratory Animal Science Association; NAVS: National Anti-Vivisection Society; NC3Rs: National Centre for the 3Rs; LAVA: Laboratory Animal Veterinary Association; RDS: Research Defence Society; SPCA: Society for the Prevention of Cruelty to Animals; UFAW: Universities Federation for Animal Welfare; UAR: Understanding Animal Research.

Hall (1790–1857)

Marshall Hall, an English neurophysiologist and contemporary of Magendie, pioneered animal welfare from within science. In 1831, he proposed that physiologic procedures be regulated in a way that took into consideration the suffering of animals [19]. He called for the formation of a society for physiological research, which would regulate animal experimentation, for he said, "every experiment … is necessarily attended by pain or suffering of a bodily or mental kind". In 1835, in his "Principles of Investigation in Physiology," he outlined five principles to govern animal experimentation, the fourth of which stated (in the days before anaesthesia) that "justifiable experiments should be carried out with the least possible infliction of suffering, often through the use of lower, less sentient animals, such as frogs and fish or even newly dead animals". (The other points were that: (1) an experiment should never be performed if the necessary information could be obtained by observation; (2) no experiment should be performed without a clearly defined and obtainable objective; (3) scientists should be well-informed about the work of their predecessors and peers to avoid unnecessary experimental repetition; (4) *vide supra*; and (5) every experiment should be performed under circumstances that would provide the clearest possible results, thereby diminishing the need for repetition of experiments.)

Morton (1819–1868), Glover (1815–1859), Simpson (1811–1870) and Snow (1813–1858)

The birth of medical anaesthesia based upon N_2O and ether in the United States did not depend on animal experimentation, although Morton tested ether on animals [23] and then patients, before administering a convincing anaesthetic at the Massachusetts General Hospital on 16 October 1846.

In the UK, Simpson's eventual "discovery" and promotion of chloroform in 1847 [24] relied as much on what could be remembered after self-administration as it did upon his own animal studies. That said, Simpson's pioneering had been eased by others establishing the compound's pharmacological properties through animal testing—namely Robert Glover, whose studies of bromine, iodine and chlorine compounds, including chloroform and dichlorethane culminated in a thesis entitled 'On the Physiological and Medicinal Properties of Bromine and its Compounds'. Published in 1842, this was the first to describe chloroform's anaesthetic effects. Among various experiments, Glover injected 30 and 60 minims (1.6 and 3.6 mL) of chloroform into the jugular vein of two dogs, causing immediate unconsciousness, loss of the eyelid reflex, insensitivity of the paws to painful stimuli and marked motor weakness, a state from which the dogs quickly recovered. Glover did not test chloroform by inhalation, although he described its smell on his animals' breath [25].

Despite Simpson's inclination to test newly discovered halogenated organic molecules on himself, it is probable that animal testing saved his life. Having recently synthesized ethylene dibromide, Wemyss Reid (the Professor of Chemistry at the University of Edinburgh) suggested to Simpson that it should be tested on rabbits first. Simpson had intended to inhale ethylene dibromide himself but surrendered the opportunity on discovering the two rabbits so treated had died overnight. It was subsequently discovered that the inhalation of ethylene dibromide causes pulmonary congestion; it is currently used as a soil fumigant to kill nematode worms [26].

John Snow—Simpson's Yorkshire-born contemporary—was described by Waters (1936) as "the first anaesthetist" because his research spanned basic science and clinical medicine and he answered fundamental questions with respect to anaesthetic safety [27]. Snow also recognised the importance of accuracy in anaesthetic delivery and patient monitoring. Investigating ether (and later chloroform), Snow performed numerous studies on animals, including birds and fish, in which his renowned attention to detail—at least on one occasion—lapsed. Whilst demonstrating ether anaesthesia on a thrush to an audience of Military Surgeons, he became distracted and killed the bird [28]. Snow's published description of the event reveals the legitimacy of Waters' accolade.

"This thrush was only in the vapour for about a minute, and it is dead. It had ceased to breathe before I took it out of the jar. It is a result I did not intend, and it has arisen from my going on with the lecture, and looking at my notes, instead of directing my whole attention to the animal".

Wakley (1795–1862)

On the 19th March 1847, *The Times* reported the death of Ann Parkinson two days after ether administration [29]. Chloroform's first UK victim was Hannah Greener, who died during induction on 28th January 1848. To assuage growing public concern with the safety of anaesthetics, large-scale animal studies were conducted [30]. In 1848, Wakley reported a study comparing the effects of ether and chloroform on 100 animals—predominantly dogs, but also cats, rabbits, rats, mice, guinea pigs, a hedgehog, a pig, two sheep, a donkey, two mares and some pigeons. From the findings that death resulted in 11 out of 32 animals (34%) anaesthetized with ether and 30 out of 67 animals (44%) receiving chloroform, Wakley concluded:

" . . . assuredly the more dangerous one of the two would be found in the vapour of chloroform". [31]

However, Youngson (1979) emphasized the studies' hopelessly unscientific character [22], e.g., of the 18 dogs studied, only one received ether (and of the 17 dogs given chloroform, only 4 died (24%)). Whilst Wakley's conclusions were based on unsound data, they were nevertheless to prove correct.

Bernard (1813–1878)

Claude Bernard, the "father of modern experimental medicine", died a revered scientist. His attitude to the use of animals and anaesthetics in physiological research was influenced by Magendie under whom he studied from 1834 to 1843. Unlike Magendie, Bernard could not claim ignorance to justify the exclusion of anaesthetics from most—but not all—of his animal studies [19]; nor could he claim adherence to the vitalist principle which he described as being, "entirely contrary to science itself". Bernard's considerable contributions to anaesthesia [32] were anthropocentric and achieved at cost to the animals involved. He studied vasomotion—the nervous control of blood vessel diameter—by cutting projections from the stellate ganglion in unanaesthetized rabbits (so discovering cervical sympathetic block and its effects on the eye). His most influential and controversial work—on curare—began in 1844. (Curare is a drug derived from tree bark and applied to blow-dart tips used by South American Indians to paralyse and capture animals. Curare prevents muscle contraction and stops all movement, including breathing. However, it does not affect the central nervous system and so darted animals suffocate whilst being fully conscious. If breathing is artificially supported, a curarised animal can be subjected to surgery and, although fully aware of the process, is unable to move—or vocalize—in agonized protestation.) Having demonstrated its neuromuscular blocking effect in frogs, he used it in dogs to control strychnine-induced convulsions. After further ranine studies, Bernard suspected that death following curare was caused by asphyxia and not loss of consciousness [33]. The realisation that it did not render the animal insentient prompted him to reflect:

"in the animals, one can judge sensitivity only by motor manifestations. Man alone, on recovering from poisoning by curare, would be able to say, supposing that he had retained the memory, whether or not he had suffered".

He wrote eloquently, and more than once, on his impression of death following curare administration:

"Within the motionless body, behind the staring eye, with all the appearance of death, feeling and intelligence persist in all their force. Can we conceive of a suffering more horrible than that of intelligence present, after succumbing, one by one, of all the organs which are destined to find themselves imprisoned alive within a cadaver?"

This concern did not prevent Bernard from continuing to use curare rather than anaesthetics in his experiments. Furthermore, he promoted its use in experiments to other physiologists. An 1864 essay asserts that the 'poison becomes an instrument which dissociates and analyses the most delicate phenomena of the living machine' [34]. Consequently, vivisectors discovered that used in the right quantities, curare made invasive procedures much easier and the drug became a common experimental tool in physiological practice.

Alfort (1846–1863)

For some 17 years, complaints and petitions had been made to various French authorities about the vivisection of horses at the Veterinary College at Alfort, Paris. *The Times* of 8 August 1863 reported that,

" ... at Alfort a wretched horse is periodically given up to a group of students to experimentalize upon. They tie him down and torture him for hours, the operations being graduated in such a manner that sixty and even more may be performed before death ensues".

British Veterinary Surgeons joined elements of the French Press to demand reform. The attack of the *British Medical Journal* was based on the exclusion of anaesthetics:

"It has never appeared clear to us that we are justified in destroying animals for mere experimental research under any circumstances; but now that we possess the means of removing sensation during experiments, the man who puts an animal to torture ought, in our opinion, to be prosecuted"

The beneficial role of anaesthetics during noxious animal experiments was being increasingly recognized. Responding to the Alfort affair, the RSPCA offered a £50 prize for the best essay received on the subject of vivisection. Dr Markham, Physician to Saint Mary's Hospital London, and Mr Fleming, Veterinary Surgeon to the Third Hussars, won prizes for essays recommending the use of anaesthetics in experiments [15].

Cobbe (1822–1904) and Schiff (1823–1896)

As one of the 19th century's most effective anti-vivisectionists, the Anglo-Irish reformer Frances Power Cobbe had a major effect on ensuring anaesthetics became a legal requirement in animal experiments in the UK. Prompted by the Alfort scandal, she published, *"The Rights of Man and the Claims of Brutes"* in 1863. In the same year, she petitioned Moritz Schiff "to spare his animals as much pain" as possible. In response, Schiff offered that all his animals had anaesthetics. This may have been true: unlike most of his French and German peers in physiology, Schiff recognized that there

"was a real problem in reconciling the needs of science with the most refined humanitarian sentiment"

and that

"Should the physiologist make use of live animals, he had to suspend their sensitivity, by means of opium, ether or chloroform, depending on which vital functions he wanted to examine and thus to maintain as normal" [35]

Burdon-Sanderson (1828–1905) [35]

The concern of other British physiologists with the suffering experimentation caused eventually prompted the British Association for the Advancement of Science to publish four recommendations in 1871, two of which promoted the use of anaesthetics, i.e.,

(1)	No experiment that can be performed under the influence of an anaesthetic ought to be done without it.
(2)	No painful experiment is justifiable for the mere purpose of illustrating a law or fact already demonstrated; in other words, experimentation without the employment of anaesthetics is not a fitting exhibition for teaching purposes.

This view was not entirely consensual. In 1873, John Burdon-Sanderson's Handbook for the Physiological Laboratory, "a practical description of experimental procedures" was published, providing Cobbe's anti-vivisectionists with evidence that scientists were indifferent to animal suffering. The handbook made no explicit reference to anaesthesia or how and when to use it in experiments. However, it did index curare, as several contributing authors recommended its use to keep animals still as an alternative to tying them down. The drug's inability to affect consciousness had been known for approximately 20 years.

Magnan (1847)

Eugene Magnan, another Magendian student, enraged delegates at the 1874 British Medical Association's congress by injecting alcohol and absinthe into two dogs. The first animal became "dead drunk" while the second had an epileptic seizure. Both later died [36]. The event had at least three consequences. It: (1) made members of the medical profession prosecutable under an amended form of Martin's 1822 Act; (2) revealed to both anti-vivisectionists and scientists that a law dealing specifically with experimental animal protection was required; (3) re-animated Cobbe's anti-vivisectionism [15]. It also involved an attempted demonstration of intravenous anaesthesia which failed.

Henniker and Playfair (1875)

Cobbe, working with the RSPCA and indirect support from Queen Victoria, lobbied the Government to legislate against vivisection. In May 1875, the Henniker Bill for Regulating the Practice of Vivisection was introduced to Parliament [37]. The Bill proposed—amongst other things—that "recovery (from anaesthesia) experiments required special approval and that anaesthetics be used in all experiments, excepting those undertaken by individuals with a personal license". One week later a second Bill representing the scientists' views was read. This, the Playfair Bill, proposed the regulation of painful animal experiments but recommended the legalisation of painless experiments, including those conducted under anaesthesia. Licenses for painful experiments undertaken without anaesthesia could be granted on several grounds, including "when the use of anaesthesia interfered with the experiment". The two Bills were in unexpected accord over the general undesirability of painful experiments but sufficiently contradictory to prompt the appointment of a Royal Commission, the creation of the World's first anti-vivisection society and the establishment of the *Association for the Advancement of Medicine by Research*.

The subsequently appointed Royal Commission of Enquiry spent less time in discussing the undesirability of painful experiments—over which there was some consensus—than in deliberating whether curare was an anaesthetic. The final recommendation was that curare would not be considered an anaesthetic by law, but the rest of the Commission's findings satisfied neither side who set about drafting new Bills.

Hoggan (1837–1891)

George Hoggan had worked in Bernard's laboratory in Paris and published an "extraordinarily powerful" letter in the London *Morning Post* on the 2 February 1875 describing Bernard's underuse of anaesthetics and over-use of curare. He wrote: "We sacrificed daily from one to 3 dogs, besides rabbits and other animals, and after four years' experience, I am of the opinion that not one of those experiments on animals was justified or necessary" [38]. Frustrated with the Commission's rate of progress, Cobbe, George Hoggan and others formed the "Victoria Street Society for the Protection of Animals from Vivisection" on 2 December 1875 [39]. Initially a non-abolitionist society, it aimed to protect laboratory animals by regulation. Along with the seventh Earl of Shaftesbury, the new society formulated a second Bill. The Cruelty to Animals Act [40] reached the statute book on 15 August 1876 and required that:

> *"animals must during the whole of the experiment be under the influence of some anaesthetic of sufficient power to prevent the animal feeling pain"*

and that

> *"if the pain is likely to continue after the effect of the anaesthetic has ceased, or if any serious injury has been inflicted on the animal, it be killed before it recovers from the influence of the anaesthetic which has been administered";*

Despite the opposition of the scientific lobby, the new Act also clarified that "the substance known as urari or curare shall not for the purposes of this Act be deemed to be an anaesthetic".

The Act radicalized the dissatisfied anti-vivisection movement [41]. In 1878, the Victoria Street Society declared its goal of total abolition of vivisection prompting the resignation of moderates like George Hoggan. As opposition to vivisection and Victoria Street Society membership grew, the Society reconfigured and in 1897 became the National Anti-Vivisection Society (NAVS). The following year the NAVS voted to accept humane animal experimentation in the short-term whilst remaining committed to a long-term abolitionist goal. Cobbe promptly left to form the British Union for the Abolition of Vivisection (BUAV) which demanded the total and immediate abolition of animal experiments, which, as Cruelty Free International, it still does.

Koller (1857–1944) and Corning (1855–1923)

Karl Koller initially tested cocaine hydrochloride to the cornea of rabbits and dogs in 1884 and reported that, after one-half to one minute, "insensitiveness" was complete and lasted ten minutes [42]. In 1888, Koller moved to the United States and practiced ophthalmic surgery in New York. Here, the neurologist James Corning injected 20 minims (1.3 mL) of a 2% cocaine solution into the space between two inferior dorsal vertebrae of a young dog [43]. Within 5 min, he noted incoordination and later, weakness and anaesthesia of the animal's hind quarters which resolved completely in approximately 4 h. Local anaesthetic techniques had been discovered.

Hobday (1869–1939)

The veterinary surgeon Sir Frederick Hobday arguably established veterinary anaesthesia as a specialty for purposes beyond the laboratory. In 1906, he published "Surgical diseases of the dog and cat: with chapters on anaesthetics and obstetrics". Chapter 4, entitled, "The administration of anaesthetics" was 23 pages long, with five pages devoted to local anaesthesia, 16 pages to general anaesthesia and two pages to morphia and chloral (hydrate) [44]. In 1915, Hobday published the first book dedicated to the subject, "Anaesthesia & Narcosis of Animals and Birds". In describing the use of chloroform, ether, ACE mixture (alcohol, chloroform, ether) ethyl chloride and nitrous oxide for use in horses, oxen, sheep, goats, pigs, dogs, cats, monkeys, wild and semi-domesticated animals and birds, Hobday provided information of potential value for improving laboratory animal anaesthesia.

Starling, Bayliss, Dale, Coleridge, Lind-af-Hageby and the brown dog (1902–1910) [45]

Starling (Professor of Physiology at University College London) conducted pancreatic surgery on a small brown terrier in December 1902. He re-opened the dog's abdomen in February 1903 before a class of medical students. On completing the second operation, Starling handed the animal to Bayliss, who, in making a new wound, contravened the "no re-use" principle extant in the 1876 Act. During the third operation, the animal "suffered greatly" and made purposeful movements, indicating inadequate anaesthesia. The dog was finally given to Henry Dale, an unlicensed research student, who killed the animal. Louise Lind-af-Hageby and Leisa Schartau observed and recorded the events in, *"The Shambles of Science: Extracts from the Diary of Two Students of Physiology"*. Stephen Coleridge, NAVS secretary, read the book and after recognising breaches of the 1876 Act, made a public statement against Bayliss, who issued a libel suit. Bayliss won the ensuing case but Coleridge won the public: £5735 was collected and partly paid for a statue of the brown dog which was unveiled in Battersea Park on 15 September 1906. Medical students vandalized the statue and it became the focus of fighting between medical pro-vivisectionists, the police, anti-vivisectionists and the locals—who had developed a fondness for the effigy. The statue's removal on 10 March 1910 provoked further anger: nine days later, more than 3000 people marched from Hyde Park Corner to Trafalgar Square, where a public meeting was held. Politically relevant public unrest had originated from the mismanagement of the small dog's anaesthetic.

Public reaction to the brown dog affair encouraged the appointment of a second Royal Commission on Vivisection in 1906. Amongst other matters, Stephen Coleridge proposed that the use of curare should be entirely prohibited in animal experiments. The Commission eventually recommended that the use of curare in experiments required special certification, that animals in such experiments must be anaesthetized before the operation and kept anaesthetized until death. They also recommended that a Home Office appointee should be present during experiments in which curare was used.

Paget (1855–1926) [46]

The publics' perception of scientists versus anti-vivisectionists began shifting after the brown dog affair because a link between scientific growth and an improving quality and quantity of life was becoming apparent. Medical scientists, becoming sensitive to the public's concerns, were also making efforts to dispel the accusations of cruelty being made against them whilst capitalising on the role of animal experimentation in understanding—if not treating—conditions such as diabetes. This was exemplified by the publication of Stephen Paget's *Experiments on Animals* in 1900. A tome of 381 pages, 24 were devoted to the chapter on "Anæsthetics Used For Animals"—a considerable improvement from Sanderson's aforementioned handbook in which advice on anaesthesia was notable by its absence. Please see Figure 3 for the dates of the publication of material contributing to the promotion of anaesthetics and analgesics in animal experiments.

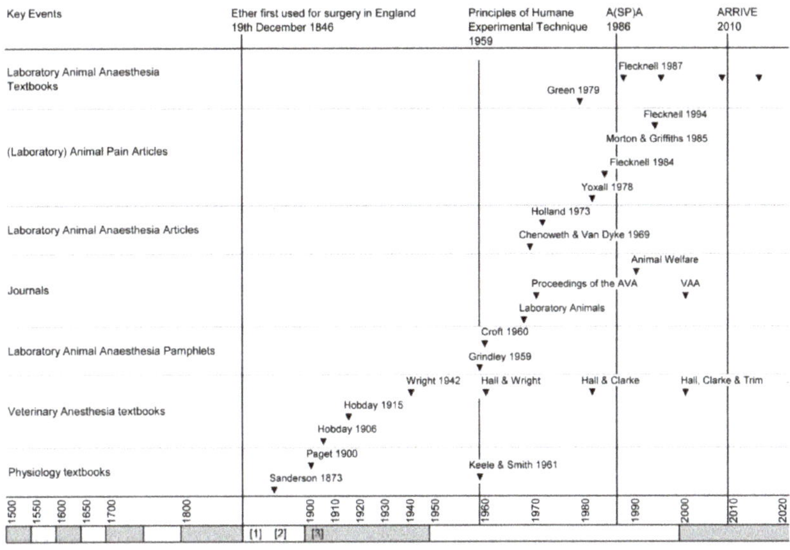

Figure 3. Dates of the publication of material contributing to promotion of anaesthetics and analgesics in animal experiments. Alternatively shaded x axis blocks each represent a 50 year epoc. A(SP)A 1986: The Animals (Scientific Procedures) Act 1986. ARRIVE: Animals in Research: Reporting in Vivo Experiments; Reporting guidelines. Additional dates for reference: [1]: 1873 Sanderson publishes, "Handbook for the Physiological Laboratory"; [2]: the Magnan "affair"; [3]: the brown dog affair. AVA: Association of Veterinary Anaesthetists; VAA: Veterinary Anaesthesia and Analgesia. Grindley [47] Keele and Smith [48] and Chenoweth and Van Dyke [49] are difficult to locate and on 12 September 2020, were unavailable online. Morton and Griffiths [50] is a landmark paper insofar that it was the first to emphasize an ethical and scientific imperative to use analgesics in laboratory animals.

Wright (1897–1971)

Hobday's publications were superseded by (J.G.) Wright's Veterinary Anaesthesia, which was first published in 1942. Its objective was to serve as a student textbook and as a reference manual for veterinarians in practice. To this end, the book in all its editions served its purpose well. However, laboratory animal anaesthesia was never prioritized, e.g., the current (11th edition) edited by Hall (deceased) Clarke and Trim, covers most of the laboratory animal species albeit subsumed in a chapter entitled "Anaesthesia of zoological species (exotic pets, zoo, aquatic, and wild animals)". An examination of Figure 3 (below) reveals that the separation of an increasingly specialized laboratory animal anaesthesia literature from that describing the common domesticated species began with Hall and Wrights 5th edition (1961).

Hume (1886–1981) Russell (1925–2006) and Burch (1926–1996) [51]

In 1926, Charles Hume founded the University of London Animal Welfare Society which became the Universities Federation for Animal Welfare (UFAW) in 1938. Its goals "were to 'tackle animal problems on a scientific basis, with a maximum of sympathy but a minimum of sentimentality". Amongst its formal aims were to enlist the influence of university men and women on behalf of animals, wild and domestic and to lessen, by methods appropriate to the special character of a university organisation, the pain and fear inflicted on animals by man. In 1954, Hume appointed William Russell and Rex Burch and inaugurated a systematic study of the ethical aspects of experimental animal techniques. Their report, published in 1959 and entitled *"The Principles of Humane Experimental Technique"*, condensed humane techniques into three categories; replacement, reduction, and refinement—the 3Rs. Replacement involved using reliable non-animal methodologies when they existed. When they did not, reduction meant using the least number of animals to achieve scientific objectives. Refinement referred to any measure improving the welfare and experiences of animals that could not evade experimentation by being replaced or surplus to study requirements. In discussing refinement and anaesthesia Russell and Burch state:

> *"the most generally important of all is that of anaesthesia, the supreme refinement procedure. This has occasioned perhaps the greatest single advance in humane technique, and has at the same time been virtually indispensable for the advance of experimental biology".* [52]

Russell and Burch's appreciation of anaesthesia can be condensed into four general themes (1) its importance, arising from its ability to eliminate pain and suffering; (2) concerns with administration, i.e., doses and timing; (3) the hazards of neuromuscular blocking agents; (4) the promotion of local anaesthetic techniques.

Croft (1919–2009)

Neither Russell nor Burch were animal anaesthetists and throughout their "Principles" refer—when necessary—to the experimental work of Phyllis Croft, a veterinary neurologist. For example,

> *"Croft has also recently (1957) discussed the condition for veterinary and experimental use of the relaxants or curariform drugs which block neuromuscular transmission among other effects and which in general should only be used in conjunction with general anaesthesia and in mammals, facilities for artificial respiration".* [53]

In 1960, Croft, in conjunction with UFAW, published, "An introduction to the anaesthetics of laboratory animals". The booklet, which was 31 pages long and written for technicians and junior graduates who had no previous experience of anaesthesia, described injection technique and the choice of anaesthetic, and contained sections on practical anaesthesia in rabbits, guinea pigs, hamsters, rats and mice. The selection of drugs was confined to ether, thiopentone and pentobarbitone, and its emphasis was on simplicity [54]. Its publication represents the beginnings of a literature devoted to laboratory animal anaesthesia.

Littlewood (1965)

By 1963, the number of animals involved in experiments was counted in millions while the Cruelty to Animals Act (1876) had remained largely unchanged [55]. A Departmental Committee on Experiments on Animals, chaired by Sir Sidney Littlewood was assigned "to consider the present control over experiments on living animals, and to consider whether, and if so what, changes are desirable in the law." Published in 1965, the Littlewood Report finally established an uncompromising recognition that, "the use of muscle relaxing drugs which, in effect, renders an animal physically helpless whilst leaving it fully conscious, opens the way to experiments of extreme cruelty."

Singer (1974)

Numerous factors, including two world wars, a major economic recession and the fear of nuclear annihilation distracted public attention from animal experimentation for much of the 20th century. However, anti-vivisectionism was re-animated in 1974 when "Animal Liberation" was published by the Australian philosopher Peter Singer. The Animal Liberation Front (ALF), founded in 1976, considered the work to be the founding philosophical statement of its *raison d'etre* and subsequently ensured—through various activities—that the subject of laboratory animal welfare became a matter of public concern. However, the chapter featuring laboratory animals and entitled "Tools for Research" tends to overstate the usefulness of in vitro and ex vivo methodologies, while understating the role of national and institutional controls. It contains harrowing and detailed descriptions of selected animal experiments, mainly conducted by the US military and psychiatric researchers delivering noxious agents and, or stimuli to conscious animals—usually primates or dogs. Importantly, the role of anaesthetics in experimental refinement is poor, or mis-represented. For example, when discussing the role of animals in traumatic shock research, Singer summarizes scientific consensus with:

> *"They (the scientists) discouraged the use of anaesthesia ... the influence of anaesthesia is controversial ... and in the reviewer's opinion prolonged anaesthesia is best avoided."* [56]

and

> *"Experimenters may consider it unnecessary to include in their reports any mention of what happens when ... animals recover consciousness in the midst of an operation because of an improperly administered anaesthetic ... ".* [56]

Singer makes no reference to "The Principle of Humane Experimental Technique", while the terms "anaesthesia", "anesthesia" and "analgesia" do not appear in the book's index. Arguably, a more balanced analysis of the benefits of anaesthetics and analgesics in animal experiments might have reduced the number of criminal acts subsequently committed by the ALF against scientists and their work-places while offsetting the public's increasing tendency to view scientists as cruel and uncompassionate.

Holland and Yoxall (1973–1978)

Concerns with unfeeling science may have been allayed had Holland's 1973 article in the Canadian Anaesthetists Society Journal [57] been more widely read. Worried about the paucity of information available for laboratory animal anaesthesia, e.g.,

> *" ... the relatively low standard of veterinary anaesthesia practised, together with the wide variety of animals which are now being anaesthetized in laboratories and veterinary hospitals"*

Holland asserted that:

> *"vertebrate animals (and perhaps some invertebrates too) have similar pain pathways and similar perceptions of pain as man—their lack of ability to communicate does not indicate a lack of awareness of pain and does not condone inhumane treatment".*

Holland goes on to describe her personal recommendations for anaesthetising dogs, cats, rabbits, guinea pigs, infrahuman primates [sic], rats, mice, pigs, ruminants and birds. She also argues that:

"In spite of their low phylogenetic position, common humanity at the very least dictates that they should be anaesthetized prior to surgical procedures",

so goes on to describe anaesthesia in fish, amphibians, reptiles and invertebrates". The beliefs expressed in this article in relation to animal pain are remarkable insofar that Holland was a medical anaesthetist with an interest in child health. Furthermore, her sentiments on animal pain were published 5 years before those of Yoxall's.

Yoxall [58] was an early member of the Association of Veterinary Anaesthetists and although he held the Royal College Diploma in Veterinary Anaesthesia, his major influence was in veterinary clinical pharmacology. His paper, "Pain in small animals - its recognition and control" (1978) was the first paper in the veterinary medical literature to highlight the importance of appreciating pain—albeit in dogs and cats. The article outlined the then current concepts in pain physiology and described the clinical use of analgesics. Taken with Holland's sentiments, the publication of Yoxall's paper represents a point in time at which some began recognising the importance of eliminating pain rather than just consciousness in animals undergoing noxious procedures.

Green (1979) [59]

Colin Green, a veterinarian working for the Medical Research Council authored the first book devoted to laboratory animal anaesthesia in 1979. Accurately entitled "Animal Anaesthesia" (the book covered all common laboratory, domestic, wild, and zoological specimens), the book represented a major advance because it condensed a mass of disparate information on the subject. Green was also in a position at the Clinical Research Centre to conduct anaesthetic trials (under the 1876 Act) on species in which little information existed. The book benefitted from Green's association with Richard Medd, who, as a veterinarian and experimental surgeon working at Huntingdon Life Sciences, commented extensively on the manuscript and provided additional information from his own experiences. "Animal Anaesthesia" did not run to further editions so the numerous references it cites are now out-dated—albeit of considerable historic interest. The book also required some understanding of physiology and pharmacology. The messages Green espoused in his book concerning safe anaesthetic practice remain applicable today.

Flecknell

Paul Flecknell was influenced by Yoxall's advocacy of analgesic use in animals and in 1984 authored "The Relief of Pain in Laboratory Animals" [60]. This work reviewed the analgesic drugs then available and the experimental evidence for their efficacy in laboratory animals. The information was then extrapolated to the clinical situation to provide guidance as to methods of achieving effective analgesia. The publication was the first to emphasize that anaesthesia did not necessarily ensure freedom from post-operative pain and suffering and that analgesics, as distinct from, and in addition to anaesthetics were required for the optimal refinement of noxious procedures.

This message was continued in a 1994 publication, "Refinement of animal use—assessment and alleviation of pain and distress" which also focused on pain recognition [61]. Warning against the dangers of uncritical anthropomorphism, Flecknell complained that the methods used for the assessment of pain and distress then available were unsatisfactory and appealed for more objective methods. This appeal led to a measureable increase in laboratory animal pain research, not least from the Comparative Biology Centre at the University of Newcastle, where Flecknell was Director.

In recognition of the increasing vintage of Green's contribution, Flecknell's "Laboratory Animal Anaesthesia: An Introduction for Research Workers and Technicians" was published by the Academic Press in 1987 [62]. As the full title suggests, these "introductions" were intended to be, and have remained, relatively uncomplicated guides for staff with limited training in veterinary anaesthesia.

However, they provide enough information for laboratory workers to be able to safely anaesthetize commonly encountered laboratory animal species. That subsequent editions have appeared in 1996, 2009 and 2015 attests to the book's appeal.

The advances being made in laboratory animal anaesthesia and analgesia at this time meant that the Animal (Scientific Procedures) Act 1986 (A[SP]A) did not justify *major* changes to the general use of anaesthetics and analgesics in laboratories. Beyond reviewing the licensing system, providing more specific definitions for "regulated procedures" and "protected animals", and introducing named personnel to ensure compliance at establishment level, the Act required additional personal and project license fulfilments for those intending to use neuromuscular blocking agents. Personal licensees were required to provide evidence that they were competent in managing anaesthetics in the species they intend to paralyse and were conversant with Appendix H (Guidance on the use of neuromuscular blocking agents [NMBAs]) [63].

Between 2005 and 2011, Flecknell and colleague conducted four structured reviews of the biomedical literature [64–67] which, in recording the reported use of analgesics in laboratory animals in the biomedical literature, acted as an estimate of experimental refinement—at least in terms of analgesic use (see Table 1 for details of these, and related studies) [68]. All studies intimated that analgesics were not being extensively used in noxious experimental procedures, but aware that discrepancies may have existed between the *reported* and *actual* use of analgesics, Richardson and Flecknell (2005) made retrospective contact with the corresponding authors of papers in which analgesics had not been mentioned [64]. Of these, 71% replied that analgesic use had not been reported because they had not been used, revealing the existence of a widespread problem.

Table 1. The Extent of anaesthetic and analgesic administration in animal experimentations as determined by literature analysis. NA: not applicable.

Publication Year	Species	Sampling Period	n Articles	% Analgesic Use Reported	% Anaesthetic Use Reported; Ranked Use of Different Techniques	Citation
2005	Rodents	1990–1992	112	2.7	NA; ether > "other" > pentobarbital > injectable combinations	[64]
		2000–2002	126	19.8	NA; pentobarbital > injectable combinations > "other" > isoflurane	
2009	Mice	2000–2001	9	7	NA; xylazine/ketamine > pentobarbital > halothane	[65]
		2005–2006	69	9	NA; pentobarbital = xylazine/ketamine > isoflurane	
	Rats	2000–2001	77	31	NA; pentobarbital > xylazine/ketamine > ether > injectable combinations	
		2005–2006	17	37	NA; isoflurane > pentobarbital > injectable combination	
2009	Rabbits	2000–2001	15	40	Anaesthetics not described	[66]
		2005–2006	15	53		
	Pigs	2000–2001	15	67		
		2005–2006	15	67		
	Sheep	2000–2001	15	64		
		2005–2006	15	73		
	Dogs	2000–2001	15	40		
		2005–2006	15	53		
	NHP	2000–2001	15	40		
		2005–2006	15	67		
2011	Rabbits	1995–1997	64	16	Anaesthetics not described	[67]
		2005–2007	63	50		
2016	Pigs	2012–2014	233	83	Isoflurane > not described > propofol	[68]

Flecknell's contribution to practical training in laboratory animal anaesthesia continues in the form of e-learning modules which have been reformulated to meet the learning outcomes set out by the EU Expert Working group on Education and Training Framework [69]. The latter was established by the European Commission to develop a common education and training framework for the EU to fulfil the requirements under Articles 23, and 24 of Directive 2010/63/EU on the protection of animals used for scientific purposes.

The burgeoning availability of educational resources and information over the last two decades may be exemplified by the NORECOPA website (https://norecopa.no) which has extensive information on laboratory animal anaesthesia and analgesia as well as a collection of links to international guidelines, information on textbooks, and links to resources for education and training. The website currently runs to some 10,000 pages and cites global resources of relevance to the 3Rs.

Kilkenny, Parsons, Kadyszewski, Festing, Cuthill, Browne, Emerson, Altman and Smith (2010–2020)

Flecknell's findings [64–67] supported a growing view that biomedical reportage was inadequate, causing the National Centre for the Replacement, Refinement and Reduction of Animals in Research (NC3Rs), a UK Government-sponsored scientific organisation, to commission a survey to review standards of description in the animal research literature. The subsequent report indicated that serious omissions were present in the way research involving animals was reported [70].

After widespread consultation the ARRIVE guidelines were published. These consisted of 20 items describing the minimum information that scientific publications reporting animal research should include. With respect to anaesthetics and analgesics, three explicit and four potential requirements were listed indicating the relative importance attributed to these in reporting animal experiments [71].

However, despite widespread uptake of the ARRIVE guidelines by journals, funding bodies, universities, organisations and learned societies, it soon became clear that their effect on reporting animal experiments was limited. For example, an examination of 400 scientific articles describing major survival surgeries in dogs, primates, pigs, mice, rats and other rodents for the completeness of the information they provided on anaesthesia and analgesia concluded that the current scientific literature could not be trusted to present full details on the use of animal anaesthetics and analgesics [72]. Another compared the extent to which all elements of the guidelines had been fulfilled in Journals which supported, or did not support the guidelines at time-points before and after guideline introduction, and found that journal support for the ARRIVE guidelines had not resulted in, "a meaningful improvement in reporting quality, contributing to ongoing waste in animal research" [73]. The reported use of NMBs in laboratory pigs in the literature was examined [74] and, based on (1) the absence of information confirming that animals were adequately anaesthetized, and (2) the affirmation by corresponding authors that reported information reflected the actuality, prompted the conclusion that a proportion of laboratory pigs undergoing noxious procedures are likely to be aware under general anaesthesia.

The ARRIVE 2.0 guidelines were introduced in 2020 in an attempt to resolve poor reporting tendencies [75]. However, it had already (2018) been pointed out that the complete description of badly planned and managed experiments would not increase study reproducibility, refine the conditions for, nor reduce the number of animals wasted in such a study [76]. Consequently, the PREPARE (Planning Research and Experimental Procedures on Animals: Recommendations for Excellence) guidelines were developed as one of Norecopa's (https://norecopa.no/PREPARE) contributions to tackling the reproducibility crisis. It is to be seen whether these and, or ARRIVE 2.0 will affect the implementation and reporting of refinement in terms of providing and describing the use of anaesthetics and analgesics in future animal experiments.

2. Conclusions

This historical review of vivisection reveals that anaesthetics and analgesics have had a major beneficial effect on laboratory animal welfare. Scientific concerns that anaesthetic and analgesic drugs may temporarily disperse data and affect scientific conclusions continue to be countered by demonstrating the similar, though adverse and oft-prevailing effects of pain. In any case, the use of analgesics increases the external validity of animal models used in the study of noxious human conditions: that animal models of painful human conditions should be deprived of the care that the modelled human could expect is fundamentally unjust and scientifically unsound.

Persistent philosophical objections to the use of animals in research are accommodated by the replacement principle, which aims to achieve the eventual exclusion of all animals from all experiments. The funding and promotional activities of Fund for the Replacement of Animals in Medical Experimentation (FRAME) and the NC3Rs attest to the sincerity of this aim. However, public objections—which have historically been based on the argument that animal research is cruel (and pointless)—are immediately addressed by the refinement afforded by terminal anaesthetics, or appropriate anaesthetic and analgesic techniques in recovery studies. Repeated Ipsos Mori polls reveal that societies' majority acceptance of animal experimentation—at least in the UK—is based on assurances that it is for medical research purposes, that there is no alternative and that there is no unnecessary suffering [77]. The latest poll also revealed a public belief that laboratory animal veterinarians continue to be the most trusted source of information on animal research. It follows that the publics' general concerns with the use of animals in research may be further allayed were qualified veterinary specialists, e.g., Diplomates of the European (or American) Colleges of Veterinary Anaesthesia and Analgesia, to become increasingly involved in assisting their colleagues in laboratory animal medicine when noxious animal experiments are planned.

Despite *some* success in replacing and reducing the numbers of animals used in research, the constant introduction of new research methodologies, e.g., gene editing, imaging, et cetera, and the requirement for new diagnostic or therapeutic devices to undergo pre-clinical safety and efficacy testing in animals, makes continuous demands on those concerned with experimental refinement. Arguably, this may apply less to anaesthesia. The compulsory training of laboratory animal personnel in the safe delivery of modern anaesthetic techniques, combined with the widespread availability of information, has reduced the need for any urgent change—at least in the management of non-invasive, non-painful and straightforward procedures. This does not apply to the use of analgesics, particularly when noxious recovery procedures are involved, because considerable gaps persist in the field of laboratory animal pain recognition and treatment. Some have suggested that this justifies the need for prospective pain studies [78]. However, the point has already been made [79] that refinement principles are better served by improved reporting. The original ARRIVE guidelines (2010) requested that articles describing animal experiments provide: (1) precise details of how anaesthetics and analgesics are used; (2) a description of welfare-related assessments and interventions; and (3) that all details of adverse events, e.g., unmitigated pain, be described, and that any modifications made to experimental protocols in order to reduce adverse events are revealed, e.g., the provision of effective analgesia. Given that these requirements have not been forthcoming, it has been suggested that amongst ARRIVE-subscribing journals (and others) a more determined editorial enforcement of the guidelines would go a long way to increasing the information available on pain management in laboratory animals [79]. Unfortunately, the ARRIVE 2.0 guidelines have relegated, "interventions taken in experimental protocols to reduce pain, suffering, and distress" and "reports of expected or unexpected adverse events" to the "Recommended Set" rather than the obligatory "Essential 10". It remains to be seen what effect this will have on the role of anaesthetics and analgesics in the ongoing refinement of animal experiments.

Funding: This research received no external funding.

Acknowledgments: The author is grateful to Ron Jones, Alistair McKenzie, Ngaire Dennison, Polly Taylor, David Morton, Judy MacArthur Clark, Yves Moens, Amanda Novak, Stephen Greenhalgh, Ambra Panti and Stefano Guido for providing hitherto unpublished information, and to Gabby Musk for commissioning this work.

Conflicts of Interest: The author has no financial nor personal relationship with people or organisations that could inappropriately influence or bias this submission's content.

References

1. Russell, W.M.S.; Burch, R.L. *The Principles of Humane Experimental Technique*; Methuen & Co Ltd.: London, UK, 1959.
2. Franco, N.H. Animal Experiments in Biomedical Research: A Historical Perspective. *Animals* **2013**, *1*, 238–273. [CrossRef] [PubMed]
3. Andreas-Holger, M.; Tröhler, U. Animal Experimentation from Antiquity to the End of the Eighteenth Century: Attitudes and Arguments. In *Vivisection in Historical Perspective*; Rupke, N.A., Ed.; Croom Helm: London, UK, 1987; pp. 15, 22, 23, 27.
4. A Guide to the Historiated Capitals of the 1543 Fabrica. Available online: https://www.vesaliusfabrica.com/en/original-fabrica/the-art-of-the-fabrica/historiated-capitals.html (accessed on 27 July 2020).
5. Slutsky, A.S. History of Mechanical Ventilation. From Vesalius to Ventilator-induced Lung Injury. *Am. J. Respir. Crit. Care Med.* **2015**, *191*, 1106–1115. [CrossRef] [PubMed]
6. Armstrong Davison, M.H. The Evolution of Anaesthesia. *Brit. J. Anaesth.* **1958**, *30*, 495. [CrossRef] [PubMed]
7. Weil, E. The Echo of Harvey's de Motu Cordis (1628) 1628 to 1657. *J. Hist. Med. Allied Sci.* **1957**, *12*, 167–174. [CrossRef]
8. Descartes, R. Animals are Machines. In *Animal Rights and Human Obligations*, 2nd ed.; Regan, T., Singer, P., Eds.; Prentice Hall: Upper Saddle River, NJ, USA, 1989; pp. 13–19.
9. Dorrington, K.L.; Poole, W. The first intravenous anaesthetic: How well was it managed and its potential realized? *Br. J. Anaesth.* **2013**, *110*, 7–12. [CrossRef]
10. Shugg, W. Humanitarian attitudes in the early animal experiments of the Royal Society. *Ann. Sci.* **1968**, *24*, 227–238. [CrossRef]
11. Cartwright, F.F. Humphry Davy's contribution to anaesthesia. *Proc. R. Soc. Med.* **1950**, *43*, 571–578. [CrossRef]
12. Bentham, J. *An Introduction to the Principles of Morals and Legislation*; Printed for W. Pickering: London, UK, 1823; pp. 142–143.
13. Schmitz, R. Friedrich Wilhelm Sertürner and the Discovery of Morphine. *Pharm. Hist.* **1985**, *27*, 61–74.
14. Elliott, P. Vivisection and the Emergence of Experimental Physiology in Nineteenth century France. In *Vivisection in Historical Perspective*; Rupke, N.A., Ed.; Croom Helm: London, UK, 1987; pp. 61–62.
15. Ryder, R.D. *Animal Revolution. Changing Attitudes towards Speciesism*; Berg: Oxford, UK, 2000; pp. 101–104, 106.
16. Neveu, R. The Introduction of Surgical Anesthesia in France. *Anesthesiology* **1947**, *8*, 334–335. [CrossRef]
17. Secher, O. Nikolai Ivanovich Pirogoff. *Anaesthesia* **1986**, *41*, 829–837. [CrossRef]
18. Sechzer, J.A. The Ethical Dilemma of Some Classical Animal Experiments. *Ann. N. Y. Acad. Sci.* **1983**, *406*, 5–12. [CrossRef] [PubMed]
19. Guerrini, A. *Experimenting with Humans and Animals*; The John Hopkins University Press: Baltimore, MD, USA, 2003; pp. 78, 80, 86.
20. Sykes, K.; Bunker, J. *Anaesthesia and the Practice of Medicine: Historical Persepectives*; The Royal Society of Medicine Press: London, UK, 2007; p. 8.
21. Smith, W.D.A. A History of Nitrous Oxide and Oxygen Anaesthesia Part Iv: Hickman and the "Introduction Of Certain Gases Into The Lungs. *Br. J. Anaesth.* **1966**, *38*, 58. [CrossRef] [PubMed]
22. Youngson, A.J. *The Scientific Revolution in Victorian Medicine*; Croom Helm: London, UK, 1979; pp. 47, 67, 74.
23. Snow, S.J. *Blessed Days of Anaesthesia. How Anaesthetics Changed the World*; Oxford University Press: Oxford, UK, 2008; p. 24.
24. Simpson, J.Y. Discovery of a new anaesthetic agent, more effective than sulphuric ether. *Lancet* **1847**, *2*, 549–551. [CrossRef]
25. Defalque, R.J.; Wright, A.J. The short, tragic life of Robert M. Glover. *Anaesthesia* **2004**, *59*, 394–400. [CrossRef]
26. Botting, J.H. Early Animal Experiments in Anaesthesia. In *Animals and Medicine: The Contribution of Animal Experiments to the Control of Disease*; Open Book Publishers: Cambridge, UK, 2015; p. 163. ISBN 9782821876347. Available online: http://books.openedition.org/obp/1988 (accessed on 30 September 2020).
27. Waters, R.M. John Snow, First Anesthetist. *Bios* **1936**, *7*, 25–40.
28. Snow, J. A Lecture on The Inhalation Of Vapour Of Ether In Surgical Operations. *Lancet* **1847**, *49*, 551–554. [CrossRef]
29. Connor, H. Anaesthesia and the British public, 1846–1856. Extracts from the undergraduate prize essay (1969). *Anaesthesia* **1970**, *25*, 115–120. [CrossRef]

30. Knight, P.R., 3rd; Bacon, D.R. An unexplained death: Hannah Greener and chloroform. *Anesthesiology* **2002**, *96*, 1250–1253. [CrossRef] [PubMed]

31. Wakley, T.H. A Record of One Hundred Experiments on Animals, with Ether and Chloroform. *Lancet* **1848**, *51*, 19–25.

32. Lee, J.A. Claude Bernard (1813–1878). *Anaesthesia* **1978**, *33*, 741–747. [CrossRef]

33. Shmuely, S. Curare: The Poisoned Arrow that Entered the Laboratory and Sparked a Moral Debate. *Soc. Hist. Med.* **2020**, *33*, 888. [CrossRef]

34. Bernard, C. E tudes Physiologiques sur Quelques Poisons Americains. *Revue des Deux Mondes* **1864**, *53*, 164–190.

35. Guarnieri, P. Moritz Schiff (1823–96): Experimental Physiology and Noble Sentiment in Florence. In *Vivisection in Historical Perspective*; Rupke, N.A., Ed.; Croom Helm: London, UK, 1987; pp. 111, 134.

36. Anon. Prosecution at Norwich. Experiments on Animals. *Br. Med. J.* **1874**, *2*, 751–754.

37. Hamilton, S. On the Cruelty to Animals Act, 15 August 1876. BRANCH: Britain, Representation and Nineteenth-Century History. Ed. Dino Franco Felluga. Extension of Romanticism and Victorianism on the Net. Available online: http://www.branchcollective.org/?ps_articles=susan-hamilton-on-the-cruelty-to-animals-act-15-august-1876 (accessed on 11 September 2020).

38. French, R.D. *Antivivisection and Medical Science in Victorian Society*; Princeton University Press: Princeton, NJ, USA, 1975; pp. 414–415.

39. Bates, A.W.H. *Anti-Vivisection and the Profession of Medicine in Britain: A Social History*; Palgrave Macmillan: London, UK, 2017; pp. 28–29.

40. legislation.gov.uk: Cruelty to Animals Act 1876. Available online: https://www.legislation.gov.uk/ukpga/Vict/39-40/77/enacted (accessed on 11 September 2020).

41. Lane-Petter, W. The Cruelty to Animals Act, 1876. *Med. Sci. Law* **1962**, *2*, 87–95. [CrossRef]

42. Goerig, M.; Bacon, D.; van Zundert, A. Carl Koller, cocaine, and local anesthesia: Some less known and forgotten facts. *Reg. Anesth. Pain. Med.* **2012**, *37*, 318–324. [CrossRef]

43. Gorelick, P.B.; Zych, D. James Leonard Corning and the early history of spinal puncture. *Neurology* **1987**, *37*, 672–674. [CrossRef]

44. Wellcome Collection. Surgical Diseases of the Dog and Cat: With Chapters on Anaesthetics and Obstetrics. Available online: https://wellcomecollection.org/works/p4mwqkwj (accessed on 1 August 2020).

45. Anon. The Brown Dog and His Friends. *Br. Med. J.* **1910**, *1*, 588–589.

46. Paget, S. Experiments on Animals. *Nature* **1906**, *75*, 121–122.

47. Grindley, G.C. The Sense of Pain in Animals. In *The Animal Year Book*; UFAW, 7A Lamb's Conduit Passsage: London, UK, 1959; Volume 2.

48. Keele, C.A.; Smith, R. The Assessment of Pain in Man and Animals. In *The Proceedings of an International Symposium Held under the Auspices of UFAW-the Universities Federation for Animal Welfare*; The Universities Federation for Animal Welfare: London, UK, 1961.

49. Chenoweth, M.; Van Dyke, R.A. Anaesthesia in biomedical research. *Fed. Proc.* **1969**, *28*, 1383.

50. Morton, D.; Griffiths, P. Guidelines on the recognition of pain, distress and discomfort in experimental animals and an hypothesis for assessment. *Vet. Rec.* **1985**, *116*, 431–436. [CrossRef]

51. Scott, W.N. The History and Impact of UFAW. *Int. J. Stud. Anim. Probl.* **1980**, *1*, 14–17.

52. Russell, W.M.S.; Burch, R. Refinement. In *The Principles of Humane Experimental Technique, Special Edtion*; Charles C Thomas: Springfield, IL, USA, 1992; p. 137.

53. Croft, P.G. Assessment of consciousness during curarisation. *Nature* **1953**, *171*, 261.

54. Croft, P.G. *An Introduction to the Anaesthesia of Laboratory Animals*; The Universities Federation for Animal Welfare: London, UK, 1960.

55. The National Archives: Littlewood Committee Report on Experiments on Animals: Consideration of Recommendations. Available online: https://discovery.nationalarchives.gov.uk/details/r/C2057470 (accessed on 12 September 2020).

56. Singer, P. *Animal Liberation*; Harper Collins: London, UK, 1976; pp. 41, 66.

57. Holland, A.J. Laboratory animal anaesthesia. *Can. Anaesth. Soc. J.* **1973**, *20*, 693–705.

58. Yoxall, A.T. Pain in small animals–its recognition and control. *J. Small Anim. Pract.* **1978**, *19*, 423–438.

59. Green, C.J. *Animal Anaesthesia*; Laboratory Animal Ltd.: London, UK, 1979.

60. Flecknell, P.A. The relief of pain in laboratory animals. *Lab. Anim.* **1984**, *18*, 147–160.

61. Flecknell, P.A. Refinement of animal use-assessment and alleviation of pain and distress. *Lab. Anim.* **1994**, *28*, 222–231.

62. Flecknell, P. *Laboratory Animal Anaesthesia*; Academic Press: London, UK, 1987.

63. Guidance on the Operation of the Animals (Scientific Procedures) Act. 1986. Available online: https://www.gov.uk/guidance/guidance-on-the-operation-of-the-animals-scientific-procedures-act-1986 (accessed on 11 September 2020).

64. Richardson, C.A.; Flecknell, P.A. Anaesthesia and post-operative analgesia following experimental surgery in laboratory rodents: Are we making progress? *Altern. Lab. Anim.* **2005**, *33*, 119–127.

65. Stokes, E.; Flecknell, P.; Richardson, C. Reported analgesic and anaesthetic administration to rodents undergoing experimental surgical procedures. *Lab. Anim.* **2009**, *43*, 149–154.

66. Coulter, C.; Flecknell, P.; Richardson, C. Reported analgesic administration to rabbits, pigs, sheep, dogs and non-human primates undergoing experimental surgical procedures. *Lab. Anim.* **2009**, *43*, 232–238. [PubMed]

67. Coulter, C.A.; Flecknell, P.A.; Leach, M.C.; Richardson, C.A. Reported analgesic administration to rabbits undergoing experimental surgical procedures. *BMC Vet. Res.* **2011**, *7*, 12.

68. Bradbury, A.G.; Eddleston, M.; Clutton, R.E. Pain management in pigs undergoing experimental surgery; a literature review (2012–2014). *Br. J. Anaesth.* **2016**, *116*, 37–45. [CrossRef] [PubMed]

69. On Protection of Animals Used For Scientific Purposes: Education and Training; Framework National Competent Authorities for the Implementation of Directive 2010/63/EU on the Protection of Animals Used for Scientific Purposes. Brussels. 19–20 February 2014. Available online: https://ec.europa.eu/environment/chemicals/lab_animals/pdf/guidance/education_training/en.pdf (accessed on 5 October 2020).

70. Kilkenny, C.; Parsons, N.; Kadyszewski, E.; Festing, M.F.W.; Cuthill, I.C.; Fry, D.; Hutton, J.; Altman, D.G. Survey of the Quality of Experimental Design, Statistical Analysis and Reporting of Research Using Animals. *PLoS ONE* **2009**, *4*, e7824. [CrossRef] [PubMed]

71. Kilkenny, C.; Browne, W.J.; Cuthill, I.C.; Emerson, M.; Altman, D.G. Improving Bioscience Research Reporting: The ARRIVE Guidelines for Reporting Animal Research. *PLoS Biol.* **2010**, *8*, e1000412. [CrossRef] [PubMed]

72. Carbone, L.; Austin, J. Pain and Laboratory Animals: Publication Practices for Better Data Reproducibility and Better Animal Welfare. *PLoS ONE* **2016**, *11*, e0155001. [CrossRef]

73. Leung, V.; Rousseau-Blass, F.; Beauchamp, G.; Pang, D.S.J. ARRIVE has not ARRIVEd: Support for the ARRIVE (Animal Research: Reporting of in vivo Experiments) guidelines does not improve the reporting quality of papers in animal welfare, analgesia or anesthesia. *PLoS ONE* **2018**, *13*, e0197882. [CrossRef]

74. Bradbury, A.G.; Clutton, R.E. Are neuromuscular blocking agents being misused in laboratory pigs? *Br. J. Anaesth.* **2016**, *116*, 476–485. [CrossRef]

75. Percie du Sert, N.; Hurst, V.; Ahluwalia, A.; Alam, S.; Avey, M.T.; Baker, M.; Browne, W.J.; Clark, A.; Cuthill, I.C.; Dirnagl, U.; et al. The ARRIVE guidelines 2.0: Updated guidelines for reporting animal research. *PLoS Biol.* **2020**, *18*, e3000410.

76. Smith, A.J.; Clutton, R.E.; Lilley, E.; Hansen, K.E.A.; Brattelid, T. PREPARE: Guidelines for planning animal research and testing. *Lab. Anim.* **2018**, *52*, 135–141. [CrossRef]

77. Clemence, M. Ipsos Mori 2018, Public Attitudes to Animal Research in 2018. Available online: https://www.ipsos.com/ipsos-mori/en-uk/public-attitudes-animal-research-2018 (accessed on 5 October 2020).

78. Gigliuto, C.; De Gregori, M.; Malafoglia, V.; Raffaeli, W.; Compagnone, C.; Visai, L. Pain assessment in animal models: Do we need further studies? *J. Pain Res.* **2014**, *7*, 227–236.

79. Clutton, R.E. A review of factors affecting analgesic selection in large animals undergoing translational research. *Vet. J.* **2018**, *236*, 12–22. [CrossRef]

Publisher's Note: MDPI stays neutral with regard to jurisdictional claims in published maps and institutional affiliations.

 *animals*

Review

The Impact of Common Recovery Blood Sampling Methods, in Mice (Mus Musculus), on Well-Being and Sample Quality: A Systematic Review

Alexandra L Whittaker [1],* and Timothy H Barker [2]

[1]   School of Animal and Veterinary Sciences, The University of Adelaide, Roseworthy Campus, Roseworthy, South Australia 5371, Australia

[2]   JBI, Faculty of Health and Medical Sciences, The University of Adelaide, South Australia 5005, Australia; timothy.barker@adelaide.edu.au

*   Correspondence: alexandra.whittaker@adelaide.edu.au

Received: 13 May 2020; Accepted: 2 June 2020; Published: 5 June 2020

Simple Summary: Blood sampling is often performed in laboratory mice. Whilst the techniques are likely to cause only momentary pain or distress, given their frequency of performance, it is essential that the method which best safeguards welfare is used. The small size of mice makes sampling challenging, and use of some routes is controversial due to perceived impact on animal welfare. However, to date, no summary of the evidence relating to welfare impacts arising from these techniques has been presented. This paper presents a systematic review of the literature, with quality appraisal of the studies and an assignment of certainty in the evidence. We conclude that there is not enough high-quality evidence available to make a determination on optimal blood sampling route. We provide recommendations for improving future laboratory animal welfare research through standardisation of outcome measures and enhanced adherence to experimental design and reporting guidelines.

Abstract: Blood sampling is often performed in laboratory mice. Sampling techniques have the potential to cause pain, distress and impact on lifetime cumulative experience. In spite of institutions commonly providing guidance to researchers on these methods, and the existence of published guidelines, no systematic evaluation of the evidence on this topic exists. A systematic search of Medline, Scopus, and Web of Science was performed, identifying 27 studies on the impact of recovery blood sample techniques on mouse welfare and sample quality. Studies were appraised for quality using the SYstematic Review Centre for Laboratory animal Experimentation (SYRCLE) risk of bias tool. In spite of an acceptable number of studies being located, few studies examined the same pairwise comparisons. Additionally, there was considerable heterogeneity in study design and outcomes, with many studies being at a high risk of bias. Consequently, results were synthesised using the Synthesis Without Meta-analysis (SWiM) reporting guidelines. Grading of Recommendations, Assessment, Development and Evaluation (GRADE) was utilised for assessment of certainty in the evidence. Due to the heterogeneity and GRADE findings, it was concluded that there was not enough high-quality evidence to make any recommendations on the optimal method of blood sampling. Future high-quality studies, with standardised outcome measures and large sample sizes, are required.

Keywords: mouse; blood sample; well-being; retrobulbar; submandibular; sublingual

---

1. Introduction

It is common in biomedical research for protocols to require blood collection from mice (Mus musculus) in order to measure a range of circulating products. The small size of these animals makes such procedures challenging, but a range of sampling methods are documented and widely used. Common considerations in the selection of sample techniques are their practicality and ease of use, the ability to attain the desired blood volume, sample quality and impact on animal well-being [1,2].

Retrobulbar bleeding (retroorbital) enables acquisition of larger blood volumes (e.g., 0.2–1 mL) [3], but has been controversial due to the risk of substantial tissue damage to the eye [3,4]. Anecdotally, it appears that this technique has fallen out of favour, particularly in some jurisdictions, such as Australia. This has led to the development of alternative methods. The most common alternative is facial (also commonly called submandibular) vein venepuncture [5]. Despite the rise in popularity of this method, perhaps driven by the aesthetically repugnant use of the ocular area, veins in the facial region lie beneath other important tissues such as glandular tissue [6]. This method then also poses a risk of causing secondary complications via tissue damage. Sublingual vein puncture is another alternative method which yields large-volume collections [4]. In contrast to the submandibular technique, sublingual sampling is generally performed under anaesthesia to immobilise the animal [4]. Anaesthesia, as an adjunct, has the potential to impact both positively and negatively on animal well-being, through minimisation of tissue damage [1], or 'hangover' effects from the drugs. Anaesthetic use will also influence the practicality of the technique due to equipment needed and the time taken to perform.

A range of methods are available for the attainment of smaller volume samples (approximately 0.1–0.15 mL), or for frequent repeat sampling. The tail is commonly used as a blood collection site. A range of collection techniques are described, including targeted collection from the lateral tail vein [7,8], tail tip amputation [9–11], and tail incision through cut of the veins [12,13]. Anaesthesia has been regarded as unnecessary for this site, enabling multiple repeat samples. However, warming of the animal may be required to encourage vasodilation [3]. This may add to overall impact on animal well-being. The lateral tarsal or saphenous vein is a common alternative to the tail. Similarly, it requires no anaesthetic and has the added advantage of not requiring external methods for vasodilation. Removal of the scab enables serial blood sampling [3].

Whilst pain, discomfort and physiological stress arising from blood sampling are likely to be short-lived, as one of the most common procedures performed on laboratory animals, researchers and animal ethics committees have a duty to utilise or promote the method with least impact on animal well-being. Furthermore, with the demise of retrobulbar sampling on supposed ethical grounds, it is imperative that an evidence-based approach to the selection of alternative methods is used. Whilst there have been a range of studies investigating the impact of a number of the sampling techniques on mice well-being, these studies typically only contrast a few techniques, and are practically limited in terms of sample sizes. The aim of this systematic review is to present the evidence related to common recovery blood sampling techniques in mice, with regard to animal well-being. Through identification of all relevant evidence, assimilation of study findings to increase statistical power, and study appraisal, it is our intention that this systematic review will provide increased strength of evidence to better inform researchers, ethics committees, and policy makers in their decision making.

2. Materials and Methods

A priori protocol was created for this review and has been registered on the SYstematic Review Centre for Laboratory animal Experimentation (SYRCLE) database for animal intervention studies [14].

2.1. Eligibility Criteria

Inclusive criteria were as follows: (P) studies that include post-weaning inbred or outbred laboratory mice. Neonatal/pre-weaning mice were excluded; (I) studies that evaluated recovery, non-surgical blood sampling techniques. Included techniques were: sublingual, retrobulbar sinus, facial, tail sampling methods, and saphenous vein. Studies that examined both one-off and serial sampling were eligible for inclusion; (C) studies were included that compared the intervention to no blood sample, or other included recovery sample method. Studies with no control group (observational studies) were also eligible for inclusion; (O) outcomes such as mortality, quantifiable measures of mouse well-being such as behavior change, bodyweight change, morbidity and quantifiable measures of sample quality such as hemolysis were included. Outcomes in either the immediate post-sampling period or over the longer term were considered for inclusion; and (S) experimental and quasi-experimental study designs including randomised controlled trials, non-randomised controlled trials, and before and after studies were eligible for inclusion. Observational studies were considered for inclusion.

2.2. Search Strategy

The search strategy aimed to locate published studies in English. An initial limited search of Medline was undertaken to identify articles on the topic. The text words contained in the titles and abstracts of relevant articles, and the index terms used to describe the articles were used to develop a full search strategy for Medline. The search strategy, including all identified keywords and index terms, was adapted for Scopus and Web of Science database searches. The three databases were searched in May 2019 using the developed search strategies (see Supplementary Table S1) and the search was updated in March 2020. Key concepts used for searching were "mice", "blood sample", "welfare" and "blood sample quality". Reference lists of all studies selected for critical appraisal were screened for additional studies. Contact with study authors was undertaken where necessary to clarify findings or seek further information. Studies published from database inception were eligible for inclusion.

2.3. Study Selection

Following the search, all identified citations were collated and uploaded into EndNote X8.0.1 and duplicates removed. Titles and abstracts were screened by one reviewer (A.W.) for assessment against the inclusion criteria for the review. Potentially relevant studies were retrieved in full and their citation details imported into the Joanna Briggs Institute System for the Unified Management, Assessment and Review of Information (JBI SUMARI, Joanna Briggs Institute, Adelaide, Australia) [15]. The full text of selected citations was assessed in detail against the inclusion criteria by two independent reviewers (A.W. and T.B.). Disagreements that arose between the reviewers at each stage of the study selection process were resolved through discussion. The results of the search, with reasons for study exclusions, are presented in the Preferred Reporting Items for Systematic Reviews and Meta-Analyses (PRISMA) flow diagram (Figure 1) [16].

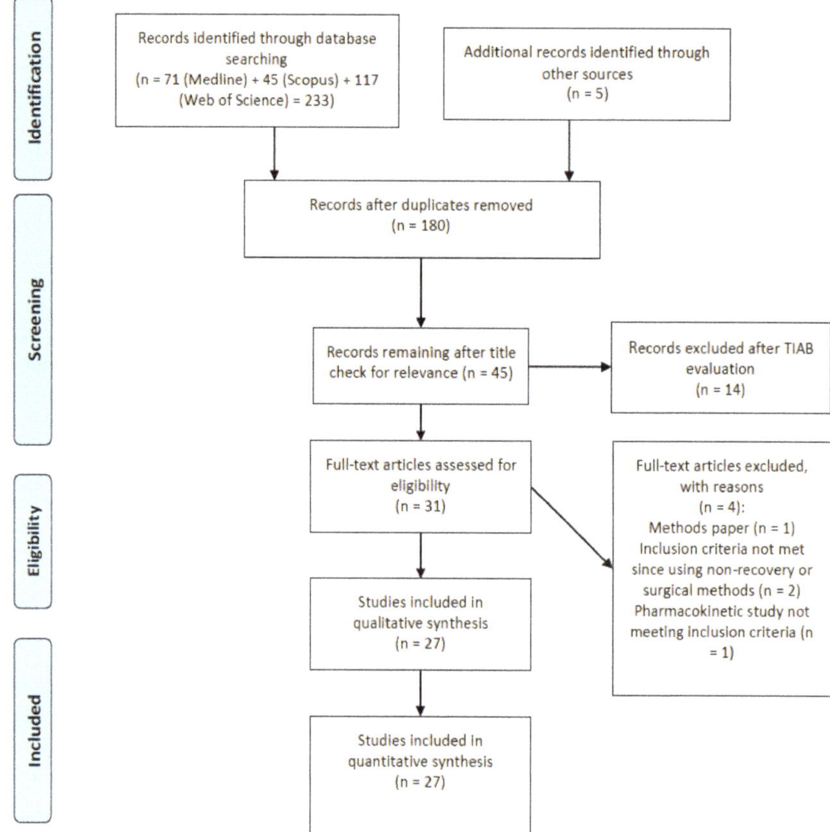

Figure 1. The PRISMA flow diagram for the systematic review detailing the database searches, the number of abstracts screened and the full texts retrieved.

2.4. Assessment of Methodological Quality

Eligible studies were critically appraised for methodological quality by two independent reviewers (A.W. and T.B.) using the SYRCLE risk of bias tool [17]. Any disagreements that arose were resolved through discussion. All studies, regardless of the results of their methodological quality, underwent data extraction and synthesis. Consideration of the methodological quality of individual studies is discussed in the narrative synthesis.

2.5. Data Extraction

Data were extracted from studies included in this review by two independent reviewers (A.W. and T.B.) using an electronic form developed by the authors (see Supplementary Table S4 for full extraction templates). Extracted data included specific details about the mice populations sampled, the study design, blood sample routes, and the outcomes of significance to the review objective, being an indicator of animal well-being. If the data were presented as figure or other form, they were extracted with the Software Getdata Graph Digitizer 2.26 0.20 (S. Federov, Moscow, Russia). Any disagreements that arose were resolved through discussion. Authors of papers were contacted to request missing or additional data, where required.

2.6. Data Synthesis

Due to the nature of the data extracted, it was decided by the review team that a meta-analysis of any format (including a pairwise meta-analysis or a network meta-analysis) was not appropriate for any data included in this review. In most cases, the studies contributing data towards a particular outcome were extremely heterogeneous. Clinical heterogeneity existed between studies in terms of the intervention timing, frequency of the blood taking procedures, the gauge of the needle used, and the amount of blood taken per procedure. There was also heterogeneity in terms of the characteristics of the mice used (strain, sex, age, etc.). In the few circumstances in which studies were homogeneous enough to facilitate an appropriate meta-analysis, the primary authors rarely provided complete reporting of data, often only reporting p-values, a statement of (non-)significance, or simply showcasing their results in the form of a figure or graph that the review team had to "digitize". The review team are cognizant that 'digitizing' data from figures is a subjective, highly variable and imprecise method in which to collect data, and are hesitant to include data collected via this method in any formal meta-analysis.

Because of these limitations and deviations from the methods as specified in the protocol, data were synthesised according to the reporting guidelines of Synthesis Without Meta-analysis (SWiM) [18] for each outcome presented. This has occurred for each outcome and is covered in detail in the results section.

2.7. Assessing Certainty in the Findings

The Grading of Recommendations, Assessment, Development and Evaluation (GRADE) approach for grading the certainty of evidence was followed [19,20] and a Summary of Findings (SoF) has been created using the GRADEPro GDT software (McMaster University, ON, Canada) [21]. The SoF reports plasma glucose concentration (mmol/L), plasma corticosterone concentrations (ng/mL), faecal corticosterone concentrations (ng/0.05 gram of faeces) and bodyweight (% change). The SoF has been presented in Table 1.

Table 1. Grading of Recommendations, Assessment, Development and Evaluation (GRADE) Summary of Findings table for primary outcomes.

The Effects of Various Phlebotomy Techniques for Animals Welfare-Related Outcomes			
Patient or Population: Laboratory Mice Used for Research Purposes **Interventions of Interest: Sublingual Sampling, Retrobulbar Sampling, Facial Vein Sampling, Tail Amputation Sampling, Tail Incision Sampling, Tail Vein Sampling and Saphenous Vein Sampling**			
Outcomes	**Impact**	**Number of Participants (Studies)**	**Certainty of the Evidenc (GRADE)**
	Plasma Glucose Concentration		
Single Sampling	The facial vein technique was involved in the most pairwise comparisons synthesised, and was found to have a beneficial effect when compared to the tail amputation and the tail incision techniques. The tail incision technique was likewise considered to be beneficial compared to the tail amputation method. The retrobulbar technique was considered beneficial to the tail vein technique, and the sublingual technique beneficial to the facial vein technique.	(3 RCTs)	⊕○○○ VERY LOW [a,b,c]
Serial Sampling	The retrobulbar technique was involved in 4 pairwise comparisons and was considered harmful compared to the tail amputation and the tail incision methods. Yet it was beneficial when compared to the facial vein method. Tail amputation was similarly beneficial over the tail incision technique, and the saphenous technique was beneficial compared to the tail vein technique.	(3 RCTs)	⊕○○○ VERY LOW [a,c,d]
	Plasma Corticosterone Concentration		
Single Sampling	The retrobulbar technique proved to be more beneficial compared to the facial vein technique in two studies, beneficial to the tail vein and saphenous methods in one study each, yet harmful when compared to the tail incision technique based on the results from one study. Neither a harmful or beneficial effect was observed comparing the facial vein and the tail vein method; the tail amputation and the saphenous technique; or the tail vein and the tail incision method.	(5 RCTs)	⊕○○○ VERY LOW [a,e,f]
Serial Sampling	The tail amputation technique was considered beneficial in two pairwise comparisons against the retrobulbar technique, and beneficial in a pairwise comparison against the facial vein technique. However, no difference was observed when compared with the saphenous vein technique. Likewise, the saphenous vein technique was considered beneficial when compared against both the retrobulbar and the facial vein technique.	(3 RCTs)	⊕○○○ VERY LOW [a,e,g]
	Faecal Corticosterone		
Serial Sampling	There appears to be a time-dependent factor for the pairwise comparison of the retrobulbar technique and the tail incision technique. While the tail vein was shown to be beneficial compared to the facial vein.	(4 RCTs)	⊕○○○ VERY LOW [a,h,i]

Table 1. *Cont.*

The Effects of Various Phlebotomy Techniques for Animals Welfare-Related Outcomes			
Patient or Population: Laboratory Mice Used for Research Purposes Interventions of Interest: Sublingual Sampling, Retrobulbar Sampling, Facial Vein Sampling, Tail Amputation Sampling, Tail Incision Sampling, Tail Vein Sampling and Saphenous Vein Sampling			
Outcomes	Impact	Number of Participants (Studies)	Certainty of the Evidenc (GRADE)
	Body Weight		
Single Sampling	The saphenous vein technique was considered 'beneficial' in pairwise comparisons with the sublingual, retrobulbar, facial, tail amputation and tail incision methods. This technique was not considered 'harmful' in any pairwise comparison synthesised. The next most beneficial technique synthesised was the sublingual technique, which was considered beneficial in two separate comparisons made to the facial vein technique, and single comparisons made to the retrobulbar; tail amputation and tail incision techniques. The tail amputation technique was beneficial over the retrobulbar; facial vein and the tail incision methods. The retrobulbar and facial vein methods were associated with the most harm, as compared through pairwise comparison.	(5 RCTs)	$\oplus$◯◯◯ VERY LOW [a,j]
Serial Sampling	The retrobulbar technique was considered beneficial in two separate pairwise comparisons made to the facial vein technique, as well as being beneficial when compared to the sublingual technique.	(3 RCTs)	$\oplus$◯◯◯ VERY LOW [k,l]

GRADE Working Group grades of evidence:

High certainty: We are very confident that the true effect lies close to that of the estimate of the effect.

Moderate certainty: We are moderately confident in the effect estimate: the true effect is likely to be close to the estimate of the effect, but there is a possibility that it is substantially different.

Low certainty: Our confidence in the effect estimate is limited: the true effect may be substantially different from the estimate of the effect.

Very low certainty: We have very little confidence in the effect estimate: the true effect is likely to be substantially different from the estimate of effect.

[a] Downgraded two levels for risk of bias as there is a high risk of performance bias across all contributing studies. [b] Downgraded two levels for inconsistency. The direction and magnitude of pairwise comparisons varied across the different studies. In addition, one comparison involved animals that were anaesthetised [c] Downgraded one level for imprecision—only 3 studies of limited size contributed data towards this outcome and no two studies included the same pairwise comparison, disallowing a synthesised pairwise comparison to be obtained. [d] Downgraded two levels for inconsistency. The direction and magnitude of pairwise comparisons varied across the different studies. [e] Downgraded one level for inconsistency. Of the one pairwise comparison that was presented in two separate studies, the same direction of effect was observed. However, multiple other comparisons included in the outcome were regarded as having no difference. [f] Downgraded one level for imprecision. Five adequately powered studies have contributed towards this outcome synthesis and there are multiple studies that provide data towards the same pairwise comparison. However due to the nature of synthesis we have borderline concerns with imprecision. [g] Downgraded one level for imprecision—only 3 studies of limited size contributed data towards this outcome and only one pairwise comparison was included across multiple studies [h] Downgraded two levels for inconsistency. Few different pairwise comparisons were made across studies, of the one comparison that was synthesised across multiple studies the result differs substantially [i] Downgraded one level for imprecision—only 3 small studies have contributed data towards this outcome synthesis [j] There are multiple pairwise comparisons made for this outcome and the direction of effect appears to be consistent. However, the review team believed that the domains of both inconsistency and imprecision were borderline calls and have downgraded one level between the two [k] Downgraded one level for inconsistency. Only one pairwise comparison was observed over multiple studies. However, the direction of that comparison occurred in the same direction across studies. Wide variation in results from other pairwise comparisons across results from included studies [l] Downgraded one level for imprecision—only 3 studies of limited size contributed data towards this outcome. Primary outcome data considered using GRADE are bolded.

3. Results

3.1. Description of Studies

Of the 31 full texts reviewed (Figure 1), 27 articles were eligible for inclusion. The date of publication of these articles ranged from 1995 to 2020, and no relevant studies were identified that were published prior to 1995. Five studies were retrieved through manual searching of the reference list of included studies. The characteristics of the included studies are summarised in Table 2. Observational studies were eligible for inclusion. However, the majority of the studies (25) adopted a randomized controlled study (RCT) design.

Out of the studies included, 78% evaluated more than one blood sampling method and tended to compare outcomes between blood sample methods. Methods evaluated included: 14 (52%) on facial vein sampling, 14 (52%) on the retrobulbar route, nine (33%) on tail incision, seven (26%) on tail tip amputation, five (19%) on the tail vein method, five (19%) using saphenous sampling, five (19%) on the sublingual, and two (7%) on a non-surgical jugular vein route. A further three studies used miscellaneous methods and uncommon routes of phlebotomy, including the use of blood-sucking bugs, a submental route, and puncture of the tail tip. For reporting, we have used consistent terminology in describing the methods. The definitions we propose are reported below.

Effects of serial blood sampling were examined in 16 (59%) of studies. Blood sampling interval varied widely across these studies, ranging from a few minutes to 8 week intervals. This finding creates challenges in comparing these studies to examine serial sampling effects. Only four studies (15%) used both male and female mice. Of these studies, 50% reported statistical analysis of sex difference and incorporated findings in data presentation.

In a number of studies there were associated conditions, which ordinarily might be considered as confounders in study interpretation. These included the use of anaesthesia for sampling, and warming methods for obtaining tail vein samples. Only four out of the 14 studies on the retrobulbar route did not use anaesthesia, whilst 3/5 on the sublingual route were performed conscious. Given that these conditions are regularly used for these methods, and may be mandated by ethics committees, they were considered a part of the method itself and were incorporated in data synthesis. However, where these conditions varied across studies, rendering comparison inappropriate, this has been reported.

Sample quality measures were reported in six (22%) of the studies. Sample quality was not a primary focus of this review and consequently it should be noted that we utilised a restricted definition of quality, mainly focusing on sample haemolysis and clotting. Furthermore, our search was restricted to studies which looked at quality in conjunction with animal welfare outcomes. We may therefore have not identified all published studies evaluating quality of samples via the different routes.

Table 2. Summary of included studies.

Author	Strain	Sex	Age at Intervention	Animals per Group	Study Design	Intervention	Accompanying Conditions	Frequency of Intervention (Occasions)	Comparator	Outcome
Aasland et al. 2010 [22]	C57BL/6JBomTac	M	NR	n = 8	RCT (crossover)	Saphenous vein Tail vein	N/A	4 by both methods at 2 weeks apart	Each other Time series	Plasma glucose Hemolysis
Abatan et al. 2008 [9]	ICR	F	NR	n = 8–13 (unclear from methods)	RCT	Saphenous vein Tail tip amputation	N/A	One-off collection Serial sample at 2–3 day intervals (4)	Each other Time series	Plasma corticosterone Behaviours noted during blood collection
Christensen et al. 2009 [23]	C57BL/6JBom	M	NR	n = 20	RCT	All sample methods were performed at 21 and 30 °C, i.e., 8 experimental groups Retrobulbar Tail incision Tail tip amputation Tail tip puncture	N/A	Serial sample at 30 min intervals (4)	Other groups	Blood glucose Haemolysis Clotting
Durschlag et al. 1996 [24]	ICR	M	9 weeks	n = 8	Case report	Tail incision	N/A	Serial sample at 2–3 day intervals (5)	Time series	Histology Plasma corticosterone
Fernandez et al. 2009 [25]	C57BL/6J	M	6 weeks	n = 10	RCT (crossover)	Retrobulbar Facial vein	Anaesthesia (retrobulbar)	Serial sample at 6–8 week intervals (3)	Each other	Blood glucose Haemolysis
Forbes et al. 2010 [26]	Balb/c	F	6–8 weeks	n = 214	Retrospective case series	Facial vein	Lancet or needle for sampling	Serial sample at 2–7 day intervals (6)	Nil	Mortality
Francisco et al. 2015 [27]	BALB/c	F	5 weeks	n = 20	RCT	Facial vein (lancet or needle)	Facial vein route with anaesthesia as one group	One-off collection	Other groups	Adverse events Gross post-mortem site evaluation
Fried et al. 2015 [28]	C57BL/6N background with a mutation in MDA5	M/F	4–6 months	n = 8	RCT	Retrobulbar	Anaesthesia	One-off collection	Time series at 0, 1, 3, 7 or 14 days after sampling	Clinical scores Histology
Frolich et al. 2018 [29]	C57BL/6NCrl	F	12–14 weeks	n = 12	RCT	Retrobulbar Facial vein	Anaesthesia (retrobulbar)	One-off collection Serial sample at 1 wk intervals (6)	Each otherSingle versus serial	Adverse events Mortality Bodyweight Histology Plasma glucose

Table 2. *Cont.*

Author	Strain	Sex	Age at Intervention	Animals per Group	Study Design	Intervention	Accompanying Conditions	Frequency of Intervention (Occasions)	Comparator	Outcome
Gjendal et al. 2020 [30]	C57BL/6	F	10 weeks	n = 30	RCT	Retrobulbar Facial vein Sublingual	Anaesthesia (retrobulbar) Facial vein and sublingual anaesthesia groups, in addition to conscious sampling	Serial sample at days 8, 9, and 10 (short protocol) and days 8, 15, and 22 (long protocol)	Eachother Single versus serial	Nest build score Faecal corticosteroid metabolites Bodyweight Haemolysis Clotting Gross post-mortem site evaluation
Harikrishnan et al. 2018 [12]	C57BL/6NTac	M/F	6 weeks	n = 12	RCT	Retrobulbar Sublingual Facial vein Tail incision	Anaesthesia (retrobulbar)	Serial sample at 24 h interval (2)	Other groups, plus isoflurane control, and behavioural test control (naïve animals)	Nest build score Elevated plus maze Open field test Faecal corticosteroid metabolites
Heimann et al. 2009 [4]	Crl: CD-1 [CR]	M/F	11 weeks	n = 30	RCT	Sublingual Retrobulbar	Anaesthesia	One-off collection	Each other	Histology
Heimann et al. 2010 [1]	CD1	F	14 weeks	n = 18	RCT	Sublingual Facial vein	Anaesthesia (for sublingual and one facial vein group)	One-off collection	Other groups, time series at 3 h, two or five days after sampling	Bodyweight Food intake Histology Blood glucose
Kim et al. 2018 [10]	CD-1 C57BL/6	M	NR	n = 4–6	RCT	Retrobulbar Tail tip amputation	Anaesthesia (retrobulbar) Restrain/unrestrained	One-off collection Serial sample tail tip groups at 30 min intervals (5)	Other groupsTime series	Plasma corticosterone
Madetoja et al. 2009 [31]	Hsdwin:NMRI	F	9 weeks	n = 10	RCT	Saphenous vein Facial vein Tail vein	Tail warming heat lamp	One-off collection	Other groups, and control with no blood samples	Plasma corticosterone Plasma ACTH
Moore et al. 2017 [32]	C57BL/6J	M	10–12 weeks	n = 8–12	RCT	Facial vein Tail tip amputation Tail incision	N/A	One-off collection	Other groups, and sham submandibular and tail tip amputation (just restraint)	Blood glucose Audible vocalisations Post-procedural epochs of inactivity Grooming behaviour Nest build score Elevated plus maze Open field test Histology

Table 2. *Cont.*

Author	Strain	Sex	Age at Intervention	Animals per Group	Study Design	Intervention	Accompanying Conditions	Frequency of Intervention (Occasions)	Comparator	Outcome
Regan et al. 2016 [33]	CD1	F	12–13 weeks	n = 15	RCT	Retrobulbar Facial vein Submental	Anaesthesia (retrobulbar)	Serial sample at 2 week intervals (3)	Other groups	Adverse events Bodyweight Extraneous blood loss Gross post-mortem site evaluation Haemolysis Clotting
Rogers et al. 1999 [34]	**Study 1** (single sample) C57BL/6	F	10–12 weeks	n = 72	RCT (crossover)	Retrobulbar	Thermostatic warming chamber	One-off collection (each method)	Each other	Plasma glucose
	Study 2 (repeated sample) C57BL/6	F	10–12 weeks	n = 48		Tail incision		Serial sample at 1 week interval (2)		Plasma insulin
Sadler et al. 2013 [7]	**Study 1** (single sample) BALB/CAnNCrl **Study 2** (repeated sample) BALB/CAnNCrl	M M	7–8 weeks 3–4 weeks	n = 5 n = 4/5	RCT	Tail incision Tail vein Tail incision	Thermostatic warming chamber Tail dipped hot water	One-off collection Serial sample at 24 h intervals (3)	Other groups Time series	Plasma corticosterone
Shirasaki et al. 2012 [35]	ICR C57BL/6N	M	5 weeks	n = 10 (unclear from methods)	RCT	Jugular Tail incision	N/A	One-off collection Serial sample at 24 h intervals (5)	Each other Time series	Plasma CRP Plasma corticosterone Plasma haemoglobin Hematocrit Plasma thrombin–antithrombin complexes
Sorenson et al. 2019 [36]	C57BL/6NTac	F	8 weeks	n = 36	RCT	Retrobulbar Saphenous Sublingual Facial vein Tail incision Tail tip amputation	Anaesthesia (retrobulbar)	One-off collection	Each other, plus isoflurane control and naïve animals (no bleeding) Time series at nine timepoints: 6 or 10 h or 1, 2, 4, 6, 8, 10, or 12 days after sampling	Bodyweight Stomach contents Plasma corticosterone Inflammatory gene expression at sample site Plasma inflammatory markers (haptoglobin and IL1ß) Histology

Table 2. *Cont.*

Author	Strain	Sex	Age at Intervention	Animals per Group	Study Design	Intervention	Accompanying Conditions	Frequency of Intervention (Occasions)	Comparator	Outcome
Tabata et al. 1998 [37]	**Study 1** (single sample) B6C3Fl/ICR **Study 2** (repeated sample) B6C3Fl	M/F M/F	8 weeks 9 weeks	n = 12 n = 20	RCT	Tail tip amputation	Tube restraint or anaesthetized with ether or pentobarbital §	One-off collection Serial sample at varied intervals for 24 h (8)	Other groups Time series	Plasma glucose
Teilmann et al. 2014a [2]	BomTac:NMRI	M	6–8 weeks	n = 8–18	RCT	Facial vein Tail vein	N/A	Two day and two night samples within 24 h (4)	Other groups, plus controls (no blood sample and naïve animals as behavioural control)	Bodyweight Plasma corticosterone Faecal corticosteroid metabolites Triple test (elevated plus maze, open field test, light–dark box)
Teilmann et al. 2014 [38]	C57BL/6J	M	5 months	n = 8–12	RCT	Retrobulbar Facial vein	N/A	One-off collection	Other group, plus control (no blood sample)	Bodyweight Plasma corticosterone Histology
Tsai et al. 2015 [8]	BALB/cO1aHsd	F	8 weeks	n = 12	RCT	Jugular Retrobulbar (with and without anaesthesia) Saphenous Facial vein Tail vein	Anaesthesia (retrobulbar group and jugular)	One-off collection	Other groups	In-cage activity Plasma corticosterone Open field test Histology
Tuli et al. 1995 [11]	**Study 1** (acute stress) BALB/c/Ola **Study 2** (tail bleeding recovery) BALB/c/Ola	M M	5–6 months 5–6 months	n = 5 n = 5	RCT	Tail tip amputation	Tail dipped hot water	One-off collection each route Serial humane killing at day 2, 4 and 8 after blood sample	Other groups and control with no tail amputation	Plasma corticosterone Adrenal weight Spleen weight
Voigt et al. 2013 [13]	C57BL/6CrlN	F	4–6 months	n = 36 (16 contributed to final results due to technical failure)	RCT (crossover)	Blood-sucking bug Retrobulbar Tail incision	Anaesthesia (retrobulbar)	One-off collection each route (note crossover design)	Other groups	Faecal corticosteroid metabolites

NR—not reported. §—The aim of this section of the study was to investigate common scenarios, such as anaesthesia, which elevated blood glucose in laboratory mice, rather than investigate the blood sampling technique per se. The reviewers still considered the study was worthy of inclusion since it could provide data on the topic but have only extracted data relevant to the review question. Note: Restraint not included as an accompanying condition unless effects of this specifically investigated as part of study design, since this would be needed for all sample routes.

3.2. Note on Terminology

A variety of sampling routes from the tail were described with little consistency in naming. For the purposes of comparison, we have defined as follows: (1) tail amputation involves the removal of the tip of the tail with a blade, (2) tail incision uses a blade to cut the tail laterally, and (3) tail vein is the targeted collection of blood by insertion of a needle directly into the lateral tail vein.

The submandibular vein, as described in [5], targets the vascular bundle in the caudal part of the jaw. This terminology is commonly used, but perhaps erroneously [39]. The preferred term based on an examination of mouse anatomy is the facial vein [39]. This is the term used in this review, in spite of some usage of 'submandibular' in reviewed studies.

During retrobulbar bleeding, also called retroorbital bleeding, a capillary tube is used to disrupt the retrobulbar venous sinus located behind the eye [8]. Some authors refer to this route as the retroorbital plexus, inferring that a plexus is present in the mouse. This anatomical nomenclature may also be incorrect, although there exists controversy in the claim [39]. We have applied the term 'retrobulbar' in summarising these studies.

The terms saphenous vein, referring to a site near the ankle [8], sublingual for beneath the tongue [4], and jugular, for accessing the vein in the craniocervical region [8] were universally referred to in the studies evaluating them, and are reported as such.

Finally, the submental vein referred to in one study [33] has also been described as a misnomer by an eminent veterinary anatomist, with a suggestion that the site actually targeted was the inferior labial vein [40]. In spite of this, given that only one study reported on this technique, we have continued to refer to this as the submental route.

3.3. Animal Welfare Outcomes

Animal welfare is an umbrella term, defined by the summation of the individual summed experiences of an individual [41]. The nature of affective states experienced by the animal, and their relative weighting over time, typically defines whether an animal has, on balance, good or poor welfare [42,43]. Affective states comprise emotions such as pain, fear and joy [44]. For laboratory animals, the term "cumulative experience" has been coined [45]. Cumulative experience has been defined as: 'the sum of all the events and effects, including their quantity, intensity, recovery between and memory thereof, that impact adversely, positively, and by way of amelioration on the welfare of an animal over its lifetime' [45]. Whilst phlebotomy techniques are generally considered to produce a short-lived response in an animal, given that they are conducted frequently, they may have a significant impact on cumulative experience that can be minimised through appropriate evidence-based selection of sampling routes.

The included studies utilised a range of measures for assessing animal affect to provide an indication as to potential welfare impact. Utilising a combination of physiological and behavioural methods is generally regarded as superior in the holistic measurement of welfare state [46,47]. A short description of the main outcomes considered in this review follows. These measures have been categorised into (1) measures of physiology, (2) clinical and pathological parameters and (3) behavioural measures and (4) blood sample quality measures.

3.4. Physiological Measures

The major physiological measures investigated were those reflecting fear or arousal, via the hypothalamic–pituitary–adrenal (HPA) axis response [2,7–11,13,24,31,35,36,38] and the associated release of stored glucose [1,22,23,25,29,32,34,37]. Stress almost invariably activates the HPA axis, which, via a sequence of steps, leads to glucocorticoid production—the principle rodent glucocorticoid being corticosterone. Corticosterone serves to regulate glucose, allowing for its release from reserves in the short term, and carrying out other metabolic actions with the goal of establishing homeostasis [48]. Typically, blood serum or plasma is used to measure corticosterone [49] or glucose concentrations [50], with an increase implying an acute stress response. Measurement of faecal corticosterone metabolites (FCM) has recently been proposed due to the advantage of being non-invasive. FCM provide a retrospective measure of the HPA axis response with lag time from peak in blood to faecal excretion being 9 h in mice [51], thus reflecting sampling method and subsequent recovery [2].

3.5. Clinical and Pathological Parameters

General health parameters such as fur condition, inactivity or dehydration status are commonly implemented in severity score sheets for rodent experiments. A change in these parameters can provide an indication of reduced welfare or disease but may not on their own be a sensitive indicator of a stress response. Studies evaluating welfare typically combine these measures with other physiological or behavioural parameters, as occurred in the included studies. Procedure-specific adverse events were important to include as outcome measures, since these may have considerable impact on individual welfare. Furthermore, an increased incidence of such events would prevent recommendation of a technique for practice. Events considered in the studies included haemorrhage from the ear and nares, ocular lesions, circling and convulsions.

Mortality rate is a commonly used indicator providing a retrospective measure of welfare, since it may be influenced by disease, trauma or environmental problems [52].

Other measures utilised are arguably less indicative of animal affective state, but may provide insight by inference—for example, the quantification of tissue damage through post-mortem or histology [1,4,8,24,27–29,32,33,36,38] may imply pain or loss of function.

3.6. Behavioural Measures

Behavioural outcomes were widely reported and ranged from evaluation of spontaneous behaviour, such as eating (ascertained through bodyweight), vocalisation or inactivity, through to the use of well-established behavioural tests of anxiety or affective state [2,8,12,30,32].

Bodyweight loss can result from a variety of causes all of relevance to well-being. These include disease, poor or lack of nutrition, as well as eating behaviour which is potentially compromised by a stress response or trauma associated with a procedure. However, stable bodyweight does not necessarily imply that well-being is not impaired, or even positive in nature [53]. Whilst the use of bodyweight alone as a measure of welfare is fairly crude and non-specific, it is a commonly used surrogate parameter for welfare [54].

Nest building is a spontaneous behaviour that has been proposed to represent a 'luxury' behaviour which is highly motivated but non-essential in the laboratory [55]. As such these behaviours are generally the first to be reduced during times of stress [44]. Therefore, poor or reduced performance in this behaviour may indicate a reduction in well-being [55].

Elevated plus maze is used as an assay for anxiety-related behaviour, and typically utilises several different outcome measures to ascertain level of anxiety behaviour, with the general presumption being that increased open arm activity is anti-anxiety behaviour [56]. Open arm activity can be measured as the number of entries or duration. A range of other parameters are frequently collected in this test, including velocity in open and closed arms and distance covered. These are, however, typically a measure of locomotor activity rather than anxiety per se [57].

The open field test is used to gather information on ambulation and emotionality [58]. Ambulation or activity can be measured using total distance travelled in the test. Thigmotaxis is used as a measure of anxiogenic behaviour, with thigmotaxis increasing as anxiety increases. This is typically measured through entries into the central zone or time spent in the centre versus the periphery [58].

3.7. Blood Sample Quality Measures

Based on our restricted definition of measures of blood sample quality, measures of hemolysis and clotting were considered in our synthesis. Hemolysis is the most common pre-analytical sources of error in clinical laboratory and generally leads to sample rejection and the need for blood re-draw.

The finding therefore has animal welfare as well as experimental implications. Furthermore, invisible hemolysis can lead to discharge of cell constituents and false results [59]. Clotting may occur where the blood is slow to fill the collection container, or when considerable manipulation of the vein by the needle has occurred. The presence of clot can therefore give a good indication of the ease with which sample can be collected via the particular route but may also be lessened by operator experience [60]. These samples are unable to be analysed for most laboratory tests.

4. Methodological Quality

Overall, the risk of bias of the included studies was high. Often, the nature of the intervention precluded adequate blinding of the operators involved in the blood taking procedures. Whilst it was discussed by the review team that this would be inherent in the included studies, it would still warrant a rating of a "high" for this particular domain. Table 3 details the risk of bias assessments for each domain for included studies, whilst Figure 2 displays the percentage of studies that achieved either a low, unclear or high risk of bias for each domain. See Supplementary Table S2 for full reviewer judgment for the assessment of methodological quality.

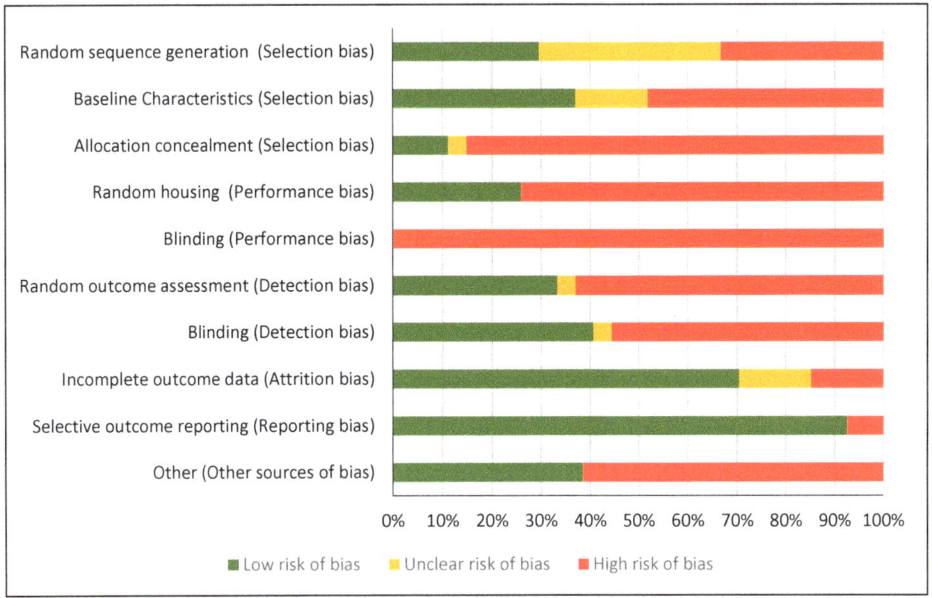

Figure 2. Percentage of studies that achieved either a low, unclear or high risk of bias for each domain using the SYstematic Review Centre for Laboratory animal Experimentation (SYRCLE) risk of bias tool. Although randomisation was mentioned in several articles, lack of reporting of the method used resulted in an unclear risk of bias for most items. Blinding was impossible to achieve due to the inherent nature of the study design.

Table 3. Risk of bias assessments for each domain for included studies. L: low risk of bias, U: unclear risk of bias, H: high risk of bias.

Study	Random Sequence Generation	Baseline Characteristics	Allocation Concealment	Random Housing	Blinding	Random Outcome Assessment	Blinding	Incomplete Outcome Data	Selective Outcome Reporting	Other
Aasland et al. 2010 [22]	L	L	H	L	H	L	L	L	L	H
Abatan et al. 2008 [9]	H	H	H	H	H	H	H	H	L	L
Christensen et al. 2009 [23]	L	L	H	H	H	H	L	H	L	H
Durschlag et al. 1996 [24]	H	H	H	H	H	H	H	H	L	H
Fernandez et al. 2009 [25]	L	H	H	L	H	L	L	H	L	L
Forbes et al. 2010 [26]	H	U	H	H	H	H	H	L	H	H
Francisco et al. 2015 [27]	L	L	L	H	H	L	H	L	L	H
Fried et al. 2015 [28]	U	H	H	H	H	H	H	L	L	L
Frolich et al. 2018 [29]	U	H	H	H	H	H	H	L	L	L
Gjendal et al. 2020 [30]	L	L	H	L	H	H	L	L	L	L
Harikishnan et al. 2018 [12]	L	L	H	L	H	L	H	L	L	L
Heimann et al. 2009 [4]	U	H	H	H	H	H	H	L	L	H
Heimann et al. 2010 [1]	U	L	H	H	H	H	H	L	L	H
Kim et al. 2018 [10]	H	U	H	H	H	L	L	U	L	L
Madetoja et al. 2009 [31]	H	L	U	H	H	H	H	L	L	H
Moore et al. 2017 [32]	U	H	H	L	H	H	H	L	L	H
Regan et al. 2016 [33]	H	H	H	H	H	H	H	U	L	H
Rogers et al. 1999 [34]	L	H	H	H	H	L	L	L	L	H
Sadler et al. 2013 [7]	H	U	H	H	H	L	L	U	L	L
Shirasaki et al. 2012 [35]	H	H	H	H	H	H	L	U	L	H
Sorenson et al. 2019 [36]	U	H	H	H	H	H	H	L	L	H
Tabata 1998 [37]	H	H	H	H	H	L	L	L	H	H
Teilmann et al. 2014a [2]	U	L	L	H	H	H	H	L	L	L
Teilmann et al. 2014b [38]	U	L	L	L	H	U	U	L	L	L
Tsai et al. 2015 [8]	U	L	H	H	H	H	H	L	L	H
Tuli et al. 1995 [11]	U	H	H	H	H	H	L	L	L	L
Voigt et al. 2013 [13]	L	U	H	L	H	L	L	L	L	H

4.1. Selection Bias

There was a high risk of selection bias for the majority of the included studies. Only one study was at a low risk of selection bias for all three signalling questions used [27]. The generation of an adequate randomisation sequence was reported in only 8 studies [12,13,22,23,25,27,30,34], while 10 studies [1,2,4,8,11,28,29,32,36,38] stated that randomisation took place but failed to report the actual methods used, thus being assigned as "unclear". For the remaining studies, randomisation did not take place, or was not mentioned. Baseline characteristics were identified to be similar between groups for only 10 included studies [1,2,8,12,22,23,27,30,31,38], and only 3 [2,27,38] studies reported adequate methodological details on the use of appropriate and proper allocation concealment.

4.2. Performance Bias

There was a high risk of performance bias across all included studies. Due to the design of the studies reviewed and the nature of the blood sampling interventions employed, blinding was not achieved, nor possible. In addition, only seven included studies [12,13,22,25,30,32,38] employed a method of random housing, or utilised a methodology which reduced the possibility of cage associated biases.

4.3. Detection Bias

Only 7 [7,10,13,22,25,34,37] of the 27 included studies were recorded as having a "low" risk for both signalling questions used to ascertain detection bias. Whilst blinding of the outcome assessor was not necessarily confirmed for these studies, the majority have been considered to be at low risk as they utilised objective, biochemically validated outcome measures. Of the studies that were of high risk due to the lack of blinding of the outcome assessor, it was mainly due to the inclusion of outcomes such as behavioural assessment that required subjective judgment on the part of the outcome assessor. These were considered to have been at a high risk.

4.4. Attrition Bias

Overall, there was a low risk of attrition bias for the included studies. Only four [9,23–25] studies were at high risk of attrition bias. One study did not analyse every animal data point [25], inappropriately removing outliers from the reported analysis. Two studies [23,24] experienced loss to follow-up due to failures in the blood collection experimental design. These data were not adequately discussed or analysed in the reported results. Finally, [9] failed to report the number of animals randomised to each group, and therefore, loss to follow-up could not be appropriately assessed. Additionally, four studies [7,10,33,35] were reported as having an unclear risk of attrition bias due to a lack of consistent reporting of animal numbers.

4.5. Reporting Bias

Only two studies [26,37] were at high risk of reporting bias. In one of these studies, behavioural observations were stated to be an outcome of interest [26]. However, no results on this outcome have been reported in relation to the bleeding technique. Meanwhile, the reporting in another study [37] was considered to be poor quality, with animal numbers not provided for each treatment group, making assessment of reporting bias difficult.

4.6. Other Bias

The only other potential source of bias was the non-disclosure of a funding or supporting body, or lack of conflict of interest. Fourteen included studies were at a high risk of bias due to this factor [1,4,8,13,22–24,27,31–35,37]. One study was also considered to be at a high risk of an additional bias, as it was a retrospective review of records from a separately reported, 3 year study [26].

5. Effects of the Interventions

Seven outcomes arising from the included studies were synthesised using vote counting based on direction of effect. These outcomes were plasma glucose (single/serial sample), plasma corticosterone (single/serial sample), facial corticosterone, bodyweight change (single/serial sample), nest building hemolysis and clotting. A summary of the results of this synthesis is provided in the Summary of Findings table (Table 1). The full analysis and reporting of these outcomes are provided in Supplementary Table S3. Remaining outcomes investigated in the studies are presented narratively below.

5.1. Mortality

Mortality, expressed as a% of study sample based on conversion from the absolute figures, was presented in two studies [26,29]. Both of these studies examined the facial vein route, with [29] additionally comparing with the retrobulbar sample route. These studies evaluated serial sampling at intervals of up to one week. Frolich et al. 2018 also made comparisons with single-sample groups. It is worth noting that the study by Forbes et al. 2010 [26] was a retrospective case-control study with some variation in blood sampling interval and a concurrent study design, potentially confounding interpretation of findings. Due to a sole pairwise comparison both single and serial sampling comparison have been synthesised narratively and are not supplemented with a table.

In a retrospective study [26] with facial vein sampling performed serially, a mortality rate of 4/214 mice was observed (≈2%). Frolich et al. 2018 [29] reported a substantially higher mortality rate of (4/12, 33%) when mice were serially sampled by the facial vein route. There was no associated mortality with the single-sample facial vein route or retrobulbar routes, or serial retrobulbar sampling [29]. Both studies utilised a similar sampling interval of approximately 1 week.

5.2. Adverse Events

Clinical signs or adverse events were considered in four studies [27,29,30,33]. These covered all three of the large sample methods, that of retrobulbar, sublingual and facial vein sample. Regan et al. 2016 also studied submental sampling [33]. There was considerable heterogeneity in the types of adverse events reported which ranged from numbers of repeat attempts at sampling to instances of haemorrage from the site. Events also ranged in severity from mild, such as corneal opacity, to life-threatening. This finding implied that simple addition of incidences of event would provide a biased picture. Furthermore, some adverse events were clearly specific to the sample location, for example ocular lesions or ear canal hemorrhage rendering direct comparison non-meaningful. For this reason, results have been summarised narratively.

The number of punctures needed to obtain a sufficient blood sample was significantly less for retrobulbar bleeding (1.03 punctures), compared to facial vein (1.45) and sublingual (1.31), ($p < 0.001$ in all comparisons) [30]. Sublingual puncture caused haemorrhage from the nares in 3.33% of mice. Interestingly this rate was increased after the use of anaesthesia to 10.56% of mice [30].

Clinical signs following facial vein bleeding were evaluated in three studies [27,29,30]. Signs included inactivity after collection and being unsteady on release. These occurred at a frequency of 1/20 animals (5%) [27]. The more serious adverse effect of haemorrhage from the ear canal occurred at a rate of 2/20 (10%) [27]. These adverse events occurred when using needle, rather than lancet puncture [26]. Frolich et al. 2018 [29] similarly reported inactivity, ear and nose haemorrhage, as well as head tilt, convulsions, circling and corneal opacity, at rates of approximately 17–25%, but only with serial facial vein samples. Rates of haemorrhage from the ear canal were 2.78%, with an approximate doubling when anaesthesia was used (5.56%) in the Gjendal et al. 2020 study [30].

The most common adverse event reported with retrobulbar sampling was corneal opacity and periocular tissue prolapse occurring at a rate of 2/12 animals (17%) [29]. The incidence of corneal opacity was increased in serial RBB to 5/12 (42%) animals [29]. However, no ocular abnormalities were observed after use of this sampling route in the Gjendal et al. 2020 study [30].

In contrast to other studies Regan et al. 2016 [33], reported no adverse signs after retrobulbar, facial or submental bleeding in a cohort of 15 per group. Minor inflammation was noted at the point of capillary tube insertion in the retrobulbar group but this resolved quickly.

5.3. Histology

Nine studies evaluated histological findings [1,4,8,24,28,29,32,36,38]. Only three of these studies evaluated small vein sample routes [24,32,36]. There was significant heterogeneity in method of reporting which ranged from narrative summary [8,24,32,38], to incidence [1,4,29,36], to a semi-subjective scoring system [28]. There were further differences since lesions were observed in different anatomic regions as would be expected based on sampling location. For example, ocular trauma was reported in retrobulbar sampling, yet was not seen after facial sampling. Similarly, foreign body steatitis is typically caused by hair shaft penetrance of the area and is less likely to occur when sampling hairless regions. In assimilating findings a judgement call has to be made as to whether the increased incidence of a histological finding implies greater welfare impact, or whether this is constituted by greater severity of lesion, or combination thereof. For these reasons, it was considered that vote counting was inappropriate and results have been summarised narratively (Figure 3).

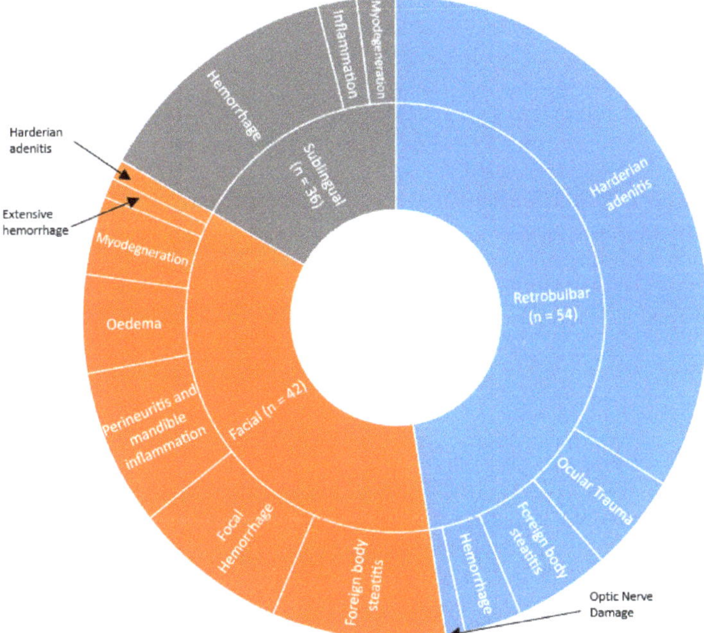

Figure 3. Histological characteristics observed in studies comparing large-volume sample sites. Size of slice represents relative incidence. Difference in reporting prevented assimilation of all findings. Compiled from [1,4,29,32,38]. Note that animals may have demonstrated more than one histological finding. *n* represents the number of animals.

Retrobulbar sampling consistently led to microscopic evidence of haemorrhage [4,8,28,36] in structures around and within the orbit, including the muscle, retroorbital sinus, harderian gland and nasolacrimal duct. Inflammatory infiltrate with constitute cells changing over time post-sample was also a key feature [4,8,28,36]. Occasional broken hair shafts setting up a foreign body reaction were reported [28,29,36,38]. Optic nerve damage appears to be a rare finding [4]. Massive necrosis, mononuclear cell infiltration, and fibroplasia of the harderian gland was observed in 2/18 animals [36].

Jugular vein sampling led to histological changes such as haemorrhage, inflammatory infiltrate, degenerative change, and oedema in muscle, subcutaneous connective and adipose tissue [8].

In facial-punctured animals, macroscopic observations were characterized by subcutaneous haemorrhage and oedema with an acute inflammatory response [1,36]. Focal muscle necrosis was also observed [1]. Signs of trauma persisted for five days [1]. When compared together, sublingual sampling led to fewer traumatic lesions than facial [1]. Lesions in the former were characterized by minimal to slight haemorrhage, with minimal acute inflammation [1,36]. Scar formation and production of granulation tissue was occurring after 5 days [1]. Trichogranuloma has been observed with both of these methods [2,36].

Anaesthesia appears to impact on severity and type of lesions noted. Comparatively fewer indicators of histological change were observed in retrobulbar sampling with anaesthesia compared to the conscious method [8]. In facial vein-sampled animals a diffuse acute neutrophilic/fibrinoid inflammatory response was noted after conscious sampling, whereas the inflammatory response was more chronic in anaesthetized animals [1].This difference in response was likely related to the presence of hair fragments deep within the puncture site in a number of animals [1].

Histology findings following tail sampling were generally mild. Mild neutrophilic inflammation was a consistent finding in mice sampled by tail incision [24,36], which also extended to the dermis in some animals [32]. In a proportion (3/5) of animals, tail amputation resulted in transection of the last caudal vertebra [32]. Additionally all of these animals had neutrophilic inflammation (generally mild) and fibrin at the tail tip [32]. In contrast to tail tip amputation, tail incision led to a shorter period of epidermal oedema, tail muscle involvement with necrosis and inflammatory infiltrate, and earlier proliferation of fibroblasts [36]. The inflammatory infiltrate progressed from neutrophilic to mononuclear cell in both groups over time [36]. Lesions following tail vein sample were primarily in the subcutaneous tissue and adipose tissue [8].

The incidence of histological change was higher after saphenous sample than tail vein (29.4% vs. 13.97% after 1 h) and lesions in the muscle were reported [8]. Alternately, minimal histological change, characterised by minor inflammatory infiltrate and bleeding into the muscle, was observed after saphenous puncture in another study [36]. The authors commented that this finding may have resulted from imperfect tissue sampling.

5.4. Behavioural Tests of Anxiety

5.4.1. Elevated Plus Maze (EPM)

Three studies evaluated EPM performance after blood sampling [2,12,32]. Harikrishnan et al. 2018 [12] investigated all three major blood sampling routes, as well as tail incision. Whilst Moore et al. 2017 [32] examined tail amputation and tail incision in comparison with facial vein sampling. Teilmann et al. 2014 [2] compared facial and tail vein sampling using an EPM as part of a triple test. A further consideration is that the study by Harikrishnan et al. 2018 [12] utilised anaesthesia for retrobulbar sampling, but all other routes were performed conscious in this study and in other studies. This factor may confound study interpretation in spite of the behavioural testing occurring 24 h after sampling. As a result of the differences in experimental design, vote counting was considered inappropriate and results have been summarised narratively. A summary findings table (Table 4) for both the EPM and OFT behavioural test is provided below.

Table 4. Summary of direction of effect for behavioural tests elevated plus maze (EPM) and open field test (OFT) when compared to sham/unmanipulated controls or baseline values.

Single-Sample Method and Reference	Behaviour Test	Measure	General Direction of Effect on Measure	Timeframe for Measure
		Retrobulbar		
Harikrishnan et al. 2018 [12]	EPM	Anxiety	↑	24 h
	OFT	Anxiety	↑	24 h
	OFT	Locomotor activity	↓	24 h
		Facial Vein		
Harikrishnan et al. 2018 [12]	EPM	Anxiety	=	24 h
	OFT	Anxiety	=	24 h
	OFT	Locomotor activity	=	24 h
Moore et al. 2017 [32]	EPM	Anxiety	=	Few hours post-procedural
	OFT	Anxiety	=	Few hours post-procedural
Teilmann et al. 2014 [2]	Triple Test	Anxiety	↑	24 h
		Sublingual		
Harikrishnan et al. 2018 [12]	EPM	Anxiety	↑	24 h
	OFT	Anxiety	=	24 h
	OFT	Locomotor activity	↓	24 h
		Tail Incision		
Harikrishnan et al. 2018 [12]	EPM	Anxiety	↑	24 h
	OFT	Anxiety	=	24 h
	OFT	Locomotor activity	=	24 h
Moore et al. 2017 [32]	EPM	Anxiety	=	Few hours post-procedural
	OFT	Anxiety	=	Few hours post-procedural
		Tail Vein		
Teilmann et al. 2014 [2]	Triple Test	Anxiety	↓	24 h
		Tail Amputation		
Moore et al. 2017 [32]	EPM	Anxiety	=	Few hours post-procedural
	OFT	Anxiety	=	Few hours post-procedural

In Harikrishnan et al. 2018 [12], the groups spent significantly different durations in the open arms ($p = 0.03$), yet there were no differences in time spent in closed arms or the number of open-arm visits. The groups also differed in the number of centre visits ($p = 0.002$). Facially vein-punctured animals did not differ from controls in these parameters, whereas retrobubular sampling caused greatest deviation from controls with reduced centre visits, least time in the centre and reduced activity. Sublingual and tail incision led to intermediate deviations from control values, frequently exhibiting more anxiety behaviour than facially sampled animals, but not differing from each other.

Moore et al. 2017 examined facial, tail amputation and tail incision routes to find that phlebotomy group did not affect performance in the elevated plus maze [32].

5.4.2. Open Field Test (OFT)

Four studies evaluated OFT performance after blood sampling [2,8,12,32]. Harikrishnan et al. 2018 [12] investigated all three major blood sampling routes, as well as tail incision. Whilst Moore et al. 2017 [32] evaluated tail amputation and tail incision, in comparison with facial vein sampling. Teilmann et al. 2014 [2] compared facial and tail vein sampling using an OFT as part of a triple test. Tsai et al. 2015 [8] studied retrobulbar (with and without anaesthesia), facial, tail vein, and saphenous bleeding.

Tsai et al. 2015 [8] measured total distance travelled in the test which is an important measure of locomotor activity. Distance travelled differed between groups being longest in mice that underwent tail vein bleeding (2.183 cm) or saphenous (2.110 cm), followed by mice that underwent retrobulbar (1.699 cm) and facial vein bleeding (1.226 cm). This difference was significant for facially vein-sampled animals, but non-significant for other pairwise comparisons. Facial vein-sampled animals also had a lower average speed. This study has not been included in the summary table because values were not compared to a sham/unmanipulated control to enable a determination on direction of effect.

The study by Harikrishnan et al. 2018 [12] utilised anaesthesia for retrobulbar sampling but all other routes were performed conscious in this and in other studies. This may be a consideration in interpretation, in spite of testing occurring 24 h after sampling.

Mice subjected to retrobulbar sinus puncture were significantly less active than control mice for all three OFT parameters in Harikrishan et al. 2018 [12]. This was not observed for mice that underwent tail incision or facial vein puncture. An effect of anaesthesia was apparent since mice subjected only to isoflurane anaesthesia showed greater activity, and a higher number of centre entries, than mice subject to retrobulbar puncture under isoflurane anaesthesia [12]. This contrasts with the Tsai et al. 2015 study [8], where mice that underwent retrobulbar bleeding with anaesthesia performed similarly in the OFT to mice sampled without anaesthesia.

There were no differences in display of anxiety behaviour between facial vein-sampled, tail incision, sublingual and retrobulbar groups, as indicated by visits and time spent in the centre of the field [12]. Alternately, mice sampled by facial vein generally avoided the open field in the triple test with the authors concluding that they expressed more anxious behaviour than tail vein-sampled animals [2].

In Moore et al. 2017 [32], the facial vein group exhibited a significantly lower average speed in the OFT compared to tail amputation and incision groups. No other between-group differences were observed.

6. Discussion

As common techniques performed in biomedical research studies involving mice, blood sampling methods may have considerable impact on cumulative experience. Furthermore, retrobulbar sampling has been considered controversial and its use is vetoed in some laboratories. Whilst summary documents or guidelines exist on this topic, a number of these are now older documents and it is not clear whether they have been based on a systematic review of the available evidence (see, e.g., [3,61]). This review represents the first systematic review on recovery blood sampling techniques in mice based on their impact on animal welfare and sample quality.

6.1. Impact of Blood Sample Route on Mouse Welfare

Whilst there is a substantial body of evidence on the impact of blood sampling on animal welfare with 27 studies sourced, the heterogeneity in terms of sampling routes compared, and outcomes measured, renders it problematic to make any recommendation. Despite the large number of studies, few performed the same pairwise comparisons, with the same outcome measures. We were also unable to perform any assimilation of the behavioural data given the heterogeneity. This is unfortunate, as arguably these measures may provide a better measure of well-being and animal impact than a measure of the short-term stress response. Given the above caveat, some general points taken from the assimilation are presented.

For serial blood sampling, as a general rule, small-volume sampling routes may be beneficial over large-volume sampling routes based on the findings for glucose, plasma corticosterone and bodyweight (Table 1). This would seem plausible based on physiological principles given the reduced blood volume lost, and therefore reduced chance of hypotension, haemorrhagic shock and impaired tissue and organ metabolism that can result [62].

Whilst it might be expected that head-focused routes would have a greater impact on bodyweight than methods focused at the tail and leg, this was only borne out for the facial and retrobulbar routes. The finding that bodyweight loss did not occur with sublingual sampling (data from two studies) is intriguing given the mouth focus, and requires further study to confirm.

It is often assumed that anaesthesia improves animal well-being following procedure performance, yet the findings are inconsistent. For example, beneficial effects of isoflurane use were found, with reduced anxiety being observed in the OFT [12] and decreased histological lesion severity after retrobulbar and facial sampling [1,8]. However, anaesthesia caused a doubling in the incidence of the serious adverse effect of ear bleeding after facial vein bleeding, and a tripling in rate of bleeding from the nares after sublingual puncture as observed in [30]. Perhaps the best current advice in this regard is to determine anaesthetic use on a case-by-case basis, dependent on the blood sampling route, and the researcher's level of technical expertise and comfort with the procedure. These apparent discrepancies also need future targeted research focus.

The difference in serial mortality rate observed for the facial vein route between the studies of [26,29] (2% vs. 33%) is striking. This is especially so since the technique is widely used in laboratory animal practice, and this mortality rate would likely raise ethical concerns for continued performance amongst many ethics committees. These results may be an artefact of the small sample sizes employed in the Frolich et al. 2018 study, with a potential effect of learning/re-familiarisation. Mortality rates may have declined as the skill was reacquired with larger numbers of animals sampled. Only one of the included studies specifically investigated the effect of experience on outcomes measured [28]. Based on pathological findings, findings from this study were that experience with retrobulbar sampling had little impact on outcome. However, it is suggested that this effect may have been overlooked in the included studies, and deserves further research attention, and an agreed criterion to standardise expertise level.

As a final point, based on the evidence available, the reason for the demise of the retrobulbar route for ethical reasons remains unclear. The synthesis implies that it is associated with zero mortality [29], none [30,33] to mild clinical ocular abnormalities [29], and similar severity of histological lesions to other large-volume sampling routes [36,38]. This change in policy direction is more surprising given that this route has generally been replaced by facial vein sampling, which arguably can lead to a similar number, if not more, potentially negative outcomes [27,29,33,38]. It is speculated that pictures, or dialogue showing the horrendous (likely rare) outcome of globe perforation after sampling, may have influenced decision making in this regard. Perhaps a more appropriate focus for policy makers should be how to best train and ensure operator competency in this technique in order to avoid the occurrence of serious adverse effects. It is certainly by no means clear whether the proposed alternative facial route is more beneficial for animal well-being and this should be a priority for future research.

6.2. Evidence Completeness and Quality and Recommendations for Future Research

This review identifies a number of factors preventing recommendations on choice of mouse blood sampling route being made. These include (1) that in spite of a reasonable number of studies on the topic, there were often few studies examining the same pairwise comparison; (2) there was a lack of standardised outcome measures relevant to well-being and timepoints for comparison; (3) many studies were at high risk of bias, either by virtue of study design or deficiencies in reporting. When considering that there have been significant and repeated recent efforts to improve the reporting standards of animal research, and that guidelines to assist animal research have been widely available since 2010 [63–66], the overall poor quality of reporting of the included studies is problematic. Simple details prescribed by the ARRIVE (Animal Research: Reporting of In Vivo Experiments) guidelines, such as description of the randomisation procedure, were only reported by three studies published after 2010 (of 16 total). This highlights how far animal-based research needs to come to be considered rigorous, transparent and transferable.

Generally, RCTs are desirable due to their placement in the hierarchy of evidence. However, given the practical and widespread nature of these interventions, and therefore the importance of external validity, it may be desirable to investigate these techniques using large-scale, well-designed observational studies. This may avoid issues alluded to earlier, where small sample sizes may lead to artefacts. Alternatively, RCTs that are multicentre in nature to increase sample size could be performed. These studies may be able to reduce confounding by issues such as technique experience as part of a dilution effect due to the use of many operators.

One of the major limitations of this systematic review was the inability to perform a meta-analysis, which is the ideal method to synthesise and present data from multiple, comparable studies. However, due to the heterogeneity that existed between studies, performing a meta-analysis was never considered to be appropriate. While this factor has severely limited the impact of our results in terms of making a determination on the effects of blood sample route on animal welfare, it does provide us some insight as to how a meta-analysis may be facilitated for future systematic reviews on this topic. Ideally, the relevance and reproducibility of outcome measures used to assess welfare in adult mice should be validated and discussed in the field, so that consensus may be reached, and experiments standardised accordingly. However, by far the most common barrier to performing a meta-analysis encountered was the underreporting of results by authors of the primary studies. Often, authors simply reported their results as a figure or a graph, and a statement of (non-)significance. While this may suit the purposes of the primary author in confirming or rejecting their null hypothesis, this underreporting greatly reduces the transferability and comparability of these data. A simple solution to this problem would be for scientific journals to start mandating that submitting authors provide complete data sets. This tactic is already being employed by international journals such as PLoS One, Springer Nature and Science [67–69] and may reduce the need to resort to alternate, less robust data synthesis strategies [70]. However, until this becomes a standard, as it has in clinical research, this issue will continue to plague animal-based research.

7. Conclusions

In spite of a substantial body of evidence investigating welfare associated with blood sampling techniques in mice, it was concluded that there was not enough, high-quality evidence to make any recommendations on the optimal method of blood sampling from the point of view of animal welfare. Future high-quality studies, with standardised outcome measures and large sample sizes, are required.

There is an urgent need, as highlighted by many authorities, to increase quality (and/or reporting) of animal research at all stages from inception to reporting. The use of guidelines such as those published by ARRIVE [64], and protocol registration, can assist in achieving this. Journal editors also need to advise researchers of guidelines and enforce provisions, which will no doubt serve as an educative as well as compliance function.

Supplementary Materials: The following are available online at http://www.mdpi.com/2076-2615/10/6/989/s1, Search strategy, Table S1; risk of bias determinations, Table S2; Results of vote counting based on direction of effect, Table S3; Full extraction templates, Table S4.

Author Contributions: Conceptualization, A.L.W.; methodology, A.L.W. and T.H.B.; data curation, A.L.W. and T.H.B.; writing—original draft preparation, A.L.W. and T.H.B.; writing—review and editing, A.L.W. and T.H.B.; project administration, A.L.W. and T.H.B. All authors have read and agreed to the published version of the manuscript.

Funding: This review received no external funding. A.W. is supported by an Australian Government, NHMRC Peter Doherty Biomedical Research Fellowship (APP1140072).

Conflicts of Interest: The author A.W. was an author on one of the included studies. [26] T.B. made the final call on decisions relating to inclusion of this article and the risk of bias determination. No other potential conflicts exist.

References

1. Heimann, M.; Roth, D.R.; Ledieu, D.; Pfister, R.; Classen, W. Sublingual and submandibular blood collection in mice: A comparison of effects on body weight, food consumption and tissue damage. *Lab. Anim.* **2010**, *44*, 352–358. [CrossRef] [PubMed]

2. Teilmann, A.; Kalliokoski, O.; Sørensen, D.B.; Hau, J.; Abelson, K.S. Manual versus automated blood sampling: Impact of repeated blood sampling on stress parameters and behavior in male nmri mice. *Lab. Anim.* **2014**, *48*, 278–291. [CrossRef] [PubMed]

3. Diehl, K.H.; Hull, R.; Morton, D.; Pfister, R.; Rabemampianina, Y.; Smith, D.; Vidal, J.M.; van de Vorstenbosch, C. A good practice guide to the administration of substances and removal of blood, including routes and volumes. *J. Appl. Toxicol.* **2001**, *21*, 15–23. [CrossRef] [PubMed]

4. Heimann, M.; Käsermann, H.P.; Pfister, R.; Roth, D.R.; Bürki, K. Blood collection from the sublingual vein in mice and hamsters: A suitable alternative to retrobulbar technique that provides large volumes and minimizes tissue damage. *Lab. Anim.* **2009**, *43*, 255–260. [CrossRef] [PubMed]

5. Golde, W.T.; Gollobin, P.; Rodriguez, L.L. A rapid, simple, and humane method for submandibular bleeding of mice using a lancet. *Lab. Anim. (New York)* **2005**, *34*, 39–43. [CrossRef] [PubMed]

6. Popesko, P.; Rajitová, V.; Horák, J. *A Colour Atlas of the Anatomy of Small Laboratory Animals*; Wolfe Publishing: London, UK, 1992.

7. Sadler, A.M.; Bailey, S.J. Validation of a refined technique for taking repeated blood samples from juvenile and adult mice. *Lab. Anim.* **2013**, *47*, 316–319. [CrossRef] [PubMed]

8. Tsai, P.-P.; Schlichtig, A.; Ziegler, E.; Ernst, H.; Haberstroh, J.; Stelzer, H.D.; Hackbarth, H. Effects of different blood collection methods on indicators of welfare in mice. *Lab. Anim. (New York)* **2015**, *44*, 301–310. [CrossRef]

9. Abatan, O.I.; Welch, K.B.; Nemzek, J.A. Evaluation of saphenous venipuncture and modified tail-clip blood collection in mice. *J. Am. Assoc. Lab. Anim. Sci.* **2008**, *47*, 8–15.

10. Kim, S.; Foong, D.; Cooper, M.S.; Seibel, M.J.; Zhou, H. Comparison of blood sampling methods for plasma corticosterone measurements in mice associated with minimal stress-related artefacts. *Steroids* **2018**, *135*, 69–72. [CrossRef]

11. Tuli, J.; Smith, J.; Morton, D. Corticosterone, adrenal and spleen weight in mice after tail bleeding, and its effect on nearby animals. *Lab. Anim.* **1995**, *29*, 90–95. [CrossRef]

12. Harikrishnan, V.; Hansen, A.K.; Abelson, K.S.; Sørensen, D.B. A comparison of various methods of blood sampling in mice and rats: Effects on animal welfare. *Lab. Anim.* **2018**, *52*, 253–264. [CrossRef] [PubMed]

13. Voigt, C.; Klockner, P.; Touma, C.; Neuschl, C.; Brockmann, G.; Goritz, F. Hormonal stress response of laboratory mice to conventional and minimally invasive bleeding techniques. *Anim. Welf.* **2013**, *22*, 449–455. [CrossRef]

14. SYstematic Review Centre for Laboratory Animal Experimentation (SYRCLE) Database for Animal Intervention Studies. Available online: https://drive.google.com/file/d/1O7KD-1vuzlM-k56wD7--XpGKBveAUhH-/view (accessed on 5 June 2020).

15. Munn, Z.; Aromataris, E.; Tufanaru, C.; Stern, C.; Porritt, K.; Farrow, J.; Lockwood, C.; Stephenson, M.; Moola, S.; Lizarondo, L.; et al. The development of software to support multiple systematic review types: The Joanna Briggs institute system for the unified management, assessment and review of information (jbi sumari). *Int. J. Evid. Based Healthc.* **2019**, *17*, 36–43. [CrossRef] [PubMed]

16. Moher, D.; Liberati, A.; Tetzlaff, J.; Altman, D.G.; The, P.G. Preferred reporting items for systematic reviews and meta-analyses: The prisma statement. *PLoS Med.* **2009**, *6*, e1000097. [CrossRef]

17. Hooijmans, C.R.; Rovers, M.M.; de Vries, R.B.; Leenaars, M.; Ritskes-Hoitinga, M.; Langendam, M.W. Syrcle's risk of bias tool for animal studies. *BMC Med. Res. Methodol.* **2014**, *14*, 43. [CrossRef]

18. Campbell, M.; McKenzie, J.E.; Sowden, A.; Katikireddi, S.V.; Brennan, S.E.; Ellis, S.; Hartmann-Boyce, J.; Ryan, R.; Shepperd, S.; Thomas, J.; et al. Synthesis without meta-analysis (swim) in systematic reviews: Reporting guideline. *BMJ* **2020**, *368*, l6890. [CrossRef]

19. Schünemann, H.J.; Oxman, A.D.; Vist, G.E.; Higgins, J.P.T.; Deeks, J.J.; Glasziou, P.; Akl, E.; Guyatt, G.H. Chapter 12: Interpreting results and drawing conclusions. In *Cochrane Handbook for Systematic Reviews of Interventions*; Version 5.1.0 (Updated March 2011); Higgins, J.P.T., Green, S., Eds.; Cochrane: London, UK, 2011.

20. Schünemann, H.; Brożek, J.; Guyatt, G.; Oxman, A. Handbook for Grading the Quality of Evidence and the Strength of Recommendations for Using the GRADE Approach (Updated October 2013). Available online: gdt.guidelinedevelopment.org/app/handbook/handbook.html (accessed on 23 November 2016).

21. GRADE Working Group. *GRADEpro GDT*; McMaster University: Hamilton, ON, USA, 2014.

22. Aasland, K.; Skjerve, E.; Smith, A. Quality of blood samples from the saphenous vein compared with the tail vein during multiple blood sampling of mice. *Lab. Anim.* **2010**, *44*, 25–29. [CrossRef]

23. Christensen, S.D.; Mikkelsen, L.; Fels, J.; Bodvarsdottir, T.; Hansen, A. Quality of plasma sampled by different methods for multiple blood sampling in mice. *Lab. Anim.* **2009**, *43*, 65–71. [CrossRef]

24. Durschlag, M.; Wurbel, H.; Stauffacher, M.; Von Holst, D. Repeated blood collection in the laboratory mouse by tail incision—Modification of an old technique. *Physiol. Behav.* **1996**, *60*, 1565–1568. [CrossRef]

25. Fernández, I.; Peña, A.; Del Teso, N.; Pérez, V.; Rodríguez-Cuesta, J. Clinical biochemistry parameters in c57bl/6j mice after blood collection from the submandibular vein and retroorbital plexus. *J. Am. Assoc. Lab. Anim. Sci.* **2010**, *49*, 202–206.

26. Forbes, N.; Brayton, C.; Grindle, S.; Shepherd, S.; Tyler, B.; Guarnieri, M. Morbidity and mortality rates associated with serial bleeding from the superficial temporal vein in mice. *Lab. Anim.* **2010**, *39*, 236–240. [CrossRef] [PubMed]

27. Francisco, C.C.; Howarth, G.S.; Whittaker, A.L. Effects on animal wellbeing and sample quality of 2 techniques for collecting blood from the facial vein of mice. *J. Am. Assoc. Lab. Anim. Sci.* **2015**, *54*, 76–80. [PubMed]

28. Fried, J.H.; Worth, D.B.; Brice, A.K.; Hankenson, F.C. Type, duration, and incidence of pathologic findings after retroorbital bleeding of mice by experienced and novice personnel. *J. Am. Assoc. Lab. Anim. Sci.* **2015**, *54*, 317–327. [PubMed]

29. Frohlich, J.R.; Alarcón, C.N.; Toarmino, C.R.; Sunseri, A.K.; Hockman, T.M. Comparison of serial blood collection by facial vein and retrobulbar methods in c57bl/6 mice. *J. Am. Assoc. Lab. Anim. Sci.* **2018**, *57*, 382–391. [CrossRef] [PubMed]

30. Gjendal, K.; Kiersgaard, M.K.; Abelson, K.; Sorensen, D.B.; Ottesen, J.L. Comparison of sublingual, facial and retro-bulbar blood sampling in mice in relation to animal welfare and blood quality. *J. Pharmacol. Toxicol. Methods* **2020**, *103*, 106680. [CrossRef]

31. Madetoja, J.; Madetoja, M.; Mäkinen, J.; Riuttala, E.; Jokinen, J. Blood sampling from the tail vein, in comparison with two other techniques, causes less stress to mice. *Scand. J. Lab. Anim. Sci.* **2009**, *36*, 215–221.

32. Moore, E.S.; Cleland, T.A.; Williams, W.O.; Peterson, C.M.; Singh, B.; Southard, T.L.; Pasch, B.; Labitt, R.N.; Daugherity, E.K. Comparing phlebotomy by tail tip amputation, facial vein puncture, and tail vein incision in c57bl/6 mice by using physiologic and behavioral metrics of pain and distress. *J. Am. Assoc. Lab. Anim. Sci.* **2017**, *56*, 307–317.

33. Regan, R.D.; Fenyk-Melody, J.E.; Tran, S.M.; Chen, G.; Stocking, K.L. Comparison of submental blood collection with the retroorbital and submandibular methods in mice (mus musculus). *J. Am. Assoc. Lab. Anim. Sci.* **2016**, *55*, 570–576.

34. Rogers, I.T.; Holder, D.J.; Mcpherson, H.E.; Acker, W.R.; Brown, E.G.; Washington, M.V.; Motzel, S.L.; Klein, H.J. Influence of blood collection sites on plasma glucose and insulin concentration in conscious c57bl/6 mice. *J. Am. Assoc. Lab. Anim. Sci.* **1999**, *38*, 25–28.

35. Shirasaki, Y.; Ito, Y.; Kikuchi, M.; Imamura, Y.; Hayashi, T. Validation studies on blood collection from the jugular vein of conscious mice. *J. Am. Assoc. Lab. Anim. Sci.* **2012**, *51*, 345–351.

36. Sørensen, D.B.; Metzdorff, S.B.; Jensen, L.K.; Andersen, K.H.; Teilmann, A.C.; Jensen, H.E.; Frøkiær, H. Time-dependent pathologic and inflammatory consequences of various blood sampling techniques in mice. *J. Am. Assoc. Lab. Anim. Sci.* **2019**, *58*, 362–372. [CrossRef] [PubMed]

37. Tabata, H.; Kitamura, T.; Nagamatsu, N. Comparison of effects of restraint, cage transportation, anaesthesia and repeated bleeding on plasma glucose levels between mice and rats. *Lab Anim.* **1998**, *32*, 143–148. [CrossRef] [PubMed]

38. Teilmann, A.C.; Madsen, A.N.; Holst, B.; Hau, J.; Rozell, B.; Abelson, K.S.P. Physiological and pathological impact of blood sampling by retro-bulbar sinus puncture and facial vein phlebotomy in laboratory mice. *PLoS ONE* **2014**, *9*, e113225. [CrossRef] [PubMed]

39. Silverman, J. (Ed.) Clinical biochemistry parameters in c57bl/6j mice after blood collection from the submandibular vein and retroorbital plexus. *J. Am. Assoc. Lab. Anim. Sci.* **2010**, *49*, 202–206.

40. Constantinescu, G.M.; Duffee, N.E. (Eds.) Comparison of submental blood collection with the retroorbital and submandibular methods in mice (mus musculus). *J. Am. Assoc. Lab. Anim. Sci.* **2017**, *56*, 711–712.

41. Mellor, D.J.; Beausoleil, N. Extending the 'five domains' model for animal welfare assessment to incorporate positive welfare states. *Anim. Welf.* **2015**, *24*, 241–253. [CrossRef]

42. Mason, G.; Mendl, M. Why is there no simple way of measuring animal welfare? *Anim. Welf.* **1993**, *2*, 301–319.

43. Whittaker, A.L.; Marsh, L.E. The role of behavioural assessment in determining 'positive' affective states in animals. *CAB Rev.* **2019**, *14*, 1–13. [CrossRef]

44. Boissy, A.; Manteuffel, G.; Jensen, M.B.; Moe, R.O.; Spruijt, B.; Keeling, L.J.; Winckler, C.; Forkman, B.; Dimitrov, I.; Langbein, J.; et al. Assessment of positive emotions in animals to improve their welfare. *Physiol. Behav.* **2007**, *92*, 375–397. [CrossRef]

45. Pickard, J. *Review of the Assessment of Cumulative Severity and Lifetime Experience in Non-Human Primates Used in Neuroscience Research*; Animal Procedures Committee: London, UK, 2013.

46. EFSA Panel on Animal Health & Welfare (AHAW). Welfare. Statement on the use of animal-based measures to assess the welfare of animals. *EFSA J.* **2012**, *10*, 2767.

47. Otovic, P.; Hutchinson, E. Limits to using hpa axis activity as an indication of animal welfare. *ALTEX* **2014**, *32*, 41–50. [CrossRef] [PubMed]

48. Ralph, C.R.; Tilbrook, A.J. Invited review: The usefulness of measuring glucocorticoids for assessing animal welfare. *J. Anim. Sci.* **2016**, *94*, 457–470. [CrossRef] [PubMed]

49. Bekhbat, M.; Glasper, E.R.; Rowson, S.A.; Kelly, S.D.; Neigh, G.N. Measuring corticosterone concentrations over a physiological dynamic range in female rats. *Physiol. Behav.* **2018**, *194*, 73–76. [CrossRef] [PubMed]

50. Brown, M.J.; Winnicker, C. Chapter 39—Animal Welfare. In *Laboratory Animal Medicine*, 3rd ed.; Fox, J.G., Anderson, L.C., Otto, G.M., Pritchett-Corning, K.R., Whary, M.T., Eds.; Academic Press: Boston, MA, USA, 2015; pp. 1653–1672.

51. Touma, C.; Palme, R.; Sachser, N. Analyzing corticosterone metabolites in fecal samples of mice: A noninvasive technique to monitor stress hormones. *Horm. Behav.* **2004**, *45*, 10–22. [CrossRef] [PubMed]

52. Clough, G. Environmental effects on animals used in biomedical research. *Biol. Rev. Camb. Philos. Soc.* **1982**, *57*, 487–523. [CrossRef] [PubMed]

53. Jirkof, P.; Rudeck, J.; Lewejohann, L. Assessing affective state in laboratory rodents to promote animal welfare—What is the progress in applied refinement research? *Animals (Basel)* **2019**, *9*, 1026. [CrossRef]

54. Dietze, S.; Lees, R.K.; Fink, H.; Brosda, J.; Voigt, J.-P. Food deprivation, body weight loss and anxiety-related behavior in rats. *Animals* **2016**, *6*, 4. [CrossRef]

55. Jirkof, P. Burrowing and nest building behavior as indicators of well-being in mice. *J. Neurosci. Methods* **2014**, *234*, 139–146. [CrossRef]

56. Walf, A.A.; Frye, C.A. The use of the elevated plus maze as an assay of anxiety-related behavior in rodents. *Nat. Protoc.* **2007**, *2*, 322–328. [CrossRef]

57. Biedermann, S.V.; Biedermann, D.G.; Wenzlaff, F.; Kurjak, T.; Nouri, S.; Auer, M.K.; Wiedemann, K.; Briken, P.; Haaker, J.; Lonsdorf, T.B.; et al. An elevated plus-maze in mixed reality for studying human anxiety-related behavior. *BMC Biol.* **2017**, *15*, 125. [CrossRef]

58. Seibenhener, M.L.; Wooten, M.C. Use of the open field maze to measure locomotor and anxiety-like behavior in mice. *J. Vis. Exp.* **2015**, e52434. [CrossRef] [PubMed]

59. Koseoglu, M.; Hur, A.; Atay, A.; Cuhadar, S. Effects of hemolysis interference on routine biochemistry parameters. *Biochem. Med. (Zagreb)* **2011**, *21*, 79–85. [CrossRef] [PubMed]

60. Magnette, A.; Chatelain, M.; Chatelain, B.; Ten Cate, H.; Mullier, F. Pre-analytical issues in the haemostasis laboratory: Guidance for the clinical laboratories. *Thromb. J.* **2016**, *14*, 49. [CrossRef] [PubMed]

61. Removal of blood from laboratory mammals and birds: First report of the bva/frame/rspca/ufaw joint working group on refinement. *Lab. Anim.* **1993**, *27*, 1–22. [CrossRef] [PubMed]

62. McGuill, M.; Rowan, A. Biological effects of blood loss: Implications for sampling volumes and techniques * commentary: H. Richard Adams. *ILAR J.* **1989**, *31*, 5–20. [CrossRef]
63. Osborne, N.; Avey, M.T.; Anestidou, L.; Ritskes-Hoitinga, M.; Griffin, G. Improving animal research reporting standards: Harrp, the first step of a unified approach by iclas to improve animal research reporting standards worldwide. *EMBO Rep.* **2018**, *19*, e46069. [CrossRef]
64. Kilkenny, C.; Browne, W.J.; Cuthill, I.C.; Emerson, M.; Altman, D.G. Improving bioscience research reporting: The arrive guidelines for reporting animal research. *PLoS Biol.* **2010**, *8*, e1000412. [CrossRef]
65. O'Connor, A.M.; Sargeant, J.M.; Gardner, I.A.; Dickson, J.S.; Torrence, M.E.; Dewey, C.E.; Dohoo, I.R.; Evans, R.B.; Gray, J.T.; Greiner, M.; et al. The reflect statement: Methods and processes of creating reporting guidelines for randomized controlled trials for livestock and food safety. *J. Vet. Intern. Med.* **2010**, *24*, 57–64. [CrossRef]
66. Sargeant, J.M.; O'Connor, A.M.; Dohoo, I.R.; Erb, H.N.; Cevallos, M.; Egger, M.; Ersbøll, A.K.; Martin, S.W.; Nielsen, L.R.; Pearl, D.L.; et al. Methods and processes of developing the strengthening the reporting of observational studies in epidemiology—Veterinary (strobe-vet) statement. *J. Vet. Intern. Med.* **2016**, *30*, 1887–1895. [CrossRef]
67. Nature Publishing Group. Availability of Data & Materials. 2017. Available online: http://www.nature.com/authors/policies/availability.html (accessed on 30 April 2020).
68. Science. Science: Editorial Policies. 2017. Available online: http://www.sciencemag.org/authors/science-editorial-policies (accessed on 30 April 2020).
69. Silva, L. PLOS' New Data Policy: Public Access to Data. EveryONE: PLOS ONE Community Blog. 2014. Available online: http://blogs.plos.org/everyone/2014/02/24/plos-new-data-policy-public-access-data-2/ (accessed on 30 April 2020).
70. McKenzie, J.; Brennan, S. Chapter 12: Synthesizing and presenting findings using other methods. In *Cochrane Handbook for Systematic Reviews of Interventions*; Version 6.0 (Updated July 2019); Higgins, J.P.T., Thomas, J., Chandler, J., Cumpston, M., Li, T., Page, M.J., Welch, V.A., Eds.; Cochrane: London, UK, 2019; Available online: www.training.cochrane.org/handbook (accessed on 30 April 2020).

Article

The Pharmacokinetics of Medetomidine Administered Subcutaneously during Isoflurane Anaesthesia in Sprague-Dawley Rats

Leila T. Kint [1], Bhedita J. Seewoo [2,3,4], Timothy H. Hyndman [5], Michael W. Clarke [6], Scott H. Edwards [7], Jennifer Rodger [3,4], Kirk W. Feindel [2,8] and Gabrielle C. Musk [1,9,*]

[1] Faculty of Health and Medical Sciences, The University of Western Australia, Perth 6009, Australia; leila.kint@uwa.edu.au
[2] Centre for Microscopy, Characterisation and Analysis, Research Infrastructure Centres, The University of Western Australia, Perth 6009, Australia; bhedita.seewoo@research.uwa.edu.au (B.J.S.); Kwfeinde@dal.ca (K.W.F.)
[3] Experimental and Regenerative Neurosciences, School of Biological Sciences, The University of Western Australia, Perth 6009, Australia; jennifer.rodger@uwa.edu.au
[4] Brain Plasticity Group, Perron Institute for Neurological and Translational Science, Perth 6009, Australia
[5] School of Veterinary Medicine, Murdoch University, Perth 6150, Australia; t.hyndman@murdoch.edu.au
[6] Metabolomics Australia, Centre for Microscopy, Characterisation and Analysis, The University of Western Australia, Perth 6009, Australia; michael.clarke@uwa.edu.au
[7] School of Animal Veterinary Sciences, Charles Sturt University, Wagga Wagga 2650, Australia; sedwards@csu.edu.au
[8] School of Biomedical Sciences, the University of Western Australia, Perth 6009, Australia
[9] Animal Care Services, the University of Western Australia, Perth 6009, Australia
*   Correspondence: gabrielle.musk@uwa.edu.au

Received: 28 May 2020; Accepted: 17 June 2020; Published: 18 June 2020

Simple Summary: Rodents, including rats, are used as animal models for research investigating neurological diseases in humans. To enable this research the animals are anaesthetized to facilitate imaging of the brain, but the anaesthetic drugs impact the results of the research. To minimize the variation between studies anaesthetic protocols should be similar. A common anaesthetic regime is the combination of two drugs (medetomidine and isoflurane); however, there is much variation in the doses of these drugs and the way in which they are administered. To provide some evidence to facilitate the standardization of anaesthetic protocols this study was performed to elucidate the details of what the body does to these drugs when they are administered in a certain way. Three groups of rats were studied to determine the desired dose of medetomidine when isoflurane is used at a low dose (approximately 0.5%). The results of the study are an evidence-based suggestion for medetomidine and isoflurane anaesthesia during functional magnetic resonance imaging (fMRI) studies.

Abstract: Anaesthetic protocols involving the combined use of a sedative agent, medetomidine, and an anaesthetic agent, isoflurane, are increasingly being used in functional magnetic resonance imaging (fMRI) studies of the rodent brain. Despite the popularity of this combination, a standardised protocol for the combined use of medetomidine and isoflurane has not been established, resulting in inconsistencies in the reported use of these drugs. This study investigated the pharmacokinetic detail required to standardise the use of medetomidine and isoflurane in rat brain fMRI studies. Using mass spectrometry, serum concentrations of medetomidine were determined in Sprague-Dawley rats during medetomidine and isoflurane anaesthesia. The serum concentration of medetomidine for administration with 0.5% (vapouriser setting) isoflurane was found to be 14.4 ng/mL (±3.0 ng/mL). The data suggests that a steady state serum concentration of medetomidine when administered with 0.5% (vapouriser setting) isoflurane can be achieved with an initial subcutaneous (SC) dose of 0.12 mg/kg of medetomidine followed by a 0.08 mg/kg/h SC infusion of medetomidine.

Consideration of these results for future studies will facilitate standardisation of medetomidine and isoflurane anaesthetic protocols during fMRI data acquisition.

Keywords: functional MRI; rat anaesthesia; refinement

---

1. Introduction

Anaesthetic protocols using a combination of medetomidine and isoflurane, are increasingly being used in functional magnetic resonance imaging (fMRI) studies of the rodent brain [1–10]. The use of medetomidine for these studies was first reported in 2002 [9], whilst the use of low dose isoflurane (<0.5% vapouriser setting) in conjunction with medetomidine as an anaesthetic regime was first reported in 2012 [3,4,8].

Medetomidine is an α_2-adrenoceptor agonist that causes sedation, hypertension, bradycardia, respiratory depression, hyperglycaemia, diuresis, muscle relaxation and analgesia [11–26]. The potency and receptor selectivity of medetomidine has led to its widespread use in veterinary anaesthesia, mostly in dogs and cats [27]. Medetomidine causes sedation through the activation of central α_2-adrenoceptors in the locus coeruleus, which prevents excitatory neurotransmitter release in the central nervous system and thereby depresses cortical arousal [13–15]. Vascular side effects of medetomidine occur due to the activation of peripheral α_2-adrenoceptors, which causes a transient and marked increase in systemic vascular resistance [17,18]. This vasoconstriction is followed by a decrease in vascular tone due to suppression of central nervous system-mediated sympathetic stimulation on blood vessels.

Isoflurane is a GABAergic fluorinated ether that causes anaesthesia, respiratory depression, bronchodilation, vasodilation, hypotension and muscle relaxation [28,29]. Isoflurane is commonly used for clinical and veterinary anaesthesia due to its rapid onset of action, short recovery time, safety and titratability [29–31]. The minimum alveolar concentration of isoflurane in adult Sprague-Dawley rats is $1.46 \pm 0.06\%$ [32].

The benefit of combining medetomidine with isoflurane specifically for fMRI studies has been described. When >0.1% isoflurane is administered with medetomidine, the epileptic activity caused by medetomidine is suppressed [4,21,22]. Furthermore, the drug combination allows for maintenance of a steady state of anaesthesia for >4 h, with consistent fMRI data [3]. In contrast, when medetomidine is administered alone via a constant rate infusion (subcutaneous (SC) or intravenous (IV)), it is not possible to maintain a steady state of sedation for >3 h. It has been reported that medetomidine administered alone can only be used in fMRI experiments >3 h if the initial infusion dose is increased three-fold after 90 min, or if medetomidine is specifically administered using an initial IV injection of at least 0.05 mg/kg medetomidine followed by a subsequent continuous SC or IV infusion of at least 0.1 mg/kg/h medetomidine, whereby the initial dose cannot be omitted, and the dose cannot be decreased [1,33].

Despite the increasing popularity of this combination of medetomidine and isoflurane, a standardised anaesthetic protocol for their combined use for rodent brain fMRI studies has not been established [4,34]. Various protocols are described with variable doses of both medetomidine and isoflurane, different routes of administration of medetomidine, and variation in the time of fMRI data collection relative to the time of medetomidine administration [3,4,6,7,34–36]. For example, in seven rodent fMRI studies employing medetomidine and isoflurane anaesthesia, the dose of isoflurane for maintenance of anaesthesia varied from 0.25–1.4% [3,4]. Furthermore, reported loading doses for medetomidine range from 0.03 to 0.15 mg/kg and the subsequent infusion doses range from 0.03 to 0.1 mg/kg/h [35,36]. In addition, the initial injection was administered via the intravenous (IV), intramuscular, intraperitoneal or subcutaneous (SC) routes and the infusion via the IV, intramuscular or SC routes. The time of fMRI data collection after the initial administration of medetomidine ranged from 15 min to 90 min [3,7,36]. This variation in the use of medetomidine and isoflurane in rodent brain fMRI studies may be attributed to a lack of comprehensive data on the pharmacokinetics and pharmacodynamics of medetomidine in rodents. Importantly, the serum concentration of medetomidine when administered with low dose isoflurane for rodent brain fMRI studies is unknown. Thus, the rationale for the administration of medetomidine alongside isoflurane for rodent brain fMRI studies is largely derived empirically [1–9]. However, there is now evidence that both resting-state and evoked blood-oxygen-level-dependent (BOLD) fMRI signals are altered by the type of anaesthetic drug(s) used and their dose [37–44]. Thus, the many aforementioned inconsistencies in the use of medetomidine and isoflurane may be hindering the interpretation, generalisation, meta-analysis and reproducibility of rodent brain fMRI studies.

Medetomidine substantially reduces the dose of isoflurane required to achieve stable anaesthesia, therefore minimising anaesthetic-induced distortions of BOLD fMRI signals. When dogs are administered a dose of 0.03 mg/kg IV medetomidine, there is a reduction of the minimum alveolar concentration of isoflurane by 47.2% [45]. Furthermore, when rodents are anaesthetised with a combined medetomidine and isoflurane dose of 0.06 mg/kg/h IV and 0.5–0.6%, respectively, they exhibit levels of anaesthesia comparable to rodents treated either medetomidine 0.1 mg/kg/h IV or isoflurane 1.3% [34]. Reducing the dose requirement of each drug is beneficial, as high doses of each drug in isolation are associated with significant drug-specific distortions of BOLD fMRI signals [34]. This artefact occurs because BOLD fMRI studies rely on the coupling between local blood flow and local neuronal activity (known as neurovascular coupling) to infer and therefore measure neural activity [44,46,47]. BOLD signals in anaesthetised rodents are considered an accurate measure of neural activity when they produce an image reflective of brain activity in the awake rodent. Conversely, BOLD signals are considered inaccurate when they produce an image reflective of fMRI-induced-stress or anaesthetic-induced changes in the BOLD effect [34,48]. Recent evidence suggests that BOLD fMRI signals obtained during medetomidine and isoflurane anaesthesia can be used to accurately measure rodent brain activity [34]. This attribute can be partially explained by the synergistic effects of the drugs on preserving neurovascular coupling [3,35]. When administered alone, medetomidine alters the BOLD effect by causing cerebral vasoconstriction, bradycardia, decreased cerebral blood flow and altered astrocyte activity [4,35,49]. In contrast, when isoflurane is administered alone, it alters the BOLD effect by inducing vasodilation in cerebral vasculature [1,50]. Accordingly, when medetomidine and isoflurane are administered together, medetomidine appears to attenuate isoflurane-induced cerebral vasodilation, leading to better preservation of neurovascular coupling [51].

To better utilise medetomidine and isoflurane anaesthesia in rodent fMRI studies, their use should be standardised. To this end, the pharmacokinetic profile of medetomidine during combined medetomidine and isoflurane anaesthesia needs to be elucidated, and the serum concentration of medetomidine in this context needs to be identified. The aim of this study was to describe the pharmacokinetics of medetomidine during isoflurane anaesthesia and determine the serum concentration of medetomidine when administered with 0.5% (vapouriser setting) isoflurane, so that an evidence-based dosing regimen of medetomidine could be determined for rat brain fMRI studies.

2. Materials and Methods

The study was approved by the University of Western Australia's Animal Ethics Committee (RA/3/100/1599) and conducted in accordance with the Australian code for the care and use of animals for scientific purposes, 8th edition [52]. The rats were housed in an AAALAC (Association for the Assessment and Accreditation of Laboratory Animal Care) facility.

2.1. Animals

Twenty-four male, eight-week-old, Sprague-Dawley rats (*Rattus norvegicus*) were imported from the Animal Resources Centre (Canning Vale, WA, Australia) as specific pathogen free rats. Rats were transported in groups to the animal care facility and held for at least three days prior to the study.

The rats were housed in a temperature-controlled environment on a 12 h light-dark cycle with food and water ad libitum at M-block in QEII Medical Centre (Nedlands, WA, Australia). The cages were individually ventilated with minimum dimensions of 38.8 cm wide, 40.6 cm long and 21 cm high on coarse aspen bedding. The rats were housed in pairs, fed a commercial rat diet (Specialty Feeds Meat Free Rat and Mouse Diet, Glen Forrest, Australia) that was autoclaved prior to introduction into the animal facility and were provided with acidified drinking water (pH 2.5–3). Food was not withheld prior to anaesthesia. On the day of the procedure, the rats were transferred to the Centre for Microscopy Characterisation and Analysis (University of Western Australia, Nedlands, Australia).

2.2. Experimental Procedure

The rats were randomly allocated to three experimental groups: Group T for determination of the target serum concentration of medetomidine when administered with low dose isoflurane for rodent brain fMRI studies ($n = 8$); Groups IV and SC for determination of the SC bioavailability of medetomidine during isoflurane anaesthesia ($n = 8$ each).

On the days of the procedures, the rats were anaesthetised with isoflurane (Isothesia™, Henry Schein Animal Health, 2000, Australia) in an induction chamber (4% isoflurane in 100% medical oxygen, 2 L/min). Once adequately anaesthetised (recumbent, no response to toe pinch) the rats were transferred onto the experimental benchtop and positioned for delivery of isoflurane throughout the experiment (0.5–2% isoflurane vapouriser setting in 100% medical oxygen, 1.5 L/min, Darvall Zero Dead Space face mask circuit, Advanced Anaesthesia Specialists) under a heat lamp. Physiological monitoring included body temperature, respiratory rate, heart rate, electrocardiography (PC-SAM Small Animal Monitor, SA Instruments Inc., 1030 System), exhaled isoflurane and CO_2 (data not shown) (ISATM Sidestream Gas Analyzer, Masimo Sweden AB and PHASEIN and Lightning Multi-Parameter Monitor Vetronic Services Ltd., Newton Abbot, UK) and blood glucose concentration (Accu-Chek Guide, Roche, Mannheim, Germany). These variables were recorded every 5 min. A single rat was studied at any one time, during the hours of 8 a.m. and 6 p.m.

Medetomidine (1 mg/mL, Ilium Medetomidine Injection, Troy Laboratories Pty. Limited, Glendenning, Australia) was administered according to the treatment group. In Group T, rats were administered an initial dose of medetomidine of 0.05 mg/kg SC over 1 s via a 29 G insulin syringe (BD Ultra-Fine Insulin Syringe, Becton Dickinson Pty Ltd., Macquarie University Research Park North Ryde, Australia), immediately followed by a continuous medetomidine infusion of 0.15 mg/kg/h SC, administered via a 25 G butterfly catheter connected to a single syringe infusion pump (Legato 100 Syringe Pump, KD Scientific Inc., Holliston, MA, USA). This protocol was developed empirically and used in our laboratory [10]. In the IV and SC groups, rats were manually administered a single dose of either IV (through a catheter placed in a lateral tail vein) or SC (under the skin over a flank) medetomidine at 0.05 mg/kg. The concentration of isoflurane was immediately reduced to 0.5% after administration of the initial dose of medetomidine and then subsequently altered to maintain an adequate depth of anaesthesia as assessed by response to toe pinch, heart rate and respiratory rate.

For serial blood sampling, a catheter was placed in the lateral tail vein (22 G, 1 IN, BD Angiocath IV Catheter, BD Australia, Seven Hills, NSW, Australia), secured with surgical tape and flushed with heparinised saline (5 IU/mL). In Group T, blood samples were collected 60 and 90 min after the initial dose of medetomidine. The conditions during anaesthesia were consistent with those observed in previous studies performed in this laboratory and were considered suitable for identification of the target concentration of medetomidine. In the IV group, blood was collected before medetomidine administration and 2, 5, 10, 20, 30, 60, 120 and 180 min afterwards. In the SC group, blood was collected before medetomidine administration and 10, 20, 30, 40, 50, 60, 120, 180 and 240 min afterwards. Following collection of the final sample, but before recovery from anaesthesia, the rats were euthanised via an intraperitoneal or IV injection of pentobarbitone (160 mg/kg, Lethabarb, Jurox, Rutherford, Australia).

2.3. Blood Sampling

Approximately 0.5 mL of blood was collected at each timepoint by inserting a 23 G butterfly catheter (SV*23BLK, Terumo Australia Pty Ltd., Macquarie Park, NSW, Australia) into the injection port of the tail vein catheter. The initial saline-diluted drops of blood were discarded before sample collection. A glucometer was used to immediately measure the blood glucose concentration (Accu-Chek Guide, Roche, BellaVista, Australia). After each sample, the catheter was flushed with 0.5 mL of heparinised saline (5 IU/mL) to prevent clot formation in the catheter and replace blood volume. In the event that sufficient blood could not be collected from the catheter, blood was drawn percutaneously from the lateral saphenous veins, medial saphenous veins or femoral arteries through a butterfly catheter.

All blood samples were collected in 3 mL Eppendorf tubes and allowed to clot at room temperature for 10 min before refrigeration. Refrigerated samples were centrifuged within 4 h of collection using an Eppendorf MiniSpin plus centrifugation at $2000\times g$ for 10 min. Approximately 0.2 mL of serum supernatant from each sample was collected and transferred into new 3 mL Eppendorf tubes. These serum samples were then frozen at $-80\,^{\circ}\text{C}$.

2.4. Serum Analysis

The analyses were performed at Metabolomics Australia (University of Western Australia, Nedlands, Australia). Medetomidine concentrations of the serum samples were analysed using a liquid chromatography-tandem mass spectroscopy (LC-MS/MS) technique. The internal standard during analysis was medetomidine-13C,d_3 hydrochloride (Sapphire Bioscience, Redfern, Australia).

To process the serum for analysis, 20 µL of serum were added to 50 µL of working internal standard (50 ng/mL labelled medetomidine-13C,d_3 in 50:50 methanol:water plus 0.1% formic acid) and vortexed for 10 s. The mixture was then vortexed with 1 mL ethyl acetate for 120 s, after which they were centrifuged at 3000 rpm for 5 min. Then, 900 µL solvent were evaporated to dryness for 30 min at $40\,^{\circ}\text{C}$ before being reconstituted in 70 µL of 50:50 methanol:water.

Processed serum extracts of 2 µL were run on an Agilent 6460 LC-MS/MS in 2D mode using isotope dilution to adjust for instrument response. Solvent A was LC-MS/MS grade water (Thermo Optima) with 0.1% formic acid (Merck). Solvent B was LC-MS/MS grade methanol (B & J) and 0.1% formic acid (Merck). Column one was an Agilent 2.1×50 mm 2.6 µm C18 Poroschell and column two was a Phenomenex Kinetex 3×150 mm 2.6 µm Biphenyl phase. The flow rate was set at 0.5 mL/min and a gradient was run from 50% B to 80% B in 10 min. The column was washed with 98% B and then returned to 50% B by 7 min. Compounds were heart cut from column one to column two between 0.4–0.9 min. Medetomidine and medetomidine-13C,d_3 were monitored with transitions 201 > 95 and 204.1 > 98, respectively, with a collision energy of 15. Assay calibration was achieved by spiking drug free matrix matched rat plasma to create a calibration curve, with the r^2 typically >0.9999. Assay precision was assessed during the project by extracting 4 samples in triplicate and the intra-assay CV ranged from 2.1–5.7%. The limit of quantitation for the assay was 0.1 ng/mL.

2.5. Pharmacokinetic and Pharmacodynamic Calculations

The maximum serum concentration (C_{max}) of medetomidine following SC administration was the highest measured concentration for each animal. The time at C_{max} (t_{max}) was also determined. The elimination rate constant (λ_z) was calculated as the negative slope of the semilogarithmic plot of each animal created from the terminal three time points (t = 120, 180 and 240 min). The elimination half-life ($t_{1/2\beta}$) was calculated as $\ln(2)/\lambda_z$. The area under the serum concentration time curve ($AUC_{0\to\infty}$) was estimated by the trapezoidal rule extrapolated to infinite time. Standard formulae were used to calculate the total body clearance (Cl = dose/AUC) [53] and volume of distribution at pseudo-equilibrium ($Vd_{area} = Cl/\lambda_z$) [54].

The target serum concentration of medetomidine (C_{target}) was obtained from the rats in Group T and was taken as the mean serum concentration of MED at t = 60 and 90 min. The loading dose (LD) was estimated from the product of Vd_{area} and C_{target}. The maintenance dose rate (MD) was calculated from the product of Cl and C_{target}.

2.6. Trial of Results

To trial the calculated drug administration regime for SC administration of medetomidine an additional two rats were administered medetomidine with isoflurane to ensure the conditions for anaesthesia were stable and uneventful. The dose of medetomidine in these two trials was an initial SC dose of 0.12 mg/kg medetomidine delivered over 5 s followed by a SC infusion of 0.08 mg/kg/h with 0.5% (vapouriser setting) isoflurane.

2.7. Statistical Analyses

Data were tested for normality using a D'Agostino and Pearson test and compared using Student's *t*-test or Mann–Whitney test (GraphPad Prism). The *p*-value used to define statistical significance was 0.05. Data are expressed as mean ± standard deviation or as otherwise stated.

3. Results

3.1. Group T

The rats weighed 333.2 ± 19.3 g (*n* = 6). Data from two rats were excluded from the study due to inaccurate weight records at the time of anaesthesia and therefore incorrect doses of medetomidine being administered. Otherwise anaesthesia was uneventful and a stable heart rate (307.9 ± 30.7 bpm), respiratory rate (52.9 ± 8.3 breaths/min) and normothermic temperature (38.1 ± 0.7 °C) were maintained. The blood glucose concentration at 60 min was 20.9 ± 3.0 mmol/L and at 90 min was 23.2 ± 2.6 mmol/L (Figure 1). The vapouriser setting for inhaled isoflurane was maintained at 0.5% after induction of anaesthesia, whereby from 5 to 90 min after the initial dose of medetomidine the exhaled isoflurane concentration was 0.49 ± 0.05% (Table 1).

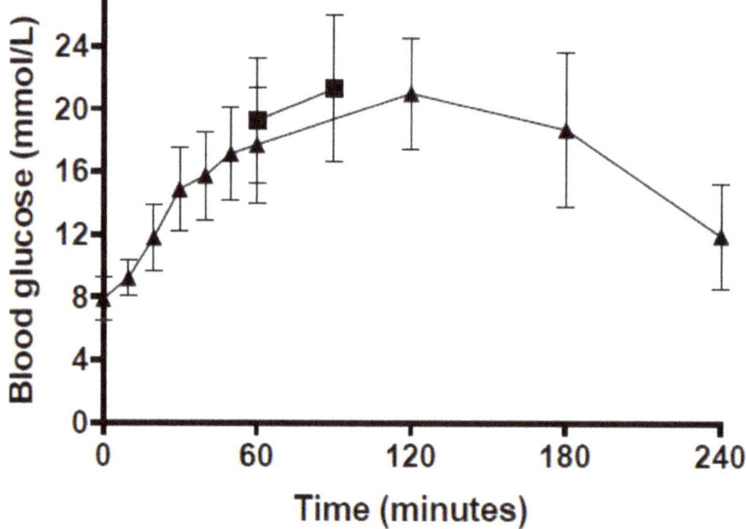

Figure 1. Time course of mean (±standard deviation) blood glucose concentration during anaesthesia of Sprague-Dawley rats in Group T (0.05 mg/kg medetomidine subcutaneous (SC) followed by a continuous infusion of 0.15 mg/kg/h SC with 0.5% isoflurane; squares) and Group SC (0.05 mg/kg medetomidine SC; triangles).

Table 1. Mean (±standard deviation) concentration of expired isoflurane during administration of isoflurane after an initial dose of medetomidine of 0.05 mg/kg SC followed by a continuous medetomidine infusion of 0.15 mg/kg/h SC (Group T) or a medetomidine dose of 0.05 mg/kg SC (Group SC). The delivery of isoflurane was adjusted as necessary to maintain an adequate depth of anaesthesia, as assessed by response to toe pinch, heart rate, respiratory rate and expired carbon dioxide concentration. Only results for the first 90 min are shown.

	Expired Isoflurane (%)							
	5 min	10 min	15 min	25 min	35 min	45 min	60 min	90 min
Group T ($n = 6$)	0.6 (± 0.2)	0.6 (± 0.4)	0.5 (± 0.3)	0.5 (± 0.04)	0.5 (± 0.1)	0.5 (± 0.1)	0.4 (± 0.1)	0.5 (± 0.1)
Group SC ($n = 7$)	1.2 (± 0.5)	1.0 (± 0.5)	1.0 (± 0.3)	1.0 (± 0.1)	1.0 (± 0.1)	1.0 (± 0.1)	1.0 (± 0.2)	1.2 (± 0.3)

The serum medetomidine concentration at 60 min after the initial medetomidine dose was 13.9 ± 3.9 ng/mL (range 9.9–20.8 ng/mL) which was similar to that at 90 min ($p = 0.329$): 15.0 ± 2.0 ng/mL (range 12.0–18.1 mg/mL). Therefore, for the purposes of identifying the serum concentration of medetomidine when administered with low dose isoflurane for rat brain fMRI studies, these data were grouped, and the target serum concentration of medetomidine was determined to be 14.4 ± 3.0 ng/mL.

3.2. Group IV

The rats weighed 333.6 ± 17.2 g ($n = 8$). In seven of the Group IV rats, respiratory arrest was observed immediately after manual administration of the IV injection of medetomidine, and gentle external chest compressions were performed. After 2 min, spontaneous ventilation resumed. Anaesthesia was otherwise uneventful. The blood glucose concentration peaked at 60 min at 16.6 ± 2.4 mmol/L and at 180 min at 11.2 ± 2.3 mmol/L.

Fifteen minutes after the administration of IV medetomidine, half of the rats required the vapouriser setting for isoflurane to be increased from 0.5% isoflurane. By 35 min, all the rats required the vapouriser setting for isoflurane to be increased from 0.5% isoflurane and maintained at approximately 1–2%.

The serum medetomidine concentration peaked at 2 min at 754.6 ± 672.5 ng/mL (range 122.1–2139.4 ng/mL). Given the variability of these data, the IV group was excluded from pharmacokinetic calculations.

3.3. Group SC

The rats weighed 317.9 ± 19.9 g ($n = 7$). Data from one rat (the first) was excluded from the study as it was administered an initial SC dose of medetomidine of 0.1 mg/kg and became apnoeic for approximately 2 min, requiring external chest compressions. The seven subsequent rats were administered a lower dose of 0.05 mg/kg SC medetomidine and anaesthesia was uneventful. The blood glucose concentration peaked at 120 min at 20.4 ± 3.4 mmol/L (Figure 1).

The inhaled isoflurane concentration required to maintain an adequate depth of anaesthesia throughout the procedure in Group SC was more variable than in the other groups (Table 1). The serum medetomidine concentration peaked at 60 min at 3.4 ± 0.9 ng/mL (Figure 2).

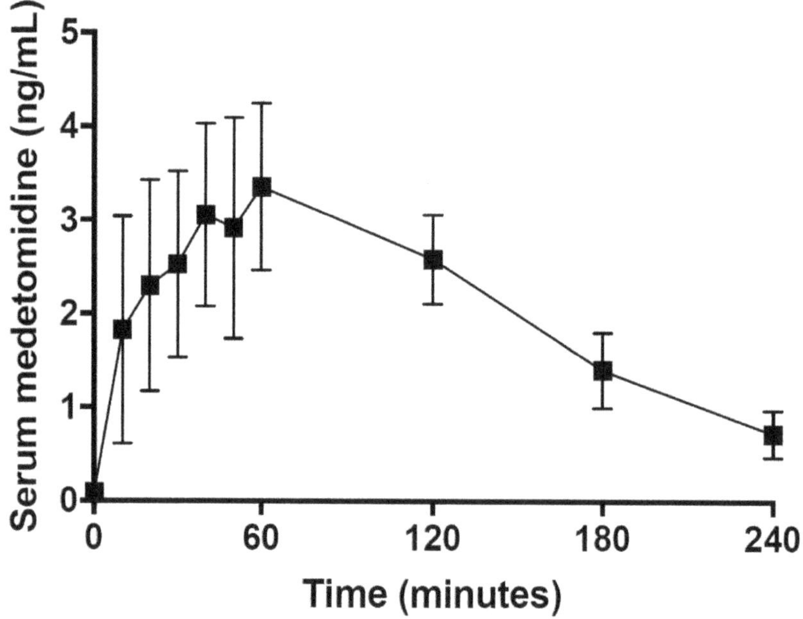

Figure 2. Time course of mean (±standard deviation) serum medetomidine concentration after subcutaneous administration of 0.05 mg/kg medetomidine in six Sprague-Dawley rats.

3.4. Pharmacokinetic Calculations

The pharmacokinetic and pharmacodynamic parameters were calculated from the mean serum medetomidine concentration data (Table 2). To achieve a target medetomidine concentration of 14.4 ± 3.0 ng/mL an initial SC dose of 0.12 mg/kg medetomidine followed by a SC infusion of 0.08 mg/kg/h medetomidine should be administered during isoflurane anaesthesia.

Table 2. Individual and mean (±standard deviation) pharmacokinetic (PK) and pharmacodynamic parameters after administration of 0.05 mg/kg SC medetomidine in seven Sprague-Dawley rats. C_{max} = maximum serum concentration; t_{max} = time of C_{max}; λ_z = elimination rate constant; $t_{1/2\beta}$ = elimination half-life; $AUC_{0\rightarrow\infty}$ = area under the serum concentration time curve from time = 0 to ∞; Cl = total body clearance; Vd_{area} = volume of distribution at pseudo-equilibrium; LD = loading dose; MD = maintenance dose.

Rat ID	C_{max} (ng/mL)	t_{max} (min)	λ_z (/min)	$t_{1/2\beta}$ (min)	$AUC_{0\rightarrow\infty}$ (ng.min/mL)	Cl (mL/kg/min)	Vd_{area} (L/kg)	LD (mg/kg)	MD (mg/kg/h)
R	4.9	60	0.0095	73.0	664.7	75.2	7.9	0.1142	0.0651
S	3.3	60	0.0118	58.7	570.8	87.6	7.4	0.1070	0.0758
T	4.4	50	0.0112	61.9	610.1	82.0	7.3	0.1055	0.0709
U	3.7	40	0.0130	53.3	485.3	103.0	7.9	0.1143	0.0891
V	2.9	120	0.0107	64.8	511.8	97.7	9.1	0.1316	0.0845
W	2.8	60	0.0112	61.9	485.6	103.0	9.2	0.1325	0.0891
X	3.3	120	0.0084	82.5	704.4	71.0	8.5	0.1218	0.0614
Mean	3.6	72.9	0.0108	65.2	576.1	88.5	8.2	0.1181	0.0765
(SD)	(0.7)	(30.6)	(0.0014)	(9.0)	(81.1)	(12.1)	(0.7)	(0.0101)	(0.0105)

3.5. Trial of Results

Two additional rats were administered medetomidine with isoflurane at the doses calculated in this study. The vapouriser setting for isoflurane could be maintained at or below 0.5% and anaesthesia was uneventful. Given the calculated initial dose was higher than that used in groups SC and IV the initial dose was administered over five seconds to mitigate the risk of apnoea (as observed in the IV group and the first rat in the SC group that was administered 1.0 mg/kg SC). Apnoea did not occur when the initial dose was delivered over five seconds.

4. Discussion

The present study shows that steady state serum concentrations of medetomidine will be achieved in male Sprague-Dawley rats if an initial SC dose of medetomidine of 0.12 mg/kg is administered in combination with continuous 0.5% (vapouriser setting) isoflurane, followed by a SC infusion of medetomidine at 0.08 mg/kg/h. This regime appears to provide suitable conditions for anaesthesia when the initial dose is delivered over five seconds. This result is within the range of doses reported in the literature [34,35].

The Group T result was used as the target serum concentration of medetomidine when administered with 0.5% (vapouriser setting) isoflurane. The anaesthetic protocol in this group was based on consultation with researchers using combined medetomidine and isoflurane anaesthesia in ongoing resting-state rodent fMRI studies. Given the apparent empirical success of the protocol in achieving strong and reproducible fMRI signals [10], rats under this protocol were hypothesised to achieve a steady state concentration of medetomidine. Data from rats in Group SC were used to determine the SC bioavailability of medetomidine during combined medetomidine and isoflurane anaesthesia. Collectively, the data from the two groups were used to inform the SC administration of medetomidine in rodents with low dose isoflurane.

The intention was to use data from both the IV and SC groups to perform pharmacokinetic calculations. However, the data from Group IV were excluded from the analysis due to considerable variation in this data set. We attribute the variation to the use of a single cannula for both IV drug administration and subsequent serial blood sampling. Issues arising from the use of a single cannula have been investigated and described by Gaud et al. [55]. They report that the use of a single cannula is not suitable for pharmacokinetic studies. Some compounds will experience non-specific binding to the cannula that may contaminate the first few blood samples taken from the cannula and lead to overestimation of serum concentrations [55]. Manually flushing the cannula with heparinised saline can help dislodge bound medetomidine, therefore reducing serum concentration overestimation. However, the flushing can also cause increased variation in measured serum medetomidine concentration due to the random error associated with repeated hand-operated techniques. This oversight likely led to inaccurate serum concentrations in the Group IV and hence exclusion of these data.

The Group T data suggest that a steady state serum medetomidine concentration of 14.4 ± 3.0 ng/mL is suitable for rats undergoing brain fMRI with 0.5% (vapouriser setting) isoflurane. This combination of drugs creates conditions suitable for prolonged anaesthesia (hours) without major anaesthetic-specific distortion of BOLD fMRI signals [10]. Future studies could consider using various doses of medetomidine and isoflurane to better define the therapeutic range for these drugs in the context of optimising the quality of fMRI images.

In the present study, the elimination half-life of medetomidine in rats was calculated to be 65.2 (±9.0) min. Similar values were calculated by Bol et al. (56.2 and 57.4 min) in a study of dexmedetomidine that was administered to Harlan-Sprague-Dawley male rats by two different IV infusion protocols [11]. In our study, and the work by Bol et al., drug concentrations were analysed for 210–240 min, and in neither study did blood medetomidine nor dexmedetomidine concentrations become undetectable. In contrast, a slower elimination half-life (1.6 h) was reported by Salonen et al. after tritium (3H)-labelled medetomidine was administered SC to male and female Sprague-Dawley rats. Furthermore, Salonen et al. detected plasma radioactivity at five and eight hours after administration of 3H- medetomidine [56]. The persistence of medetomidine at these time points (five and eight hours) may suggest that in our four-hour study, and in the study by Bol et al., the elimination rate constant was overestimated and therefore the elimination half-life was underestimated. This parameter could be explored in future studies by quantifying blood medetomidine (or dexmedetomidine) levels for several hours after administration.

The first rat in Group SC was administered a rapid initial dose (<one second, delivered manually) of 0.1 mg/kg SC medetomidine and became apnoeic for approximately two minutes. This response was not previously observed when the initial SC medetomidine dose was mechanically delivered in one second. Thus, the decision was made to alter the initial SC dose in Group SC from 0.1 mg/kg to 0.05 mg/kg for the remaining seven rats in that group. Although rats in Group IV also became apnoeic after administration of medetomidine, the dose in this group was not altered. The rationale to not alter the dose in Group IV was that transient apnoea could be managed with manual external chest compressions with the rat in sternal recumbency. To mitigate the risk of apnoea, the initial dose could be delivered over a longer time period; so during the trial of the calculated initial and infusion dose, the initial dose was administered over five seconds by the infusion pump. The conditions during anaesthesia were stable and uneventful.

Measurement of blood glucose concentrations was performed opportunistically and was not the primary aim of the project. Nevertheless, hyperglycaemia developed in all the rats in this study and although this side effect of medetomidine is described in rats its impact on experimental outcomes is not clear [57]. The mechanism of hyperglycaemia is a combination of anti-ADH (antidiuretic hormone) effects and alterations in insulin sensitivity, resulting in an osmotic diuresis [58]. This side effect of administration of medetomidine should be considered when designing anaesthetic regimens for research.

There are a number of limitations to this study which must be considered when interpreting the results. Only male, eight-week-old, Sprague-Dawley rats were used in this small study. This cohort limits the direct applicability of the results to female rats, other rat strains and mice. The age of the rats in this study is also a limitation of the model as adult animals may have a different pharmacokinetic profile for medetomidine. Future studies could expand the applicability of these results by investigating the pharmacokinetics of medetomidine in female rats, pregnant rats, obese rats, different ages and strains of rats and mice. In addition, the pharmacokinetic calculations could only be performed with serum concentrations of medetomidine that were obtained following SC administration. The data from Group IV was unfortunately excluded. Nevertheless, the data from Group SC were utilized in isolation, which meant that during the sampling period, the serum concentrations of medetomidine were assumed to be in pseudo-equilibrium. Thus, calculating the volume of distribution using the area method (Vd_{area}) was appropriate [59]. The loading dose should be calculated using the volume of distribution calculated at steady state (Vd_{ss}). Given Vd_{area} is usually only larger than Vd_{ss} by a small amount, our calculated loading dose is likely to still be a reliable estimate. Furthermore, single doses of medetomidine and isoflurane were evaluated in this study as the aim was to determine a target concentration of medetomidine based upon empirical evidence of using these doses. Future work should consider the evaluation of alternative doses and their impact on fMRI outputs. Finally, the target dose as determined by Group T was based on the premise that quality fMRI images were acquired (in previous work in the lab) with the empirical protocol. Correlation of our conclusions with the quality of fMRI images has not been performed.

For studies where multiple imaging sessions are scheduled and the animals recover from anaesthesia, the administration of atipamezole is prudent. This drug antagonises medetomidine and is routinely administered in the laboratory in which this study was performed when rats recover from anaesthesia.

The benefit of combined medetomidine and isoflurane anaesthetic protocols in rodent brain fMRI studies may be compromised by inconsistencies in these anaesthetic protocols between studies. Anaesthetics alter BOLD fMRI signals and these inconsistencies hinder the interpretation, generalisation, meta-analysis and reproducibility of rodent brain fMRI studies. Future brain fMRI studies should consider an evidence-based approach to the use of medetomidine and isoflurane anaesthetic protocols to standardise the regime between studies.

5. Conclusions

The data suggest that a serum medetomidine concentration of 14.4 ± 3.0 ng/mL is suitable for rats undergoing brain fMRI with 0.5% (vapouriser setting) isoflurane.

Author Contributions: Conceptualization, G.C.M., K.W.F. and J.R.; methodology, G.C.M., T.H.H., K.W.F. and J.R.; data collection, L.T.K., B.J.S. and G.C.M.; pharmacokinetic analysis, T.H.H. and S.H.E.; medetomidine quantification, M.W.C.; data curation, L.T.K., B.J.S. and G.C.M.; writing—original draft preparation, L.T.K., G.C.M. and B.J.S.; writing—review and editing, all authors. All authors have read and agreed to the published version of the manuscript.

Funding: This research was funded by the University of Western Australia internal research funds held by K.W.F.; K.W.F. is an Australian National Imaging Facility Fellow, a facility funded by the University, State and Commonwealth Governments. L.T.K. is supported by a University of Western Australia Winthrop Scholarship. B.J.S. is supported by a Forrest Research Foundation Scholarship, an International Postgraduate Research Scholarship, and a University Postgraduate Award. J.R. is supported by a Senior Research Fellowship from Multiple Sclerosis Western Australia and the Perron Institute for Neurological and Translational Science.

Acknowledgments: The authors thank Andrea Holme for her assistance with the serum collection process, Sandy Goodin and the team at M Block Animal Care Services for their assistance with animal care and pre-procedural monitoring. The authors acknowledge the facilities, scientific assistance and technical assistance of the National Imaging Facility, a National Collaborative Research Infrastructure Strategy (NCRIS) capability, at the Centre for Microscopy, Characterisation and Analysis, University of Western Australia.

Conflicts of Interest: The authors declare no conflicts of interest. The funders had no role in the design of the study; in the collection, analyses, or interpretation of data; in the writing of the manuscript, or in the decision to publish the results.

References

1. Pawela, C.P.; Biswal, B.B.; Hudetz, A.G.; Schulte, M.L.; Li, R.; Jones, S.R.; Cho, Y.R.; Matloub, H.S.; Hyde, J.S. A protocol for use of medetomidine anesthesia in rats for extended studies using task-induced BOLD contrast and resting-state functional connectivity. *NeuroImage* **2009**, *46*, 1137–1147. [CrossRef] [PubMed]
2. Jonckers, E.; Van Audekerke, J.; De Visscher, G.; Vander Linden, A.; Verhoye, M. Functional connectivity fMRI of the rodent brain: Comparison of functional connectivity networks in rat and mouse. *PLoS ONE* **2011**, *6*, e18876. [CrossRef]
3. Lu, H.; Zou, Q.; Gu, H.; Raichle, M.E.; Stein, E.A.; Yang, Y. Rat brains also have a default mode network. *Proc. Natl. Acad. Sci. USA* **2012**, *109*, 3979–3984. [CrossRef] [PubMed]
4. Fukuda, M.; Vazquez, A.L.; Zong, X.; Kim, S.-G. Effects of the α(2)-adrenergic receptor agonist dexmedetomidine on neurovascular responses in somatosensory cortex. *Eur. J. Neurosci.* **2013**, *37*, 80–95. [CrossRef] [PubMed]
5. Sierakowiak, A.; Monnot, C.; Aski, S.N.; Uppman, M.; Li, T.-Q.; Dambert, P.; Brene, S. Default mode network, motor network, dorsal and ventral basal ganglia networks in the rat brain: Comparison to human networks using resting State-fMRI. *PLoS ONE* **2015**, *10*, e0120345. [CrossRef]
6. Zerbi, V.; Grandjean, J.; Rudin, M.; Wenderoth, N. Mapping the mouse brain with rs-fMRI: An optimized pipeline for functional network identification. *Neuroimage* **2015**, *123*, 11–21. [CrossRef]
7. Hsu, L.M.; Liang, X.; Gu, H.; Brynildsen, J.K.; Stark, J.A.; Ash, J.A.; Lin, C.P.; Lu, H.; Rapp, P.R.; Stein, E.A.; et al. Constituents and functional implications of the rat default mode network. *Proc. Natl. Acad. Sci. USA* **2016**, *113*, E4541–E4547. [CrossRef]
8. Weber, R.; Ramos-Cabrer, P.; Wiedermann, D.; van Camp, N.; Hoehn, M. A fully noninvasive and robust experimental protocol for longitudinal fMRI studies in the rat. *Neuroimage* **2006**, *29*, 1303–1310. [CrossRef]
9. Sommers, M.G.; Pikkemaat, J.A.; Booij, L.H.D.J.; Heerschap, A. Improved Anesthesia Protocols for fMIU Studies in Rats: The Use of Medetomidine For Stable, Reversible Sedation. 2002. Available online: https://cds.ismrm.org/ismrm-2002/PDF2/0393.PDF (accessed on 20 May 2020).
10. Seewoo, B.J.; Feindel, K.W.; Etherington, S.J.; Rodger, J. Frequency-specific effects of low-intensity rTMS can persist for up to 2 weeks post-stimulation: A longitudinal rs-fMRI/MRS study in rats. *Brain Stimul.* **2019**. [CrossRef]

11. Bol, C.; Danhof, M.; Stanski, D.R.; Mandema, J.W. Pharmacokinetic-pharmacodynamic characterization of the cardiovascular, hypnotic, EEG and ventilatory responses to dexmedetomidine in the rat. *J. Pharm. Exp.* **1997**, *283*, 1051–1058.

12. Sinclair, M.D. A review of the physiological effects of $\alpha(2)$-agonists related to the clinical use of medetomidine in small animal practice. *Can. Vet. J.* **2003**, *44*, 885–897. [PubMed]

13. Doze, V.A.; Chen, B.X.; Maze, M. Dexmedetomidine produces a hypnotic-anesthetic action in rats via activation of central alpha-2 adrenoceptors. *Anesthesiology* **1989**, *71*, 75–79. [CrossRef] [PubMed]

14. Correa-Sales, C.; Rabin, B.C.; Maze, M. A hypnotic response to dexmedetomidine, an alpha 2 agonist, is mediated in the locus coeruleus in rats. *Anesthesiology* **1992**, *76*, 948–952. [CrossRef] [PubMed]

15. De Sarro, G.B.; Ascioti, C.; Froio, F.; Liberi, V.; Nistico, G. Evidence that locus coeruleus is the site where clonidine and drugs acting at alpha 1- and alpha 2-adrenoceptors affect sleep and arousal mechanisms. *Br. J. Pharm.* **1987**, *90*, 675–685. [CrossRef] [PubMed]

16. Pypendop, B.H.; Verstegen, J.P. Hemodynamic effects of medetomidine in the dog: A dose titration study. *Vet. Surg. Vs* **1998**, *27*, 612–622. [CrossRef] [PubMed]

17. Savola, J.M. Cardiovascular actions of medetomidine and their reversal by atipamezole. *Acta Vet. Scand. Suppl.* **1989**, *85*, 39–47.

18. Schmeling, W.T.; Kampine, J.P.; Roerig, D.L.; Warltier, D.C. The effects of the stereoisomers of the alpha 2-adrenergic agonist medetomidine on systemic and coronary hemodynamics in conscious dogs. *Anesthesiology* **1991**, *75*, 499–511. [CrossRef]

19. Yaksh, T.L. Pharmacology of spinal adrenergic systems which modulate spinal nociceptive processing. *Pharm. Biochem. Behav.* **1985**, *22*, 845–858. [CrossRef]

20. Stenberg, D. Physiological role of alpha 2-adrenoceptors in the regulation of vigilance and pain: Effect of medetomidine. *Acta Vet. Scand. Suppl.* **1989**, *85*, 21–28. [PubMed]

21. Mirski, M.A.; Rossell, L.A.; McPherson, R.W.; Traystman, R.J. Dexmedetomidine decreases seizure threshold in a rat model of experimental generalized epilepsy. *Anesthesiology* **1994**, *81*, 1422–1428. [CrossRef] [PubMed]

22. Miyazaki, Y.; Adachi, T.; Kurata, J.; Utsumi, J.; Shichino, T.; Segawa, H. Dexmedetomidine reduces seizure threshold during enflurane anaesthesia in cats. *Br. J. Anaesth.* **1999**, *82*, 935–937. [CrossRef] [PubMed]

23. Gellai, M. Modulation of vasopressin antidiuretic action by renal alpha 2-adrenoceptors. *Am. J. Physiol.* **1990**, *259*, F1–F8. [CrossRef]

24. Talukder, M.H.; Hikasa, Y. Diuretic effects of medetomidine compared with xylazine in healthy dogs. *Can. J. Vet. Res. Rev. Can. De Rech. Vet.* **2009**, *73*, 224–236.

25. Kanda, T.; Hikasa, Y. Effects of medetomidine and midazolam alone or in combination on the metabolic and neurohormonal responses in healthy cats. *Can. J. Vet. Res. Rev. Can. De Rech. Vet.* **2008**, *72*, 332–339.

26. Zuurbier, C.J.; Keijzers, P.J.; Koeman, A.; Van Wezel, H.B.; Hollmann, M.W. Anesthesia's effects on plasma glucose and insulin and cardiac hexokinase at similar hemodynamics and without major surgical stress in fed rats. *Anesth. Analg.* **2008**, *106*, 135–142. [CrossRef]

27. Virtanen, R.; Savola, J.M.; Saano, V.; Nyman, L. Characterization of the selectivity, specificity and potency of medetomidine as an alpha 2-adrenoceptor agonist. *Eur. J. Pharm.* **1988**, *150*, 9–14. [CrossRef]

28. Eger, E.I., 2nd. The pharmacology of isoflurane. *Br. J. Anaesth.* **1984**, *56* (Suppl. 1), 71s–99s.

29. Dohoo, S.E. Isoflurane as an inhalational anesthetic agent in clinical practice. *Can. Vet. J. = La Rev. Vet. Can.* **1990**, *31*, 847–850.

30. Paddleford, R.R. *Manual of Small Animal Anesthesia*; Churchill Livingstone: New York, NY, USA, 1988.

31. Ludders, J.W. Advantages and Guidelines for Using Isoflurane. *Vet. Clin. North Am. Small Anim. Pr.* **1992**, *22*, 328–331. [CrossRef]

32. Mazze, R.I.; Rice, S.A.; Baden, J.M. Halothane, isoflurane, and enflurane MAC in pregnant and nonpregnant female and male mice and rats. *Anesthesiology* **1985**, *62*, 339–341. [CrossRef]

33. Paasonen, J.; Stenroos, P.; Salo, R.A.; Kiviniemi, V.; Grohn, O. Functional connectivity under six anesthesia protocols and the awake condition in rat brain. *Neuroimage* **2018**, *172*, 9–20. [CrossRef] [PubMed]

34. Brynildsen, J.K.; Hsu, L.-M.; Ross, T.J.; Stein, E.A.; Yang, Y.; Lu, H. Physiological characterization of a robust survival rodent fMRI method. *Magn. Reson. Imaging* **2017**, *35*, 54–60. [CrossRef]

35. Pirttimaki, T.; Salo, R.A.; Shatillo, A.; Kettunen, M.I.; Paasonen, J.; Sierra, A.; Jokivarsi, K.; Leinonen, V.; Andrade, P.; Quittek, S.; et al. Implantable RF-coil with multiple electrodes for long-term EEG-fMRI monitoring in rodents. *J. Neurosci. Methods* **2016**, *274*, 154–163. [CrossRef] [PubMed]

36. Grandjean, J.; Schroeter, A.; Batata, I.; Rudin, M. Optimization of anesthesia protocol for resting-state fMRI in mice based on differential effects of anesthetics on functional connectivity patterns. *Neuroimage* **2014**, *102*, 838–847. [CrossRef] [PubMed]

37. Jonckers, E.; Delgado y Palacios, R.; Shah, D.; Guglielmetti, C.; Verhoye, M.; Van der Linden, A. Different anesthesia regimes modulate the functional connectivity outcome in mice. *Magn. Reson. Med.* **2014**, *72*, 1103–1112. [CrossRef] [PubMed]

38. Hamilton, C.; Ma, Y.; Zhang, N. Global reduction of information exchange during anesthetic-induced unconsciousness. *Brain Struct. Funct.* **2017**, *222*, 3205–3216. [CrossRef]

39. Masamoto, K.; Fukuda, M.; Vazquez, A.; Kim, S.G. Dose-dependent effect of isoflurane on neurovascular coupling in rat cerebral cortex. *Eur. J. Neurosci.* **2009**, *30*, 242–250. [CrossRef]

40. Masamoto, K.; Kim, T.; Fukuda, M.; Wang, P.; Kim, S.G. Relationship between neural, vascular, and BOLD signals in isoflurane-anesthetized rat somatosensory cortex. *Cerebral Cortex (New York, NY: 1991)* **2007**, *17*, 942–950. [CrossRef] [PubMed]

41. Bukhari, Q.; Schroeter, A.; Cole, D.M.; Rudin, M. Resting State fMRI in Mice Reveals Anesthesia Specific Signatures of Brain Functional Networks and Their Interactions. *Front. Neural Circuits* **2017**, *11*. [CrossRef] [PubMed]

42. Van Alst, T.M.; Wachsmuth, L.; Datunashvili, M.; Albers, F.; Just, N.; Budde, T.; Faber, C. Anesthesia differentially modulates neuronal and vascular contributions to the BOLD signal. *NeuroImage* **2019**, *195*, 89–103. [CrossRef] [PubMed]

43. Masamoto, K.; Kanno, I. Anesthesia and the Quantitative Evaluation of Neurovascular Coupling. *J. Cereb. Blood Flow Metab.* **2012**, *32*, 1233–1247. [CrossRef] [PubMed]

44. Sirmpilatze, N.; Baudewig, J.; Boretius, S. Temporal stability of fMRI in medetomidine-anesthetized rats. *bioRxiv* **2019**, 667659. [CrossRef] [PubMed]

45. Ewing, K.K.; Mohammed, H.O.; Scarlett, J.M.; Short, C.E. Reduction of isoflurane anesthetic requirement by medetomidine and its restoration by atipamezole in dogs. *Am. J. Vet. Res.* **1993**, *54*, 294–299. [PubMed]

46. Gao, Y.-R.; Ma, Y.; Zhang, Q.; Winder, W.T.; Liang, Z.; Antinori, L.; drew, P.J.; Zhang, N. Time to wake up: Studying neurovascular coupling and brain-wide circuit function in the un-anesthetized animal. *NeuroImage* **2017**, *153*, 382–398. [CrossRef]

47. Hillman, E.M.C. Coupling Mechanism and Significance of the BOLD Signal: A Status Report. *Annu. Rev. Neurosci.* **2014**, *37*, 161–181. [CrossRef] [PubMed]

48. Chuang, K.-H.; Lee, H.-L.; Li, Z.; Chang, W.; Nasrallah, F.A.; Yeowd, L.Y.; Singhd, K.K.D.R. Evaluation of nuisance removal for functional MRI of rodent brain. *NeuroImage* **2019**, *188*, 694–709. [CrossRef] [PubMed]

49. Prielipp, R.C.; Wall, M.H.; Tobin, J.R.; Groban, L.; Cannon, M.A.; Fahey, F.H.; Gage, H.D.; Stump, D.A.; James, R.L.; Bennett, J.; et al. Dexmedetomidine-induced sedation in volunteers decreases regional and global cerebral blood flow. *Anesth. Analg.* **2002**, *95*, 1052–1059.

50. Iida, H.; Ohata, H.; Iida, M.; Watanabe, Y.; Dohi, S. Isoflurane and sevoflurane induce vasodilation of cerebral vessels via ATP-sensitive K+ channel activation. *Anesthesiology* **1998**, *89*, 954–960. [CrossRef] [PubMed]

51. Ohata, H.; Iida, H.; Dohi, S.; Watanabe, Y. Intravenous dexmedetomidine inhibits cerebrovascular dilation induced by isoflurane and sevoflurane in dogs. *Anesth. Analg.* **1999**, *89*, 370–377. [PubMed]

52. National Health and Medical Research Council. *Australian Code for the Care and Use of Animals for Scientific Purposes*, 8th ed.; National Health and Medical Research Council: Canberra, Australia, 2013.

53. Toutain, P.L.; Bousquet-Melou, A. Plasma clearance. *J. Vet. Pharm.* **2004**, *27*, 415–425. [CrossRef] [PubMed]

54. Toutain, P.L.; Bousquet-Melou, A. Bioavailability and its assessment. *J. Vet. Pharm.* **2004**, *27*, 455–466. [CrossRef] [PubMed]

55. Gaud, N.; Kumar, A.; Matta, M.; Kole, P.; Sridhar, S.; Mandlekar, S.; Holenarsipur, V.K. Single jugular vein cannulated rats may not be suitable for intravenous pharmacokinetic screening of high logP compounds. *Eur. J. Pharm. Sci.* **2017**, *99*, 272–278. [CrossRef] [PubMed]

56. Salonen, J.S. Pharmacokinetics of medetomidine. *Acta Vet. Scand. Suppl.* **1989**, *85*, 49–54.

57. Callahan, L.M.; Ross, S.M.; Jones, M.L.; Musk, G.C. Mortality associated with using medetomidine and ketamine for general anesthesia in pregnant and nonpregnant Wistar rats. *Lab Anim.* **2014**, *43*, 208–214. [CrossRef] [PubMed]

58. Lemke, K.A. Pharmacology-Anticholinergics and Sedatives. In *Lumb and Jones' Veterinary Anesthesia and Analgesia 4th Edition*; Tranquilli, W.J., Thurmon, J.C., Grimm, K.A., Eds.; Blackwell Publishing: Ames, IA, USA, 2007; pp. 218–219.

59. Toutain, P.L.; Bousquet-Melou, A. Volumes of distribution. *J. Vet. Pharm.* **2004**, *27*, 441–453. [CrossRef]

Article

Humane Euthanasia of Guinea Pigs (*Cavia porcellus*) with a Penetrating Spring-Loaded Captive Bolt

Shari Cohen [1,*]**, Melody Kwok** [1] **and Joel Huang** [2]

1   Office of Research, Ethics, and Integrity, University of Melbourne, Melbourne 3010, Australia; melodyk@student.unimelb.edu.au

2   Agriculture Victoria, Attwood 3049, Australia; joel.huang@agriculture.vic.gov.au

*   Correspondence: shari.cohen@unimelb.edu.au; Tel.: +61-4-2972-5806

Received: 1 July 2020; Accepted: 29 July 2020; Published: 5 August 2020

Simple Summary: Various euthanasia methods are currently employed for guinea pigs at their experimental or humane endpoint; however, many have significant limitations or negative animal welfare implications. Captive bolt euthanasia has been used in the guinea pig meat industry but has not been explored in a research setting. This work aimed to investigate the use of a penetrating spring-loaded captive bolt gun as a refinement to guinea pig euthanasia in research. The study found that when compared to blunt force trauma, the captive bolt procedure performed well against all parameters of humane slaughter of production animals and appears to be a feasible refinement for animal welfare.

Abstract: Guinea pigs (*Cavia porcellus*) have been used in research since the 19th century to collect data due to their physiological similarities to humans. Today, animals perform a vital role in experiments and concerns for laboratory animal welfare are enshrined in the 3R framework of reduction, refinement and replacement. This case study explores a refinement in humane euthanasia of guinea pigs via the use of an irreversible penetrating spring-loaded captive bolt (CB). Penetrating spring-loaded CB stunning for euthanasia (CBE) was performed on 12 guinea pigs with the parameters for humane slaughter of production animals in order to assess the suitability of this method of euthanasia in contrast to blunt force trauma (BFT). All 12 of the guinea pigs were rendered immediately unconscious with excellent experimental tissue quality collection, high repeatability of results and operator ($n = 8$) preference over BFT. Overall, CBE in guinea pigs appears to be a feasible refinement for animal welfare, human preference and improved tissue quality for experimental collection in settings where uncontaminated tissues are required.

Keywords: guinea pig; Cavia porcellus; captive bolt; humane euthanasia; laboratory animal; humane killing

---

1. Introduction

For over 200 years, guinea pigs have played an important role as an animal model in biomedical research due to their greater physiological parallels with humans in comparison to other species of rodents and rabbits [1,2]. There is an increasing level of awareness and importance of animal welfare at each stage of research with animals which includes ensuring a humane death. A humane death or euthanasia is defined as achieving immediate stress-free insensibility and death [3–5] The very definition of humane euthanasia supports the 3R's framework of replacement, reduction, refinement as proposed by Russell and Burch in 1959 [6]. The full replacement of animals in some areas of research can be difficult, and at times animal use may still be required. Whenever the use of animals cannot be replaced, it is incumbent on the research community to apply the principle of refinement at all stages of research to improve animal welfare and care [6].

Captive bolt (CB) is a technique that can be applied in the humane euthanasia of animals. Despite its benefits during humane slaughter, it is yet to be applied in the research setting for the collection of uncontaminated non-brain tissue. The CB devices used can be hand-held with a retractable bolt powered by either a spring, compressed air, gun powder or a blank cartridge, and placed on a specific area on the cranium of the animal [7]. Once triggered, it will lead to either stunning (reversible loss of consciousness) or death [7]. CB placements and their effectiveness differs between various animals. [4,7–10] There are two types of CB: non-penetrating and penetrating [7]. Non-penetrating CB leads to stunning via the transfer of kinetic energy into the cranium and brain and is usually followed by a secondary euthanasia method [4,5,7,11]. In contrast, penetrating CB results in death due to the bolt fracturing the skull and entering the brain, subsequently causing irreversible physical damage to the cerebral cortex, brain stem and the thalamus [10,12–14]. The five indicators of successful stunning are lack of corneal blink reflex, rhythmic breathing, righting reflexes, vocalization and head/neck tension [5,15,16]. These indicators are reliable to assess the effectiveness of the shot as CB damages the region of the brain responsible for these responses [12,15,16]. The technique has not been tested in a research setting using a non-commercial captive bolt device and a methodology for laboratory use is yet to be developed. The application of this technique is thought to be transferable and potentially valuable in animal research models requiring chemically uncontaminated and intact non-brain tissues.

The use of a captive bolt to perform humane euthanasia (CBE) has been demonstrated through numerous experiments to be efficacious and humane across a wide range of species, including: guinea pigs [13], rabbits [17], piglets [8,18–22], joeys [11,23], ruminants [24–27], and avian species [28]. Some of these results showed CBE produced immediate death in 100% of animals [28]. The effective use of this method has also been documented in guinea pigs for humane slaughter in the meat industry, which demonstrates its feasibility in this species [22].

While blunt force trauma (BFT) is a common technique to euthanize guinea pigs in research requiring chemically uncontaminated tissues, there are no known published studies assessing the use of BFT in this species. In other vertebrates, this technique has been compared to CB. In a study based on 170 rabbits by Walsh et al., CBE was shown to be more effective than BFT, with BFT failure rates of up to 23% due to insufficient force [17]. Likewise, studies on piglets, sheep/lambs, and kids have also shown that CBE will produce extensive brain damage before pain can be perceived and caused instantaneous death. In these studies, the failure rate was between 0–6% in animals [8,18,19,22,25–27]. This is in contrast to a study on piglets which determined that BFT had a similar failure rate of up to 24%, hence the need for repeated blows [22].

Compassion fatigue is a form of post-traumatic stress disorder. It is defined as exhaustion from the stress of feeling empathetic towards animals and feeling unable to help them in any way [29,30]. This important concept is now readily recognizable in staff working with animals. Compassion fatigue is believed to be exacerbated by performing euthanasia on a regular basis and is partly dependent on the type of method used [29]. A study performed by Rohlf et al. on 148 animal workers in animal shelters, veterinary clinics and research laboratories reported that they experienced mild and moderate stress symptoms (39% and 11%, respectively) from performing euthanasia [31]. Similarly, a systematic review conducted by Scotney et al. demonstrated staff involved in performing euthanasia (animal shelters, veterinary clinics or research laboratories) was correlated with increased work stress and was hence a contributing factor to developing compassion fatigue [32]. In particular, laboratory staff may experience guilt when euthanizing research animals [33]. Therefore, if the euthanasia method has high animal welfare and acceptable aesthetics, then the staff member or "operator" may be less likely to experience compassion or emotional fatigue.

As part of a study requiring the collection of chemically uncontaminated tracheal tissues from guinea pigs (University of Melbourne Animal Ethics ID: 1814500.3), researchers would often use cranial blunt force trauma (BFT) followed by exsanguination or decapitation [34,35]. However, facility staff, animal welfare veterinarians and researchers all expressed concerns with BFT. It was found to be a method operators wished to avoid and has been shown in other studies to potentially have a risk of low repeatability due to inappropriate force or technique, and it is very dependent on the individual operator [5,35,36]. Furthermore, there are known concerns that this method of euthanasia can lead to physical, emotional and compassion fatigue in personnel due to the human–animal bond [5,18,37–41]. Prior to identifying CBE as an alternative technique to BFT, other methods of humane euthanasia were considered. Table 1 outlines the currently available experimental methods of guinea pig euthanasia with their known respective advantages and disadvantages.

After a literature review of possible known options, it was extrapolated that CBE could be suitable for non-brain tissue collection with no chemical residues and possible improvements in experimental, animal and human welfare outcomes when compared to the BFT method [5]. The aim of this case study was to develop and trial a protocol to refine the method of humane euthanasia of guinea pigs for research requiring the collection of uncontaminated non-brain tissues. This paper introduces an alternative application of an existing method of humane euthanasia in a novel setting using CB and the criteria for humane slaughter to ascertain both the practicality and humaneness (animal and human) of using CBE on guinea pigs [15] for research.

Table 1. Advantages and disadvantages of alternative euthanasia methods [5,16,34–36,42–44].

Euthanasia Method	Disadvantages	Advantages
Anesthetic Gas Overdose	<ul><li>Time consuming (slow accumulation in the lungs) [5,16,35,43].</li><li>Special equipment needed [5,35,42,44].</li><li>Exposure to agents may be harmful to operators [5,35,42,44].</li><li>Contaminated tissue [45].</li><li>Restricted substance (Schedule 4) [35].</li></ul>	<ul><li>Useful in small animals or when veins are inaccessible [5,35].</li><li>Different methods available e.g., face mask, containers, etc. [5].</li><li>Non-flammable, non-explosive gases [5].</li><li>Both a stunning and euthanasia method [5].</li></ul>
Barbiturates	<ul><li>Increased alkalinity and concentration can cause pain if injected intraperitoneal [34,35,45].</li><li>Intra-cardiac only in unconscious animals [34,35,42].</li><li>Contaminated tissue [5,34,45].</li><li>Tissue artefact [5,34,45].</li><li>Intravenous ideal but greater training required [5,34,36].</li><li>Restricted substance (Schedule 4) [35].</li></ul>	<ul><li>High speed of action (dose, concentration, route, rate dependent) [5,34].</li><li>Inexpensive [5].</li><li>Smooth euthanasia pending route [5].</li></ul>
Carbon Dioxide Stunning	<ul><li>Time consuming, especially in immature animals [5,8,16,44,45].</li><li>Onset of unconsciousness not always immediate [5,43].</li><li>High CO_2 concentrations can lead to distress (vocalization, dyspnea) and/or mucosal pain and post-mortem pulmonary and upper respiratory tract lesions [16,34,35,43,45–50].</li><li>Guinea pigs must be <600 g in Victoria, Australia [42].</li><li>Harmful to operators if significant amounts inhaled [35,43].</li></ul>	<ul><li>Not a restricted substance and readily available in cylinders [5,35].</li><li>Inexpensive, non-flammable, non-explosive [5,34,35,43].</li><li>Minimal tissue contamination [5].</li><li>Can potentially induce rapid unconsciousness [5,34].</li></ul>
Cervical Dislocation	<ul><li>Aesthetically displeasing [5,34,45].</li><li>Guinea pigs must be <150 g in Victoria, Australia [42].</li><li>Extensive training needed [5,34,35,42].</li><li>Challenging due to anatomical differences to rats/mice (shorter neck with more muscle covering) [36,45].</li><li>Electrical activity may persist for 13 s after the procedure, which may indicate consciousness and pain perception [5,16,35].</li><li>Force dependent [17].</li><li>Damage to brain tissues [5].</li></ul>	<ul><li>Inexpensive.</li><li>Can result in rapid loss of sensibility [5,16,34,36].</li><li>No tissue contamination [5,16].</li></ul>
Decapitation	<ul><li>Aesthetically displeasing [5,16,45].</li><li>Human safety risk and special training needed [16,36,43].</li><li>Special equipment is needed (guillotine) [16,36,43].</li><li>Challenging due to anatomical differences to rats/mice (shorter neck with more muscle covering) [45].</li></ul>	<ul><li>Inexpensive.</li><li>Can result in rapid loss of sensibility [5,36].</li><li>No tissue contamination [5,43].</li><li>Quick method [34].</li></ul>
Blunt Force Trauma	<ul><li>Aesthetically displeasing [5,42,45].</li><li>Force and accuracy of blow dependent [5,8,35].</li><li>Issue with repeatability due to fatigue [5,8].</li><li>Damage to brain tissues [5].</li></ul>	<ul><li>Inexpensive.</li><li>Effective when applied accurately and with sufficient force [5,35,36].</li><li>No tissue contamination [42].</li><li>When accurate with sufficient force, can be a quick method [34,35].</li></ul>

2. Materials and Methods

Equipment: A penetrating spring-loaded CB called 'The Ballista' (Bunny Rancher, Shapleigh, ME, USA), was used in this research (see Figures 1 and 2 below). This hand-held device was placed mid-forehead on the guinea pigs (see Figure 2 below). The device delivered 6.7 joules with each shot [51].

Figure 1. 'The Ballista': a penetrative spring-powered captive bolt (CB) gun used for the euthanasia of guinea pigs in this research. Image used with permission [51]

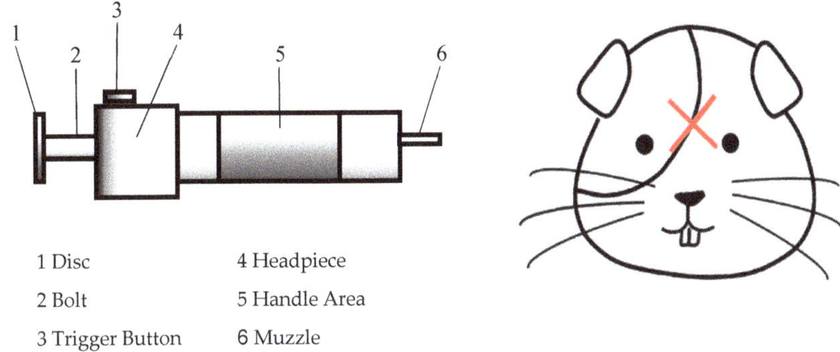

Figure 2. (**Left**): '*The Ballista*' spring-powered non-penetrating captive bolt. (**Right**): The red cross denotes where the captive bolt was positioned on the guinea pigs.

Parameters to determine humane euthanasia: There are currently no specific recommendations for CB use in guinea pigs. Hence, the livestock parameters of an absence of corneal blink reflex, rhythmic breathing, righting reflexes, vocalization and head/neck tension were utilized for this study.

Experiment: Operators first trialed the device on oranges to gain familiarity with the device. Excess breeding stock of male rat cadavers weighing 250 g or more were used to trial the device prior to use in guinea pigs. Twelve tri-color short-haired American female and male guinea pigs aged 16–22 weeks, weighing 500 g or more, were sourced as excess stock. Guinea pigs were euthanized using CBE. Accurate placement of the CB was first practiced on euthanized rats retrieved from other experiments to decrease the number of guinea pigs needed. All guinea pigs used were part of another animal ethics approved research project (University of Melbourne Animal Ethics ID: 1814500.3) as the researchers agreed to use this method as their humane euthanasia technique. Two personnel were required, with a total of 8 operators performing this method. Prior to the procedure being performed, each guinea pig was moved to a quiet place away from the others. One operator wrapped and held the guinea pig in a towel with its head resting on a rolled-up towel to elevate the head, increase comfort and decrease stress. The second operator pre-loaded the spring prior to positioning the penetrating spring-loaded CB on its head at the intersection between lines drawn from the base of the ears with the contralateral eyes (see Figure 2). When triggered, a retractable bolt was fired with the aim of producing effective and instantaneous unconsciousness. Animals were unwrapped to assess the 5 livestock humane euthanasia parameters. This was determined using five indicators (absence of corneal blink, rhythmic breathing, audible vocalization, righting reflex and head/neck tension) in each individual. Exsanguination was performed within the next 20–30 s. The results were recorded in and reviewed.

3. Results

All 12 guinea pigs failed to show signs of corneal blink, righting reflex, audible vocalization, head/neck tension and rhythmic breathing immediately after the initial shot. Based on these criteria, all guinea pigs were considered to have been stunned and rendered unconscious immediately. No additional shots were required for any of the guinea pigs. Exsanguination via severance of the lower abdominal aortic vessel was performed within 20–30 s of the initial shot to ensure death. A few of the guinea pigs were seen to display signs of involuntary muscle fasciculations and slow hindlimb pedaling motions, which lasted for less than 30 s. The bolt produced cranial skull fractures with associated subcutaneous hemorrhages, but the overlaying skin remained intact. While there was no evidence of puncture or bleeding overlying the site where the bolt penetrated the skull (Figures 3 and 4), there was a discrete palpable, circular depression in the cranium the size of the captive bolt diameter. Mild epistaxis ($n = 3$) was seen post-CBE. Informal debriefing with the operators ($n = 8$) and researchers ($n = 4$) suggested they all favored CBE over the BFT method. They stated CBE had a greater yield of higher quality uncontaminated tissue and was less physically, emotionally and compassionately fatiguing than BFT [52].

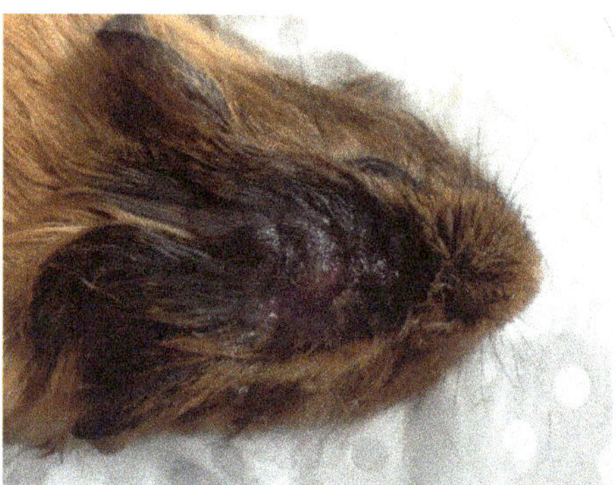

Figure 3. Skin intact from penetrative CB site.

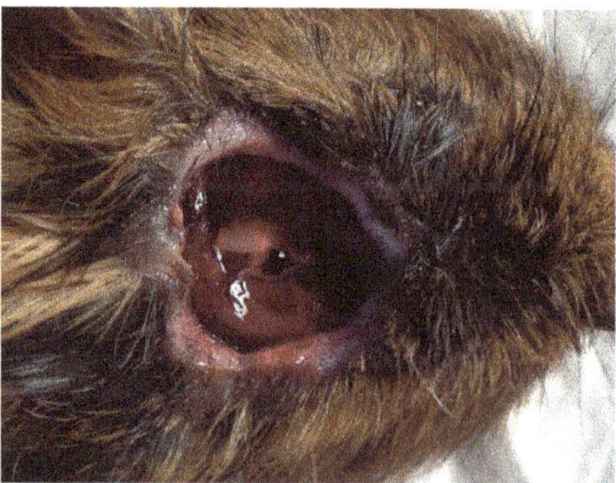

Figure 4. Skin removed from penetrative CB site.

4. Discussion

This study demonstrated that the penetrating spring-loaded CBE device can be a useful technique in the humane euthanasia of experimental animals by successfully producing instant and irreversible unconsciousness in 100% of the guinea pigs. This assessment of loss of consciousness is based on the absence of reflexes and the significant palpable physical damage to the skull, indicating sufficient damage to the vital regions of the brain [5,16,53]. The small but consistent result in this study exceeds the accepted animal welfare standard proposed by Temple Grandin for livestock in abattoirs, whereby 95% of animals should be immediately stunned with a single shot [54]. It also further supports the results from the study conducted by Limon et al. on South American commercial guinea pigs and is comparable to findings for other vertebrate species. In this study, some guinea pigs displayed involuntary muscle fasciculations and slow hindlimb pedaling motions after the initial shot. Both pedaling and head/neck movements are known autonomic movements controlled by reflex circuits in the spinal cord; thus, they commonly occur even after spinal severance due to the residual activity of these circuits [7,53,55].

The CB device used in this study was purchased from a commercial supplier which advertised it as a penetrating CB device. From the above results, all the skulls had extensive fractures, which is characteristic of a penetrating CB, although the skin remained intact after the procedure. This appearance is more characteristic of a non-penetrating CB which has a concussive effect without the bolt entering the cranium or penetrating the skin. Hence, while the skin remained intact, this paper refers to the device as a penetrating CB based on the photographs post-euthanasia as the bolt has clearly entered the cranium. An unexpected outcome of the experiment was that the overlying cranial skin of all guinea pigs was not compromised and remained intact. This observation was different to what was anticipated, especially given the extent of the fractures observed, and is not a typical characteristic of a penetrating CB. The intact skin may potentially be the result of the increased elasticity of the skin around the head in a guinea pig compared to the skin of other animals, such as cattle and rabbits, where the skin is more tightly adhered to their skulls. The cranial fractures associated with the position of the CB on the skull was expected given the nature of the device. The combination of the absence of measured parameters and post-mortem examination (hemorrhages and fractures) demonstrated that CBE resulted in effective stunning and insensibility in all the guinea pigs studied, even though the overlying skin was not compromised.

4.1. Effectiveness of Captive Bolt in Animals

Currently, there is only one published review of CBE in guinea pigs. This study was performed on South American commercial guinea pigs and four different slaughter methods were compared (cervical dislocation, electrical head-only stunning, carbon dioxide stunning and penetrating CB) [13]. Similarly to this paper, a small sample size of guinea pigs was used (10) with one guinea pig requiring a second shot. This could be argued as a 10% fail rate which is above the Temple Grandin recommendations for humane euthanasia with CB. Possible reasons for this guinea pig requiring a second shot may have been related to handling and restraint, incorrect placement of the CB, differences in devices or device failure. These problems have been described in other studies investigating CB in other species, in which insufficient pressure and/or penetration of the device, inaccurate placement of the device, inappropriate CB chosen for the intended species or prior head injury in animals were associated with failure of CBE or the need for multiple attempts [10,20,23,26,28]. These are all aspects to be managed and explored appropriately when using CBE. It is noteworthy that the use of CBE in our study for experimental tissue collection did not result in failure for any animal involved, despite the relative inexperience of the researchers with this technique.

The findings from these papers, along with this present study, support the conclusion that CBE can be an effective and humane method of euthanasia across many vertebrate species. In contrast, BFT has been shown to have a comparatively high failure rate. BFT also has the potential to more frequently fail the international standard of having at least 95% of animals being stunned after an initial blow/strike to preserve animal welfare. Therefore, when compared to CBE, BFT has a higher potential to negatively impact the welfare of guinea pigs and other vertebrate animals. The use of CBE appears to support the 3R framework principle of refinement by utilizing a more humane and preferable method to BFT in animals used for research.

4.2. Uncontaminated Tissues

At times, high quality chemically uncontaminated non-brain tissues are required for medical or biomedical research purposes. According to the operators in this study, the tracheal tissues collected were of a higher quality compared with tissues obtained from previous experimental work utilizing the BFT method. In these types of experiments, the majority of other euthanasia options available for guinea pigs (see Table 1) are not ideal as they have the potential to contaminate tissues, decrease tissue quality and affect metabolic serum biomarkers [45,46]. These issues can be potential confounders that can affect or compromise experimental outcomes, which could mean the information obtained may be incorrect or inconclusive. As such, the use of CBE over BFT supports the 3R principle of reduction as fewer animals may be required when higher quality tissues are used with better results. By utilizing high quality tissues, fewer animals may be used as research outcomes which may be more robust with a lower rate of rejected tissues. In regards to guinea pigs raised for human consumption, there is a legal and ethical requirement to ensure that meat products do not have any chemical residues unsuitable for human consumption and that animals are humanely killed. The use of CBE can be seen as a preferred method to support refinement due to potential improvements in animal welfare and research outcomes whenever intact brain or nasal tissues are not required [5].

4.3. Physical, Emotional and Compassion Fatigue

Operators anecdotally reported increased satisfaction with using a CBE compared to BFT due to the decreased variability of the effectiveness and easier application of the technique. This could potentially reduce the likelihood of physical, emotional and compassion fatigue being experienced by the operators. This is similar to statements made by other operators in different experiments which also rated CBE as the most aesthetically pleasing method of euthanasia when compared to BFT [8,17,22]. Aspects of operator fatigue include physical, emotional and compassion fatigue. Physical fatigue can be reduced when the CB is correctly placed on the cranium as the kinetic energy is delivered by the device [56,57]. This is different to the use of BFT which relies on the force applied by operators.

Emotional and compassion fatigue can also be linked to the human–animal bond. This bond is thought to be a contributing factor in the compassion fatigue experienced by people working in the animal profession [29]. Thus, by reducing the strain on the human–animal bond, both emotional and compassion fatigue can possibly be reduced. In this study, the operators using the CB reported this method to be more aesthetically appealing, as the bolt did not appear to significantly compromise the skin (Figures 3 and 4). They also felt the method of euthanasia was more humane. As a result, this method appeared to be less confronting and overall more appealing to operators, and potentially less detrimental to their wellbeing.

While the human element is an important aspect of humane euthanasia, the psychological aspect of this euthanasia technique was not objectively assessed in this paper. However, the initial anecdotal evidence appears to be positive. In future studies, a survey could be used to assess the emotional effect of CBE use to further investigate the human welfare aspect of various techniques of humane euthanasia. Overall, there was a verbalized positive impact on the physical and psychological well-being of operators when using this method in this case study. However, for CBE to be successful in future experimental studies or settings, adequate training and correct restraint of the animals must be ensured, and an appropriately sized CB device must be used and maintained. Other causes for CBE failure can arise if personnel are not adequately trained, which can lead to incorrect restraint of animals and placement of the device. If any of the above are performed incorrectly, higher levels of physical and emotional fatigue can occur due to failure of the device or technique [5,7–9,20]. Therefore, correct training and device usage is essential to decrease the chances for physical, emotional and compassion fatigue, as well as reducing risks to personnel and animals [3,36,58].

4.4. Limitations and the Future

Due to the small sample size of guinea pigs, aspects such as device fatigue and physical fatigue experienced by operators from loading the spring with each guinea pig were not evaluated. Other potential parameters that determine insensibility and success of stunning include measuring auditory evoked potential (AEP) and electroencephalogram (EEG) activity with electrodes. Both correlate to the cessation of convulsions, loss of reflexes and brain death [8,59,60]. These were not used in this study as the five parameters for humane livestock slaughter were deemed adequate to indicate insensibility. The use of EEG in CBE also risks damaging the electrodes and was deemed inappropriate for this study [8].

Additional parameters to measure could include the absence or presence of a heartbeat. However, cardiac contractions can continue in unconscious or brain-dead animals and thus this parameter was not used [8,18,61]. Whilst histological changes or scoring of the damage to the cranium and brain could have been assessed (skull fractures were easily observed and palpated), the degree of brain and cranial damage was not considered essential in this study [8,13,17,18]. An area that would benefit from additional investigation would be to explore alternate types of restraint and equipment. The current technique described requires a handler and a CB operator. It would be beneficial to develop or modify a device to enable the task to be performed by only one CB operator.

Finally, a different type of CB device could be tested. The use of an industrial CB (gunpowder-powered, compressed air or CO_2 blank cartridges) may be more suitable than the spring-powered CB in larger cohorts or in an abattoir setting for animals bred for human consumption. This is because the spring-loaded CB is designed to be used for <1000 animals per year and can experience device fatigue more quickly compared to an industrial CB [51]. The downside of an industrial CB is that they are more expensive and require more comprehensive training to use. However, the use of an industrial CB can result in less physical fatigue compared to a spring-loaded CB as there is no requirement to physically pull the spring with each use. Thus, the use of industrial CBs could be considered if a greater number of animals are to be used in a study due to larger cohorts or in other settings (i.e., an abattoir). It should also be noted that as of 2020, within Australia, CB is an accessible humane euthanasia technique as there are no legal requirements to hold a firearms license when using a CB in six out of the seven Australian states and territories. The authors are aware of similar legislation in other countries, which makes CBE a potentially accessible and humane euthanasia technique internationally in research and other settings.

From this study and other studies that have been undertaken, it can be tentatively proposed that CBE can be used as a suitable alternative for the collection of non-brain uncontaminated tissues. A drawback of this study is the small sample size available at the time. Further research should be undertaken to review the potential failure rate on a larger population to determine potential complications. Future studies in refinement could explore different handling/restraint techniques or devices, and the effects of the use of different CBs. Additional work can be performed to ensure consistency of the results from this study. As compassion and emotional fatigue are so prevalent, any future studies should incorporate formal surveys of operator preferences regarding their levels of physical, emotional and compassion fatigue and how this changes when performing euthanasia in different conditions.

5. Conclusions

In conclusion, this small study has shown CBE can be an ideal and humane method compared to other techniques (i.e., BFT) when uncontaminated tissues are needed. Thus far, it has been shown to have favorable experimental, animal and human welfare outcomes in guinea pigs where high quality, uncontaminated non-brain tissue collection is required. Future research could increase the sample size and use alternative parameters (e.g., AEP, EEG, post-mortem and histological changes) to corroborate the evidence for the effectiveness of CBE. Other areas to be explored could be the use of an industrial CB, alternative restraining methods or devices, and the inclusion of formal surveys to evaluate operator attitudes and experiences in performing CBE.

Author Contributions: Conceptualization, S.C. and J.H.; methodology, S.C. and J.H.; formal analysis, S.C.; investigation, S.C.; resources, S.C.; data curation, S.C.; validation, S.C.; writing—original draft preparation, M.K. and S.C.; writing—review and editing, S.C. and J.H.; visualization, S.C. and M.K.; supervision, S.C.; project administration, S.C. All authors have read and agreed to the published version of the manuscript.

Funding: This research received no external funding.

Acknowledgments: This work was supported by the University of Melbourne (UoM) Animal Welfare Team in the Office of Research Ethics and Integrity. The authors would like to thank the UoM Bioresources Platform facility staff, Maya Kesar and Keith Anderson. We would also like to acknowledge Gary Anderson and Andrew Jarnicki, from the Mazzone Laboratory of Respiratory Sensory Neuroscience from the UoM Department of Pharmacology and Therapeutics, as this work could not have been completed without their support.

Conflicts of Interest: The authors declare no conflict of interest.

References

1. Padilla-Carlin, D.J.; McMurray, D.N.; Hickey, A.J. The guinea pig as a model of infectious diseases. *Comp. Med.* **2008**, *58*, 324–340. [PubMed]

2. Taylor, D.K.; Lee, V.K. Chapter 25—Guinea Pigs as Experimental Models. In *The Laboratory Rabbit, Guinea Pig, Hamster, and Other Rodents*; Suckow, M.A., Stevens, K.A., Wilson, R.P., Eds.; Academic Press: Boston, MA, USA, 2012; pp. 705–744.

3. National Research Council. Chapter 5—Humane Endpoints for Animals in Pain. In *Recognition and Alleviation of Pain and Distress in Laboratory Animals*; The National Academies Press: Washington, DC, USA, 1992; pp. 102–116.

4. Grandin, T. Euthanasia and slaughter of livestock. *J. Am. Vet. Med. Assoc.* **1994**, *204*, 1354–1360. [PubMed]

5. Leary, S.L.; Underwood, W.; Anthony, R.; Gwaltney-Brant, S.; Poison, A.; Meyer, R. AVMA guidelines for the euthanasia of animals: 2013 edition. In *Proceedings of American Veterinary Medical Association*; American Veterinary Medical Association: Schaumburg, IL, USA, 2013; pp. 18–50. Available online: https://www.avma.org/sites/default/files/2020-01/2020-Euthanasia-Final-1-17-20.pdf (accessed on 29 July 2020).

6. Russell, W.M.S.; Burch, R.L. *The Principles of Humane Experimental Technique*; Methuen: London, UK, 1959.

7. Grandin, T. Recommended Captive Bolt Stunning Techniques for Cattle. Available online: https://www.grandin.com/humane/cap.bolt.tips.html (accessed on 24 January 2019).

8. Casey-Trott, T.; Millman, S.; Turner, P.; Nykamp, S.; Widowski, T. Effectiveness of a nonpenetrating captive bolt for euthanasia of piglets less than 3 d of age. *J. Anim. Sci.* **2013**, *91*, 5477–5484. [CrossRef] [PubMed]

9. Grandin, T. Captive Bolt Stunning of Cattle, Sheep, and Pigs. Available online: http://www.grandin.com/humane/captive.bolt.html (accessed on 20 January 2019).

10. Finnie, J.; Manavis, J.; Summersides, G.; Blumbergs, P. Brain damage in pigs produced by impact with a nonpenetrating captive bolt pistol. *Aust. Vet. J.* **2003**, *81*, 153–155. [CrossRef] [PubMed]

11. Hampton, J.O. Gunpowder-powered captive bolts for the euthanasia of kangaroo pouch young. *Aust. Mammal.* **2018**, *41*, 250–254. [CrossRef]

12. Terlouw, C.; Bourguet, C.; Deiss, V. Consciousness, unconsciousness and death in the context of slaughter. Part I. Neurobiological mechanisms underlying stunning and killing. *Meat Sci.* **2016**, *118*, 133–146. [CrossRef]

13. Limon, G.; Gonzales-Gustavson, E.A.; Gibson, T.J. Investigation Into the Humaneness of Slaughter Methods for Guinea Pigs (Cavia porcelus) in the Andean Region. *J. Appl. Anim. Welf. Sci. Jaaws* **2016**, *19*, 280–293. [CrossRef]

14. Finnie, J.; Manavis, J.; Blumbergs, P.; Summersides, G. Brain damage in sheep from penetrating captive bolt stunning. *Aust. Vet. J.* **2002**, *80*, 67–69. [CrossRef]

15. Grandin, T. How to Determine Insensibility (Unconsciousness) in Cattle, Pigs, and Sheep in Slaughter Plants. Available online: https://www.grandin.com/humane/insensibility.html (accessed on 24 January 2019).

16. Leary, S.L.; Underwood, W.; Anthony, R.; Gwaltney-Brant, S.; Grandin, T.; Meyer, R. AVMA Guidelines for the Humane Slaughter of Animals: 2016 Edition. In *Proceedings of the American Veterinary Medical Association*; AVMA: Schaumburg, IL, USA, 2016; pp. 24–39. Available online: https://www.avma.org/sites/default/files/resources/Humane-Slaughter-Guidelines.pdf (accessed on 29 July 2020).

17. Walsh, J.L.; Percival, A.; Turner, P.V. Efficacy of Blunt Force Trauma, a Novel Mechanical Cervical Dislocation Device, and a Non-Penetrating Captive Bolt Device for On-Farm Euthanasia of Pre-Weaned Kits, Growers, and Adult Commercial Meat Rabbits. *Animals* **2017**, *7*, 100. [CrossRef]

18. Grist, A.; Lines, J.A.; Knowles, T.G.; Mason, C.W.; Wotton, S.B. The Use of a Non-Penetrating Captive Bolt for the Euthanasia of Neonate Piglets. *Animals* **2018**, *8*, 48. [CrossRef]

19. Casey-Trott, T.M.; Millman, S.T.; Turner, P.V.; Nykamp, S.G.; Lawlis, P.C.; Widowski, T.M. Effectiveness of a nonpenetrating captive bolt for euthanasia of 3 kg to 9 kg pigs. *J. Anim. Sci.* **2014**, *92*, 5166–5174. [CrossRef]

20. Widowski, T.; Elgie, R.; Lawlis, P. Assessing the effectiveness of a non-penetrating captive bolt for euthanasia of newborn piglets. In Proceedings of the Allen D. Leman Swine Conference, Saint Paul, MN, USA, 22 September 2020; pp. 101–111.

21. Chevillon, P.; Mircovich, C.; Dubroca, S.; Fleho, J.-Y. Comparison of different pig euthanasia methods available to the farmers. *Proc. Int. Soc. Anim. Hyg.* **2004**, *11*, 45–46.

22. Whiting, T.; G Steele, G.; Wamnes, S.; Green, C. Evaluation of methods of rapid mass killing of segregated early weaned piglets. *Can. Vet. J.* **2011**, *52*, 753–758. [PubMed]

23. Sharp, T.M.; McLeod, S.R.; Leggett, K.E.A.; Gibson, T.J. Evaluation of a spring-powered captive bolt gun for killing kangaroo pouch young. *Wildl. Res.* **2015**, *41*, 623–632. [CrossRef]

24. Finnie, J.; Blumbergs, P.; Manavis, J.; Summersides, G.; Davies, R. Evaluation of brain damage resulting from penetrating and non-penetrating captive bolt stunning using lambs. *Aust. Vet. J.* **2000**, *78*, 775–778. [CrossRef] [PubMed]

25. Gibson, T.; Ridler, A.; Lamb, C.; Williams, A.; Giles, S.; Gregory, N. Preliminary evaluation of the effectiveness of captive-bolt guns as a killing method without exsanguination for horned and unhorned sheep. *Anim. Welf.* **2012**, *21*, 35–42. [CrossRef]

26. Grist, A.; Lines, J.A.; Knowles, T.G.; Mason, C.W.; Wotton, S.B. The Use of a Mechanical Non-Penetrating Captive Bolt Device for the Euthanasia of Neonate Lambs. *Animals* **2018**, *8*, 49. [CrossRef]

27. Grist, A.; Lines, J.A.; Knowles, T.G.; Mason, C.W.; Wotton, S.B. Use of a Non-Penetrating Captive Bolt for Euthanasia of Neonate Goats. *Animals* **2018**, *8*, 58. [CrossRef]

28. Raj, A.; O'Callaghan, M. Evaluation of a pneumatically operated captive bolt for stunning/killing broiler chickens. *Br. Poult. Sci.* **2001**, *42*, 295–299. [CrossRef]

29. Figley, C. *Compassion Fatigue in the Animal-Care Community*; The Humane Society of the United States: Washington, DC, USA, 2006.

30. Cohen, S.P. Compassion Fatigue and the Veterinary Health Team. *Vet. Clin. North Am. Small Anim. Pract.* **2007**, *37*, 123–134. [CrossRef]

31. Rohlf, V.; Bennett, P. Perpetration-induced Traumatic Stress in Persons Who Euthanize Nonhuman Animals in Surgeries, Animal Shelters, and Laboratories. *Soc. Anim. Soc. Sci. Stud. Hum. Exp. Other Anim.* **2005**, *13*, 201–219. [CrossRef]

32. Scotney, R.; McLaughlin, D.; Keates, H. A systematic review of the effects of euthanasia and occupational stress in personnel working with animals in animal shelters, veterinary clinics, and biomedical research facilities. *J. Am. Vet. Med Assoc.* **2015**, *247*, 1121–1130. [CrossRef] [PubMed]

33. Arluke, A. Trapped in a guilt cage. *New Sci.* **1992**, *134*, 33–35. [PubMed]

34. Close, B.; Banister, K.; Baumans, V.; Bernoth, E.M.; Bromage, N.; Bunyan, J.; Erhardt, W.; Flecknell, P.; Gregory, N.; Hackbarth, H.; et al. Recommendations for euthanasia of experimental animals: Part 1. *Lab. Anim.* **1996**, *30*, 298–305. [CrossRef]

35. Sharp, T.; Saunders, G. GEN001 Methods of Euthanasia. Available online: https://pdfs.semanticscholar.org/53fd/4a5923b24bd0696a4140e42ef715b3524d83.pdf?_ga=2.130248399.190541707.1566051564-314654895.1566051564 (accessed on 5 January 2019).

36. Close, B.; Banister, K.; Baumans, V.; Bernoth, E.-M.; Bromage, N.; Bunyan, J.; Erhardt, W.; Flecknell, P.; Gregory, N.; Hackbarth, H. Recommendations for euthanasia of experimental animals: Part 2. *Lab. Anim.* **1997**, *31*, 14–17. [CrossRef]

37. Casey-Trott, T. *Effectiveness of a Non-Penetrating Captive Bolt for Euthanasia of Suckling and Weaned Piglets*; University of Guelph: Guelph, ON, Canada, 2012.

38. Sivula, C.P.; Suckow, M.A. Chapter 35-Euthanasia. In *Management of Animal Care and Use Programs in Research, Education, and Testing*; Weichbrod, R.H., Thompson, G.A.H., Norton, J.N., Eds.; CRC Press: Boca Raton, FL, USA, 2018; pp. 869–882.

39. Reeve, C.L.; Rogelberg, S.G.; Spitzmüller, C.; Digiacomo, N. The Caring-Killing Paradox: Euthanasia-Related Strain Among Animal-Shelter Workers. *J. Appl. Soc. Psychol.* **2005**, *35*, 119–143. [CrossRef]

40. Chang, F.T.; Hard, L.A. Human-animal bonds in the laboratory: How animal behavior affects the perspective of caregivers. *ILAR J.* **2002**, *43*, 10–18. [CrossRef]

41. Whiting, T.L.; Marion, C.R. Perpetration-induced traumatic stress—A risk for veterinarians involved in the destruction of healthy animals. *Can. Vet. J.* **2011**, *52*, 794–796.

42. Agriculture Victoria. Appendix 2: Methods Of Euthanasia for Post-Neonatal Laboratory Mice, Rats, Guinea Pigs and Rabbits. Available online: http://agriculture.vic.gov.au/agriculture/animal-health-and-welfare/animal-welfare/animal-welfare-legislation/victorian-codes-of-practice-for-animal-welfare/code-of-practice-for-the-housing-and-care-of-laboratory-mice,-rats,-guinea-pigs-and-rabbits/appendix-2-methods-of-euthanasia-for-post-neonatal-laboratory-mice,-rats,-guinea-pigs-and-rabbits (accessed on 13 February 2019).

43. Gagea-Iurascu, M.; Craig, S. Chapter 4—Euthanasia and Necropsy. In *The Laboratory Rabbit, Guinea Pig, Hamster, and Other Rodents*; Suckow, M.A., Stevens, K.A., Wilson, R.P., Eds.; Academic Press: Boston, MA, USA, 2012; pp. 117–121.

44. Feldman, S.H.; Gografe, S.I.; Kohn, D.F.; Swindle, M.M. Methods of Euthanasia. In *Laboratory Mouse Handbook*; Gografe, S.I., Ed.; American Association for Laboratory Animal Science: Memphis, TN, USA, 2010; p. 56.

45. Hau, J.; Schapiro, S.J. Laboratory Animal Analgesia, Anesthesia, and Euthanasia. In *Handbook of Laboratory Animal Science: Volume I—Essential Principles and Practices*, 3rd ed.; CRC press: Boca Raton, FL, USA, 2010; Volume 1, pp. 526–530.

46. Conlee, K.; Stephens, M.; Rowan, A.; Kin, L. Carbon dioxide for euthanasia: Concerns regarding pain and distress, with special reference to mice and rats. *Lab. Anim.* **2005**, *39*, 137–161. [CrossRef]

47. Chisholm, J.M.; Pang, D.S.J. Assessment of Carbon Dioxide, Carbon Dioxide/Oxygen, Isoflurane and Pentobarbital Killing Methods in Adult Female Sprague-Dawley Rats. *PLoS ONE* **2016**. [CrossRef]

48. Raj, A.B.M.; Johnson, S.P.; Wotton, S.B.; McInstry, J.L. Welfare implications of gas stunning pigs: The time to loss of somatosensory evoked potential and spontaneous electrocorticogram of pigs during exposure to gases. *Vet. J.* **1997**, *153*, 329–339. [CrossRef]

49. Leach, M.C.; Bowell, V.A.; Allan, T.F.; Morton, D.B. Aversion to gaseous euthanasia agents in rats and mice. *Comp. Med.* **2002**, *52*, 249–257. [PubMed]

50. Anton, F.; Euchner, I.; Handwerker, H.O. Psychophysical examination of pain induced by defined CO2 pulses applied to the nasal mucosa. *Pain* **1992**, *49*, 53–60. [CrossRef]

51. Bunnyrancher. Which Bolt Gun Is Best for Me? Available online: https://www.bunnyrancher.com/why-bolt-guns.html (accessed on 6 June 2019).

52. Stewart, A.; University of Melbourne Department of Pharmacology and Therapeutics, Melbourne, Australia. Personal Communication, 2019.

53. Claudia Terlouw, E.M.; Bourguet, C.; Deiss, V.; Mallet, C. Origins of movements following stunning and during bleeding in cattle. *Meat Sci.* **2015**, *110*, 135–144. [CrossRef] [PubMed]

54. Grandin, T. Auditing animal welfare at slaughter plants. *Meat Sci.* **2010**, *86*, 56–65. [CrossRef]

55. Guertin, P.A. The mammalian central pattern generator for locomotion. *Brain Res. Rev.* **2009**, *62*, 45–56. [CrossRef]

56. Pierozan, P.; Jernerén, F.; Ransome, Y.; Karlsson, O. The Choice of Euthanasia Method Affects Metabolic Serum Biomarkers. *Basic Clin. Pharm. Toxicol.* **2017**, *121*, 113–118. [CrossRef]

57. Hampton, J.O.; Adams, P.J.; Forsyth, D.M.; Cowled, B.D.; Stuart, I.G.; Hyndman, T.H.; Collins, T. Improving animal welfare in wildlife shooting: The importance of projectile energy: Projectile Energy and Wildlife Shooting. *Wildl. Soc. Bull.* **2016**. [CrossRef]

58. Grandin, T. *Improving Animal Welfare: A Practical Approach*; Colorado State University Fort Collins: Fort Collins, CO, USA, 2015.

59. Dawson, M.D.; Lombardi, M.E.; Benson, E.R.; Alphin, R.L.; Malone, G.W. Using Accelerometers to Determine the Cessation of Activity of Broilers. *J. Appl. Poult. Res.* **2007**, *16*, 583–591. [CrossRef]
60. Dawson, M.D.; Johnson, K.J.; Benson, E.R.; Alphin, R.L.; Seta, S.; Malone, G.W. Determining cessation of brain activity during depopulation or euthanasia of broilers using accelerometers. *J. Appl. Poult. Res.* **2009**, *18*, 135–142. [CrossRef]
61. Sade, R.M. Brain death, cardiac death, and the dead donor rule. *J. South Carol. Med. Assoc.* **2011**, *107*, 146–149.

Communication

Survey on Sheep Usage in Biomedical Research

Corina Mihaela Berset [1,2], **Urban Lanker** [1] and **Stephan Zeiter** [1,*]

[1]   AO Research Institute Davos, Clavadelerstrasse 8, CH-7270 Davos, Switzerland; corina.berset@yahoo.com (C.M.B.); urban.lanker@aofoundation.org (U.L.)

[2]   Animal Welfare and 3R Department, University of Zurich, Winterthurerstrasse 190, CH-8057 Zürich, Switzerland

*   Correspondence: stephan.zeiter@aofoundation.org; Tel.: +41-81-414-23-11

Received: 31 July 2020; Accepted: 27 August 2020; Published: 30 August 2020

Simple Summary: Sheep are used in biomedical research. A European survey was conducted with the goal of identifying the need for improvement in the use of sheep. Most participants were veterinarians working at academic institutions. Two thirds have been working with sheep for more than 5 years, and their answers emphasized the importance of healthy sheep to be used in biomedical research, as about 60% have encountered health problems not related to study protocol. Other important points were sheep availability and the trust into and experience of the sheep supplier. This survey identified important points for refinement in the use of sheep in biomedical research with health status and monitoring as possible starting points.

Abstract: Currently, there is a lack of detailed information about sheep used for biomedical research. Therefore, a European survey was conducted among sheep users gathering information on the current situation, with emphasis on animal selection criteria and issues encountered in practice. The ultimate goal was to identify needs for improvement, which will subsequently lead to a refinement and reduction of the total number of animals used for experimental studies. From the 84 respondents, 77.4% were veterinarians, 71.4% were employed at academic institutions and 63.1% had worked with sheep as research animals for more than 5 years. The majority of the respondents were using females (79.8%) with no clear age preference, mainly for surgical procedures and testing medical devices. The main criteria for choosing a sheep supplier were the animals' health status, their availability, the trust and experience in the sheep provider and the animals' uniformity. Approximately 60% of the respondents had encountered problems in their sheep not related to the experimental protocol and almost half of them did not have a health monitoring program for their animals. In conclusion, there is definitely a need for refinement in selecting sheep used in biomedical research, with their health status as possible starting point.

Keywords: sheep; survey; preclinical research; health

---

1. Introduction

According to European Union (EU) reports [1], approximately 20,000 sheep are enrolled in research projects each year in EU member countries. Article 10 of the EU Directive 2010/63 states that a member state shall ensure that inter alia laboratory rodents, rabbits, dogs and cats, but not sheep or other farm animals, may only be used in studies where those animals have been bred for use in studies [2]. Moreover, to the authors' knowledge, there are currently no publications regarding the detailed criteria for selecting sheep for biomedical research. The health status of the animals is often only briefly described (i.e., clinically healthy) and there are no reports of compliance to the recommendations for the health monitoring of ruminants of the Federation of European Laboratory Animal Science Associations (FELASA) published in 2000 [3].

The goal of this European survey among sheep users was to gather information on the current situation, with emphasis on animal selection criteria and issues encountered in practice. Our motivation was to identify needs for improvement which will subsequently lead to a refinement and reduction of the total number of sheep used for experimental studies.

2. Materials and Methods

A survey was conducted between September 2016 and January 2017 among sheep users in biomedical research, with a focus on EU countries and Switzerland. The aim of the questionnaire was to gain more in-depth knowledge about the field of research in which sheep are used, the level of experience of the users, the animal characteristics and selection criteria as well as the issues encountered in practice.

The survey comprised 16 questions and was initially distributed as hard copies at the Swiss Laboratory Animal Science Association (SGV) Annual Meeting, Basel, Switzerland (13–14 September 2016). At the beginning of October 2016, an online version of the survey was created using surveymonkey.com in English (Figure S1) and in French (Figure S2) and distributed with the aid of several European Societies and networks: the Swiss Laboratory Animal Science Association (SGV), the Swiss Animal Welfare Officers Network, the European Society of Laboratory Animal Veterinarians (ESLAV), the European College of Laboratory Animal Medicine (ECLAM), the French Association for Laboratory Animal Science & Techniques (AFSTAL) and the Email for Vets in Laboratory Animal Medicine List (VOLE). A total of 84 responses were collected before the end of January 2017, when the survey was closed.

All answers were anonymous; the participants could write comments while answering some of the questions and at the end of the survey. After collecting all the answers, the data were exported to an Excel workbook.

3. Results

Among the respondents, 77.4% were veterinarians and 6% were animal care takers or technicians. From the total number of participants, 31% were conducting research activities and 15.5% were animal facility managers (multiple roles possible). Most of the respondents were working at academic institutions when the survey was conducted (71.4%), 15.5% were working at private companies, while 13.1% were working at other types of institutions.

The personnel's experience, together with education and training, being a very important factor when performing in vivo research, the participants were asked about their previous experience working with sheep. According to their answers, at the time that they filled out the questionnaire, 63.1% had worked with sheep for at least 5 years, 28.6% for 1–5 years and 8.3% for less than 1 year. The majority of institutions or companies where they were employed (86.9%) had more than 5 years of experience in working with sheep, 8.3% had used sheep for 1–5 years and 4.8% for less than 1 year.

When asked about the number of animals used/year, 29.8% of the respondents reported between 1 and 20 animals, 26.2% were using between 21 and 50 animals, 14.3%, 51–100 animals and 27.4% more than 100 sheep/year.

In order to have a more precise overview of the sheep biomedical studies, the survey also aimed to identify the experimental field in which the animals were used; the answers are illustrated in Figure 1.

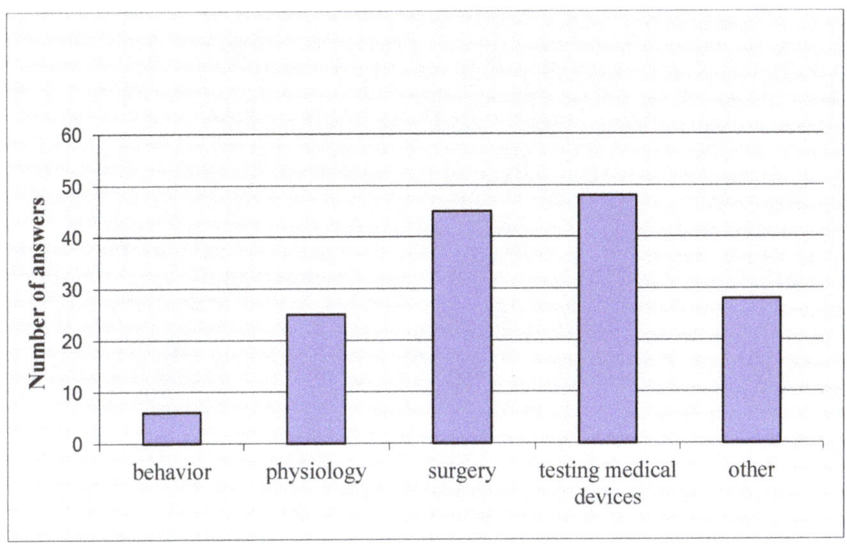

Figure 1. Main field of research using sheep (multiple answers possible).

When questioned about the origin of the sheep they were using, the respondents reported purchasing the sheep either from local farms (59.5%), commercial breeders (19%) or had their own flock (21.5%).

As the sex bias in biomedical research is an important topic that has been frequently addressed lately [4], the sheep users were also asked about the sex of the animals they used for their studies. The majority were using females (79.8%), 19% used neutered males, 16.7% intact males, while, for 19%, either the sex of the animals was not important or animals were selected based on availability. For this question, multiple answers were possible and thus the percentages add up a total of 134%, indicating both female and male (both intact and neutered) were used by some institutions.

Regarding the age range of the sheep at the beginning of the study, 46.4% of the respondents were using animals that were 1–2 years old, 33.3% older than 2 years, 32.2% younger than 1 year old, while 13.1% did not consider this to be important or used the animals that were available. Multiple answers were possible, and, based on the total percentage (125%), sheep of different ages are used by some respondents. Approximately half of the respondents (51.2%) did not have any preference for the sheep breed.

The main criteria for choosing a sheep supplier has not been reported in biomedical publications so far. The respondents were asked to rate several criteria in the order of importance. The animals' health status ranked first, followed by their availability, the trust and experience in the sheep provider and the animal homogeneity/uniformity (Figure 2).

Only 51.2% of the participants had a health monitoring program for their sheep. Among the diseases for which the animals had been screened, Q Fever was the most frequently cited. Half of the respondents reported that they were vaccinating their sheep (overall, almost one third had vaccinated against clostridia).

When asked if they had encountered problems in their sheep not related to the experimental protocol, 57.1% answered positively and, except one participant that reported problems at lambing, all the others mentioned health problems in the comments section for this question.

More than half of the respondents did not know the current purchase cost for their sheep; 67.9% could not estimate how much they would be willing to pay for a sheep with a controlled health status. The other 32.1% proposed prices ranging from EUR 50 to 1000.

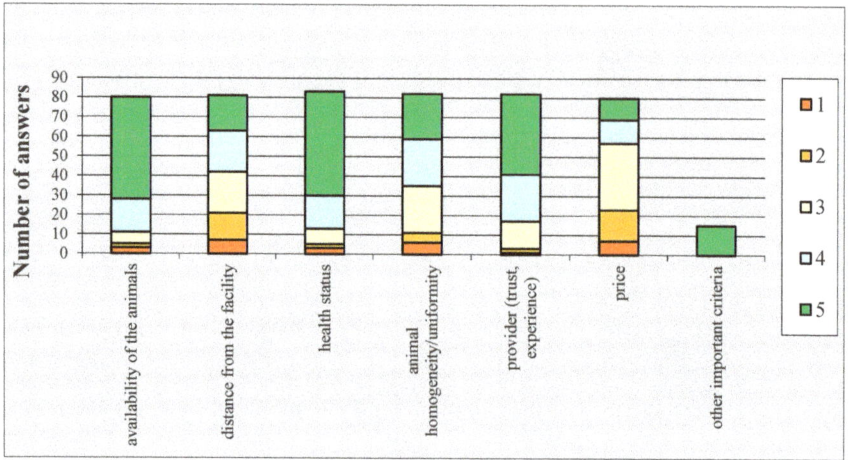

Figure 2. Main criteria for choosing a sheep supplier (several options were possible, rated from 5 (very important) to 1 (not important)).

4. Discussion

The goal of this survey among sheep users was to gather information on the current situation, with emphasis on animal selection criteria and issues encountered in practice if sheep are used for biomedical research. This survey has revealed that, currently, there is a strong need for improvement regarding sheep selection for biomedical research. The major issue identified was the animals' health status, which was, unfortunately, often insufficiently known when the animals were enrolled in experiments. With about 60% of the participants reporting on having encountered health issues unrelated to the experimental protocol, this seems to be a common problem. This may have a great impact on the experimental results, on the reproducibility of the findings and engender unnecessary ethical and scientific costs.

In this survey, only half of the participants followed a health monitoring program in sheep, whereas such programs are considered to be state of the art in rodent facilities. Furthermore, if there are sheep monitoring program in place, they differ substantially from institution to institution and are focused on only a few pathogens. It has to pointed out that the EU Directive 2010/63 states Annex III (requirements for establishments and for the care and accommodation of animals) in Section 3.1 that "establishments shall have a strategy in place to ensure that a health status of the animals is maintained that safeguards animal welfare and meets scientific requirements. This strategy shall include regular health monitoring, a microbiological surveillance program and plans for dealing with health breakdowns and shall define health parameters and procedures for the introduction of new animals" [2]. Beside the legal requirement, in the authors' opinion, health and welfare monitoring is refinement. The definition of "refinement" in animal experimentation has evolved significantly in recent decades. Russel and Burch, the creators of the 3Rs concept, described as refinement "simply to reduce to an absolute minimum the amount of stress imposed on those animals that are still used" [5]. They further characterized refinement as "an art or an ability to improvise", mentioning that "the greatest experimenters have been artists in this sense". Several decades later, Buchanan-Smith et al. [6] proposed a "harmonised progressive definition" of refinement, "in line with changes in animal ethics and animal welfare science" that included "health and welfare monitoring". Unfortunately, in the 21st century, despite guidelines and recommendations from experts [3,7–9], there are still no reports of the implementation of this harmonized definition in terms of health monitoring for sheep used for research purposes. The recently published recommendations of best practices for the health management of ruminants and pigs used for scientific and educational purposes of the FELASAmight be a first step in establishing best health and welfare management practices (i.e., refinement) at institutions [10].

Considering that the animals' availability ranked second among the criteria for selecting a sheep provider, it could be useful in the future to have specialized sheep breeders for biomedical research that ensure a good level of control and knowledge of the animals' health status.

5. Conclusions

This survey identifies health status, health monitoring and sheep provider as starting points for refinement in the use of sheep for biomedical research. This is in line with the harmonized, progressive definition of refinement [6].

Supplementary Materials: The following are available online at http://www.mdpi.com/2076-2615/10/9/1528/s1, Figure S1: Survey Sheep English version, Figure S2: Survey sheep French version.

Author Contributions: Authors were involved as following in the different parts of this study: conceptualization: C.M.B., U.L. and S.Z., methodology: C.M.B., U.L. and S.Z., formal analysis: C.M.B., writing—original draft preparation, C.M.B., writing—review and editing: C.M.B., U.L. and S.Z., supervision: U.L. and S.Z., project administration: C.M.B., funding acquisition: S.Z. All authors have read and agreed to the published version of the manuscript.

Funding: This investigation was financially supported with the assistance of the AO Foundation via the AO Strategy Fund Project number 13. No external funds were received in support of this work.

Acknowledgments: The authors would like to thank the SGV, the ECLAM, the ESLAV, the Swiss Animal Welfare Officers Network, the AFSTAL and the VOLE list for kindly accepting to distribute the questionnaire, all the participants and Jean Pierre Cabassu for his great support.

Conflicts of Interest: The authors declare no conflict of interest. The funders had no role in the design, execution, interpretation, or writing of the study.

References

1. 2019 report on the statistics on the use of animals for scientific purposes in the Member States of the European Union in 2015–2017. Available online: https://eur-lex.europa.eu/legal-content/EN/TXT/?qid=1581689520921&uri=CELEX:52020DC0016 (accessed on 30 July 2020).

2. Directive 2010/63/EU of the European Parliament and of the Council. Available online: https://eur-lex.europa.eu/legal-content/EN/TXT/?uri=CELEX:02010L0063-20190626 (accessed on 21 August 2020).

3. Rehbinder, C.; Alenius, S.; Bures, J.; de las Heras, M.; Greko, C.; Kroon, P.S.; Gutzwiller, A. Working party report. FELASA recommendations for the health monitoring of experimental units of calves, sheep and goats. *Lab. Anim.* **2000**, *34*, 329–350. [CrossRef] [PubMed]

4. Clayton, J.A.; Collins, F.S. Policy: NIH to balance sex in cell and animal studies. *Nat. News* **2012**, *509*, 282–283. [CrossRef] [PubMed]

5. Russell, W.M.S.; Burch, R.L. The Principles of Humane Experimental Technique. 1959 (reprinted 1992). Available online: https://caat.jhsph.edu/principles/chap7a (accessed on 30 July 2020).

6. Buchanan-Smith, H.M.; Rennie, H.M. Harmonising the definition of refinement. *Anim. Welf.* **2005**, *14*, 379–384.

7. Nicklas, W. International Harmonization of Health Monitoring. *ILAR J.* **2008**, *49*, 338–346. [CrossRef] [PubMed]

8. Kilkenny, C.; Browne, W.J.; Cuthill, I.C.; Emerson, M.; Altman, D.G. Improving Bioscience Research Reporting: The ARRIVE Guidelines for Reporting Animal Research. *PLoS Biol.* **2010**, *8*, e1000412. [CrossRef] [PubMed]

9. Cox, R.J.; Nol, P.; Ellis, C.K.; Palmer, M.V. Research with Agricultural Animals and Wildlife. *ILAR J.* **2019**, *60*, 66–73. [CrossRef] [PubMed]

10. Berset, C.M.; Caristo, M.E.; Ferrara, F.; Hardy, P.; Oropeza-Moe, M.; Waters, R. Federation of European Laboratory Animal Science Associations recommendations of best practices for the health management of ruminants and pigs used for scientific and educational purposes. *Lab. Anim.* **2020**. online ahead of print. [CrossRef] [PubMed]

Article

Remote Controlled Nociceptive Threshold Testing Systems in Large Animals

Polly Taylor

Taylor Monroe, Little Downham, Ely, Cambridgeshire CB6 2TY, UK; polly@taylormonroe.co.uk

Received: 31 July 2020; Accepted: 27 August 2020; Published: 2 September 2020

Simple Summary: Measurement of the nociceptive threshold (NT) is widely used in the study of pain and its alleviation. This records the intensity of a stimulus that causes pain to the test subject. The end point of the test that indicates when the subject experiences pain, the NT, is a behavioural escape response. Detection of a reliable and repeatable response depends on the animal behaving normally throughout testing. Restraint and an unfamiliar environment may prevent the animal from displaying normal behaviour and impede acquisition of robust NTs. Remotely controlled testing enables NT data to be collected from unrestrained animals behaving normally. Development of a remote controlled system for measurement of thermal and mechanical NTs in a range of large animal species is described. Normal "baseline" thermal and mechanical NTs from untreated animals are reported. This information can be used to improve both the welfare of the animals under investigation and the quality of the data collected. Remote controlled systems are now in use worldwide in both the study of pain physiology and in developing new pharmaceutical and non-drug-based methods of pain relief.

Abstract: Nociceptive threshold (NT) testing is widely used for the study of pain and its alleviation. The end point is a normal behavioural response, which may be affected by restraint or unfamiliar surroundings, leading to erroneous data. Remotely controlled thermal and mechanical NT testing systems were developed to allow free movement during testing and were evaluated in cats, dogs, sheep, horses and camels. Thermal threshold (TT) testing incorporated a heater and temperature sensor held against the animal's shaved skin. Mechanical threshold (MT) testing incorporated a pneumatic actuator attached to a limb containing a 1–2 mm radiused pin pushed against the skin. Both stimuli were driven from battery powered control units attached on the animal's back, controlled remotely via infra-red radiation from a handheld component. Threshold reading was held automatically and displayed digitally on the unit. The system was failsafe with a safety cut-out at a preset temperature or force as appropriate. The animals accepted the equipment and behaved normally in their home environment, enabling recording of reproducible TT (38.5–49.8 °C) and MT (2.7–10.1 N); precise values depended on the species, the individual and the stimulus characteristics. Remote controlled NT threshold testing appears to be a viable refinement for pain research.

Keywords: refinement; pain; nociceptive threshold; horse; cat; dog; sheep; camel

---

1. Introduction

Nociceptive threshold (NT) testing is widely used in studies of pain, analgesia, hyperalgesia and allodynia. Such testing records the intensity of a stimulus that causes pain to the test subject. Ethical, humane and repeatable methods for applying the stimuli and evaluating the responses are essential for producing robust data and maximizing animal welfare. A ramped noxious stimulus, usually thermal, mechanical or electrical, is applied, and the intensity of the stimulus at which a clear aversive response occurs is taken as the nociceptive threshold [1]. The threshold is expected to increase after analgesic treatment and to decrease with hyperalgesia and allodynia. Nociceptive threshold testing is very widely used for translational studies in rodents, both for drug development as well as for fundamental study of the physiology of pain. This is exemplified in, for example, Gugliandolo and colleagues' investigation into neuropathic pain in mice [2]. They used latency of the response to a thermal stimulus applied to the underside of the foot (Hargreaves test) and the withdrawal threshold for mechanical pressure on the plantar area of the foot (electronic von Frey filament test) to evaluate the hypersensitivity induced by sciatic nerve crush and its prevention by the treatment under investigation. Although rodents are not physically restrained for these tests, they must be contained in a special cage and be in close proximity to the tester, which may affect their behaviour [3]. Other tests such as the tail flick [1] and the Randall-Sellito method [4] used in rodents may involve actual physical restraint with even greater impact on behaviour. Larger animals are used less commonly for translational studies but their inclusion is essential for investigation into species-specific physiology and pharmacodynamics. The greater size of these species enables the equipment supplying the stimulus to be attached to the animal, for which rodents are too small. Remote control of this equipment without restraint or close proximity of the tester allows the animal to display natural behaviour.

There are a number of essential criteria for NT testing [5]. The stimulus must be repeatable over a few hours and reproducible over the weeks or months of a longitudinal study. The stimuli must not cause tissue damage, both for animal welfare and also so that the stimulated site is not sensitized to subsequent tests. It is essential that the stimulus is stopped as soon as the animal responds, and that there is a cut-out to prevent tissue damage if no response occurs. The end point of the test that indicates when the subject experiences pain, the NT, is a behavioural escape response and must be clearly identified. For instance, this includes stamping the foot or lifting the leg when a limb site is used. Turning to nose or bite at, and bending or jumping away from a site on the body is also seen. Detection of this response depends on the animal behaving normally throughout testing. Restraint and an unfamiliar environment may prevent the animal from displaying normal behaviour and impede acquisition of robust NTs. A relaxed, undisturbed animal ensures that responses are not suppressed by fear or distraction. It is often a requirement of studies investigating the effects of both analgesics (increased NT) and increased sensitivity (decreased NT) that the same animal can be tested at intervals over a period of at least 24 h. Testing over several days may be needed for investigation of, for instance, inflammatory pain and its treatment.

Thermal and mechanical testing has been used for many years in large (non-rodent) animals, particularly for analgesic drug development in the target species [6,7]. A thermal system designed specifically for cats was reported in 2002 [8] and has subsequently been widely used for analgesic drug evaluation in this species [9,10]. The thermal probe applied to the body was connected to the control unit by a cable. The cat was able to move around an individual cage while attached, but the cable was vulnerable to damage from the cat and caused some restriction of movement. The inherent disadvantages of a wired system were underlined by experience of its use in the cat. All the behavioural criteria described above apply to this species, but its unique temperament and desire to chase and bite the cables made a wireless control system highly desirable.

In horses and farm animals, the need for a cable connection generally led to the animal being restrained in stocks. Mechanical NT testing has also commonly been carried out in these larger species by exerting noxious pressure from a blunt ended pin pushed onto the surface of a limb [11–13]. These systems have all been driven by a variously sized and often noisy control box, again necessitating restraint in stocks. The animal is close to potentially fear-inducing or distracting machinery and human contact, which both prevent normal behaviour and limit the duration of a continuous study.

The potential advantages of testing unrestrained animals with a small and silent system were apparent. A wireless, remote controlled thermal NT testing system was first developed for cats and subsequently extended to larger animals, and then followed by a mechanical system for larger animals. Remote control of NT testing presents an additional challenge to purely telemetric recording as the power to drive the noxious stimulus must also be remotely controlled and attached to the animal.

This report describes the development and early validation of remote controlled systems for thermal and mechanical NT testing in a number of large (non-rodent) animal species. The data presented were collected during preparation for subsequent research protocols from animals tested in Australia, Brazil, Germany, Norway, Saudi Arabia, Switzerland, the UK and the USA under the authorization of the appropriate animal use ethical committee for each research establishment. Some of the data were presented at the World Congress of Veterinary Anaesthesia in 2006 [14] and to the Association of Veterinary Anaesthetists in 2008 [15].

2. Materials and Methods

2.1. Animals

All animals were housed, handled and fed according to the local institutional guidelines. All were confirmed as healthy from a clinical examination conducted during acclimatization. According to the species and their familiarity with the handling procedures, all animals were acclimatized to the testing environment for at least 30 min up to several days. Again according to the species, they were accustomed to wearing equipment for at least 30 min before testing. Naïve research animals wore dummy equipment for a few hours a day for several days before testing.

Cats: Group DEV was studied during the original development of the remote system. This group comprised 24 purpose-bred neutered (17 f, 7 m) domestic short haired (DSH) cats aged 1–4 years, weighing approximately 3–6 kg. They were group-housed and brought in pairs to the testing environment at least 2 h before any testing. Testing took place with the cats roaming freely in adjacent large wire-sided cages. Group CNSW comprised 2 adult neutered DSH cats housed at night and ranging freely outside during the day (6–8 kg). They were brought to the testing area at least 2 h before testing and were studied roaming freely in a large wire-sided cage. Group CWS comprised 2 young adult neutered DSH cats kept as household pets (3–4 kg). They were brought to the testing area at least 2 h before testing and were studied roaming freely in the room.

Dogs: Three groups of dogs were studied. Group DNSW comprised two mixed breed adult neutered male dogs (weighing 24 and 28 kg bodyweight). Group COL included 3 male and 3 female 8–9-month-old Walker hounds weighing 20–24 kg. Group WS comprised a Labrador, a spaniel and a mixed breed (15–25 kg). Groups DNSW and COL were kennel-housed in pairs or groups with lead and free range exercise daily. Group DWS dogs were kept as domestic pets. All dogs were brought to the study area at least two hours before any testing and studied whilst freely moving in large open-top wire cages or in the entire room.

Horses: Six groups of horses were studied. Group HPEN comprised four adult standardbreds, two mares and two geldings weighing 450–550 kg. They were turned out to pasture at night and stabled during the day. Testing took place in their own stall. Group HBR comprised 10 mixed breed young adult gelding horses weighing 268–460 kg. The horses were turned out on pasture at night and housed in part-covered yards for the duration of any testing schedule. They were brought to the testing stall and allowed at least 30 min acclimatization with the equipment in place before any testing. They were lightly restrained with a headcollar and long rope. Group HWS comprised two young Arab mares weighing 450 kg. They were kept at pasture and brought to the large open testing barn at least one hour before testing. They were also lightly restrained with a headcollar and long rope. Group HNSW comprised one adult Thoroughbred gelding weighing 475 kg. He was kept at pasture and allowed 30 min acclimatization before any testing, lightly restrained with a headcollar and rope in a small paddock. Group HNO comprised two adult Norwegian Trotter mares (450–500 kg). They were stabled throughout the study, and testing took place with them freely moving in their own stall. Group HHAN comprised seven adult warmblood horses weighing 650–700 kg, three geldings and four mares. They were also stabled throughout the study, and testing took place with them freely moving in their own stall. None of the groups was fed during testing.

Sheep: Two groups of sheep were studied. Group SWA comprised three calm, pregnant Merino sheep during preparation for a larger project investigating post-operative hyperalgesia. They were housed and tested in large, raised, wire-sided open-top pens in full view of their companions. They were allowed at least 30 min with the equipment in place before any testing. Group SZU comprised eight young adult Swiss Alpine sheep (castrated males and females, around 50 kg bodyweight). They were group-housed in large wire-sided open-top pens; testing was performed in a smaller area partitioned off with wire-sided hurdles, without separating the subject from its companions. At least 30 min acclimatization was allowed with the equipment in place before any testing started.

Camels: Three young adult dromedary camels were studied (group DCSA). One male and two females, weighing 6–700 kg, were included. They were housed in open yards but brought into individual horse stalls for the duration of the study. They were acclimatized to the testing environment for two days before any testing, which took place in their individual stalls.

2.2. Thermal system

The first remote controlled thermal NT testing system was aimed specifically for use in cats to replace the wired system developed in 2002 [8]. The remote system incorporated a similar thermal probe with the heating element adjusted to give the same heating rate (0.6 °C/s). The thermal probe comprised a heater and temperature sensor mounted together in heat conducting epoxy as a flat 10×10 mm probe weighing 5 gm [14] using battery power and an infra-red (IR) link (Maplins, UK). The probe was held against the cat's shaved thorax with an elasticated band. Constant and repeatable pressure against the skin was maintained by inflating a modified blood pressure bladder behind the probe. At each test, the heater was activated, and when the cat responded by twitching the skin, jumping, turning, flinching or occasionally vocalizing, the stimulus was terminated immediately and the threshold temperature recorded. A maximum cut-off temperature of 55 °C was set to minimize skin damage if the cat did not react at lower temperatures.

For remote control, the equipment was miniaturized so that the circuit board and batteries could be carried by the cat on a 50 cm wide back pack positioned on the dorsal thorax (Topcat Metrology Ltd. WTT1) (Figure 1). The back pack was held onto the cat using an elasticated band, and the circuit board and batteries were secured using Velcro®, with vulnerable cables under a soft, flexible cover. The underside of the band held the bladder and temperature probe against the thorax. The pressure transducer for control of the bladder pressure was attached to the circuit board, and a window between 30 and 70 mmHg was indicated by illumination of red and green LEDs on the circuit board. Tubing between bladder, transducer and a non-return valve (used for manual bladder inflation with a 20 mL syringe) was housed under the soft covering. The sensor output was displayed on a 3-digit display with peak hold, also fixed onto the band. The whole system weighed 320 g (Figure 2). Heating was controlled by IR signal operated manually and was activated only when the remote control handset button switch was depressed, so was failsafe if operator or IR contact was lost. For each test, the heater was switched on and held on by the operator who released the button, thereby stopping the heating, when the cat reacted. The display unit held the peak reading at the point the heater was switched off, and this temperature was recorded as the threshold. Peak hold was overridden by a second IR control via the handset in order to read skin temperature before the start of each test. Mirrors mounted on the walls of the cage allowed the display to be read whichever way the cat was facing.

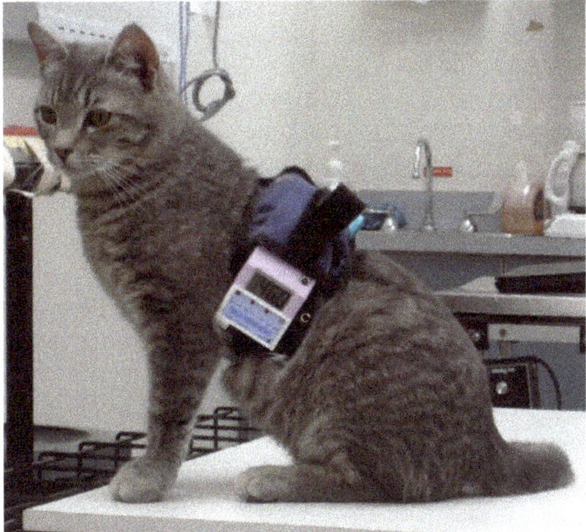

Figure 1. Remote controlled thermal unit worn by a small cat (3 kg) free roaming in a room.

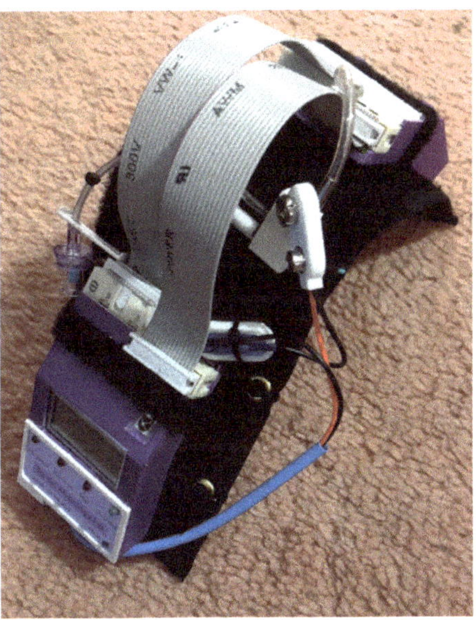

Figure 2. Remote controlled cat thermal band with cover removed.

The remote controlled system was also used with dogs. It was intended to replace the wired system previously reported [16]. The feline equipment was modified only by using a longer thoracic strap to secure the back pack on the larger species. Heating rates of 0.6–0.8 °C/s were used, and the cut-out temperature was 55–60 °C. Testing on the thorax produced similar responses, including skin twitching, turning to the probe site, biting at the band, flinching and moving away, but rarely vocalization. Testing on a leg site was also used as described below for larger animals.

For larger animals, the electronics and batteries were mounted in a shaped box on the animal's back, held in place by a larger version of the band used in cats and dogs (Topcat Metrology Ltd. WTT2) (Figure 3). Two digital temperature displays were used, mounted on each side of the control box. Two IR receivers were also used, mounted on each side. These modifications enabled IR control in the much larger accommodation space needed for larger animals and for the display to be read from outside the stall whichever way the animal was facing. In camels, a single cuboid box mounted on the side of the hump was adopted as the top of the hump was too high and rounded for the shaped box (Figure 4). A further development was to place the transmitter high on the wall of the stall, wired to the handset, and the receiver mounted on an aerial attached to the animal's back (Figure 5a,b). A similar thermal probe as in cats was used, with a heating rate of 0.8 °C/s. Constant contact between probe and skin was assured with the same pressurized bladder. The temperature probe was positioned either on the thorax, under the control unit, or it was placed on the dorsal metacarpus or metatarsus using a smaller elasticated band and connected to the control box by a longer ribbon cable. On dogs, sheep, small horses and camels, the probe was secured on a limb without a pressure bladder; consistent contact was assured by careful adjustment of the elastic strap (Figure 6). A limb site was not tested on cats. The response to thoracic stimuli was a skin twitch, turning to the site, bending away from the site or becoming agitated. Limb stimuli evoked stamping, a snatched leg lift or nosing at the site. Cut-off was set to 55–60 °C.

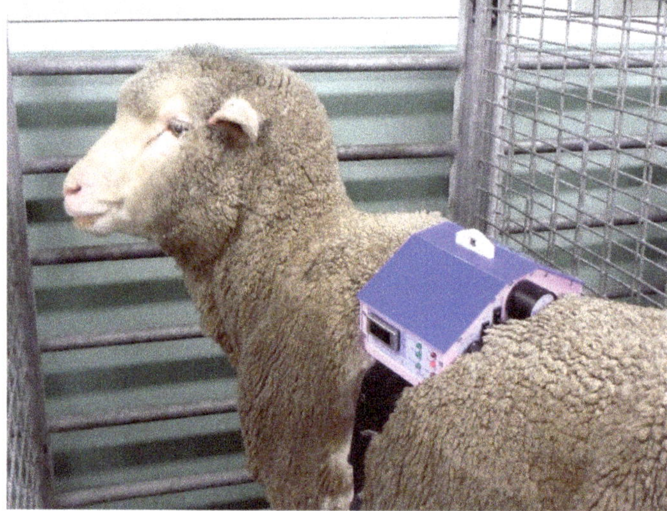

Figure 3. Remote controlled thermal unit on an unrestrained sheep in its pen.

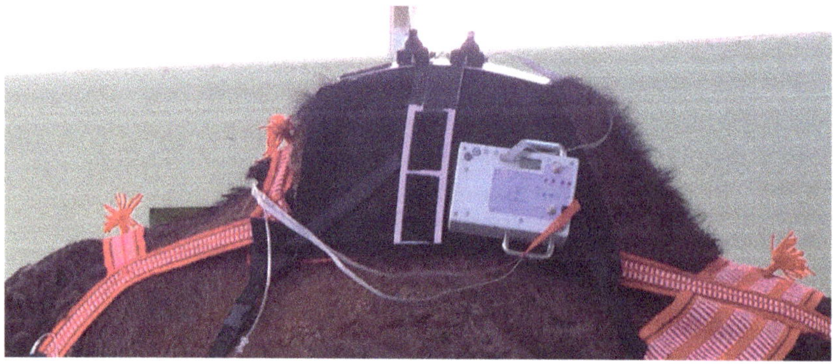

Figure 4. Cuboid thermal unit mounted on a camel's hump.

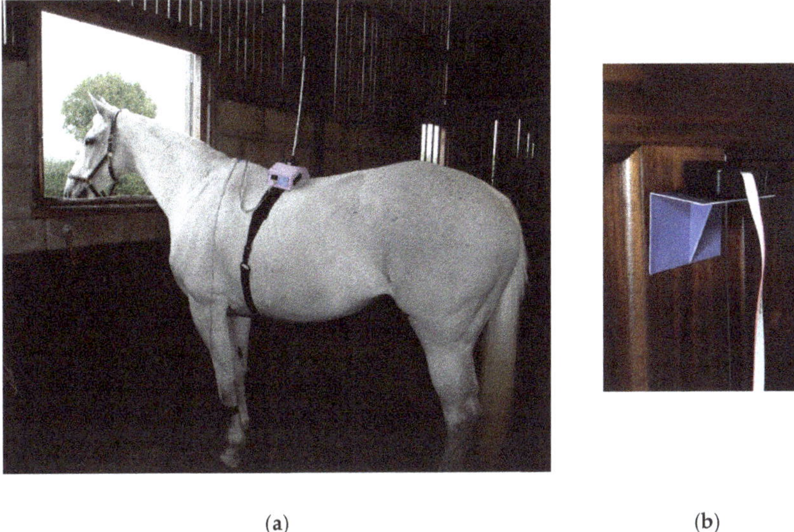

(a)   (b)

Figure 5. (**a**) Thermal unit with receiver mounted on aerial on a horse's back. (**b**) Transmitter placed high on stable wall.

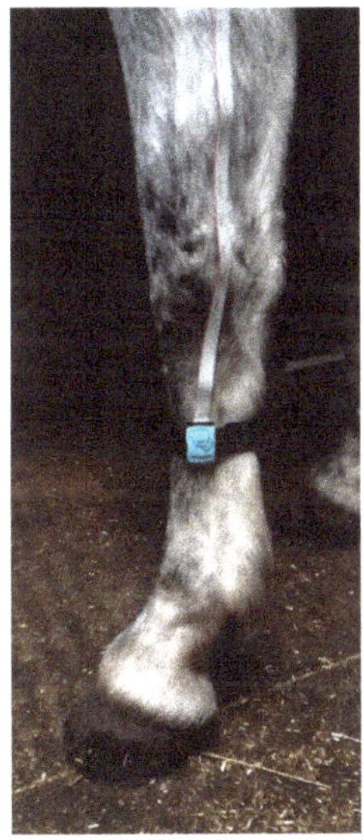

Figure 6. Thermal probe positioned on the limb with a carefully tensioned elastic and Velcro® band.

A modified probe [17] was incorporated into the system for all species from 2013. The style of heating probe used in each group is indicated in Table 1. Heating rates of 0.6 °C/s were used in cats and dogs, 0.8 °C/s in horses and sheep, and around 2 °C/sec in camels.

Thermal Testing Schedule (see Table 1)

Baseline thermal nociceptive thresholds (TTs) were collected from conscious animals who had not received any medication. The TT for each animal was recorded as the mean of at least 3 tests recorded at 10 min intervals. When first applied, familiarization and training for both tester and subject comprised a series of up to 7 tests at no less than 10 min intervals in order to establish the reaction of the individual subject. This was repeated several hours later and usually again over the following few days to allow complete familiarization. Once training was completed, baseline TT was taken as the mean of 3–5 tests within 10%. Tests were conducted at ambient temperatures of around 21 °C in dogs and cats but ranging from 11 to 30 °C in the larger species.

During development in cats, 12 of group CDEV (CDEV1) were tested with both the original wired and the remote controlled systems in order to confirm that the remote system produced stimuli and TTs similar to the original. These cats had been previously familiarized with the original wired system. The remaining 12 cats were tested with only the new system (Group CDEV2). Four DEV1 cats (DEV1a) were tested without any drug treatment with both systems on separate occasions either on consecutive days or with not more than 12 months between testing days. Two cats were tested with the wired system first and two with the wireless first. Each cat was tested 5–13 times at 15 min intervals with each system. A further four CDEV1 cats (DEV1b) were also treated with opioids to examine the performance of the wireless system when thermal thresholds were raised above normal. Two cats received intramuscular (IM) butorphanol 0.4 mg/kg and two received sub-lingual (SL) buprenorphine 20 µg/kg. Three tests at 15 min intervals were made before treatment, and at intervals not less than 15 min post treatment for up to 24 h. The mean of the pretreatment tests was taken as the baseline TT for that individual (see Table 2).

Historical data from CDEV1 cats using the wired system [6] and CDEV2 cats using the wireless system [18] were compared before, during and after treatment with buprenorphine (20–80 µg/kg). Three to five tests were performed before treatment, and at intervals not less than 30 min for 24 h after treatment.

Thoracic baseline TTs were collected from cats in groups CNSW and CWS and from dogs in DNSW, DCOL and DWS. The CNSW group cats and the DNSW dogs each received 1.0 mg/kg methadone intramuscularly (IM) after baseline TT had been recorded. The TT was then recorded at 15 min intervals until it returned to baseline.

Baseline TTs were collected from horses in groups HPEN (thorax and legs), HBR (thorax and leg), HWS (thorax and leg), HNSW (thorax), HNO (thorax) and three from HHAN (thorax). Horses in groups HBR ($n = 5$), HNSW, HNO, HWS and HPEN were treated with intravenous (IV) 0.2 mg/kg methadone, 0.15 mg/kg methadone and 0.25 mg/kg xylazine, 0.2 mg/kg methadone, 0.1 mg/kg butorphanol and 0.03 mg/kg acepromazine or 0.5 mg/kg xylazine and 0.025 mg/kg butorphanol, respectively, after baseline TT had been recorded, and NTs were measured at 15–30 min intervals for 3 h and on the following day, 16–18 h later.

In sheep, the responses to thoracic stimuli were difficult to detect and this site was abandoned. Baseline limb TTs were collected in groups SWA (metatarsal site) and SZU (stifle and metatarsal area). The thoracic site was not attempted in camels, and only baseline leg TTs were recorded.

2.3. Mechanical System

The mechanical stimulus was produced by a single blunt ended pin driven onto the surface of the skin by a pneumatic actuator [19] positioned on the dorsal surface of the metacarpus, midway between the carpus and the metacarpo-phalangeal joint (Figure 7). Increasing pressure in the actuator drives the pin. The system is calibrated to give force (in N) as the stimulus intensity. The responses to stimulation of the limb are similar to those seen with thermal stimulation at the same site.

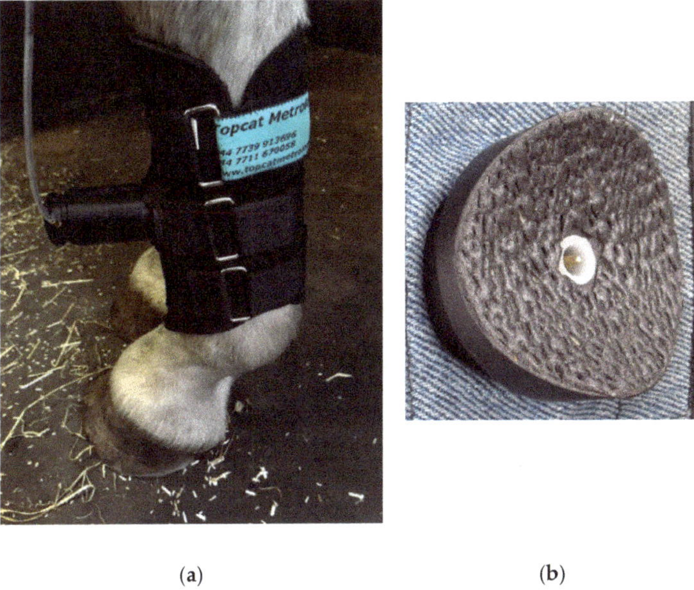

(a)   (b)

Figure 7. Pneumatic actuator: (**a**) secured with boot and carefully tensioned band on the forelimb of a horse standing free in its stall, (**b**) the leg side of the actuator showing probe tip in resting position.

The remote controlled device (Topcat Metrology Ltd. WMT2) was first developed in horses to replace the wired system by modifying the manually operated system previously described [20–22], where pressure was produced in a syringe compressed by hand. Indicator lights were used to keep the force rise rate within a predefined window. The system was silent and allowed rapid removal of the stimulus at threshold via a vent valve. The remote controlled system was operated via IR signal from a handset activated by the assessor in the same way as for the thermal system. The modification to include the aerial on the animal's back was also used for the mechanical system. The shaped unit was positioned on the horse's back and secured with Velcro® to a fly-sheet or surcingle. A pressure reservoir, recharged between tests, was mounted alongside the control unit (Figure 8). This supplied a miniature, silent, solenoid metering valve (Kinesis, UK) to increase the cuff pressure at a predetermined rate. The valve was controlled via a feedback circuit from a pressure transducer in the force actuator. The final vent valve configuration was "normally shut", but opened once a second when in standby mode. This required less power, therefore improving battery life. It also ensured that the subject was familiarized with the regular click of valve operation, which although insulated to a low volume, could act as an audible clue if only present during the test. At the start of the test, the click simply changed from the vent valve to the pressure valve.

Threshold reading was held automatically and displayed on a digital display on both sides of the control unit. The system was failsafe with a safety cut-out at a preset force and a pressure relief valve with electronic interlock to remove the stimulus if the remote signal was lost. The system was validated by mounting the probe onto a force transducer (Kenwood, UK) and recording force rate rise (FRR) during calibration. Human thresholds were measured at 5 min intervals on the dorsal metatarsals of two of the researchers (PT, MD) to evaluate an initial pressure in the supply reservoir of 32, 43 and 49 kPa (240, 320 and 370 mmHg).

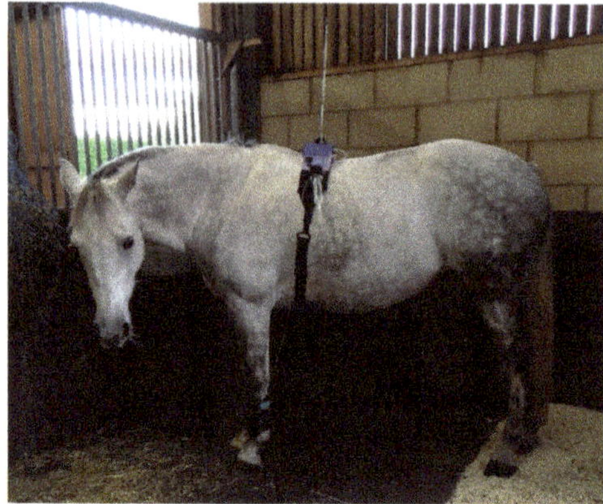

Figure 8. Horse with mechanical system in place: control unit and reservoir mounted on the back, with pressure line to the actuator on the foreleg.

The mechanical system was too large for use in cats and dogs and was not tested in sheep, although at least those of over 40 kg would be suitable. The control box for camels was a cuboid design fitted to the side of the hump in the same way as the thermal system.

Mechanical Testing Schedule

Baseline mechanical nociceptive thresholds (MTs) were collected from conscious animals who had not received any medication. The MT for each animal was recorded as the mean of at least three tests recorded at five minute intervals. When first applied, familiarization and training for both tester and subject comprised a series of up to 10 tests at no less than 5 min intervals in order to establish the reaction of the individual subject. This was repeated several hours later and usually again over the following few days to allow complete familiarization. Once training was completed, baseline MT was taken as the mean of 3–5 tests within 10%.

Baseline data were collected from horses in groups HPEN, HBR and four from HHAN (see Table 3). Horses in group HPEN were treated with 0.5 mg/kg xylazine and 0.025 mg/kg butorphanol IV after baseline MT had been recorded, and NTs were measured at 10–15 min intervals for one hour and on the following day, 20 h later. A set of baseline data was also collected from the four horses in group HHAN, each test within 2 min of the same time points tested with the wired system previously described [21] and attached to the same actuator. In this case, the tester stood close to the horse and manually operated a syringe to generate the stimulus.

Baseline data were collected from the three DCSA camels using a 0.5 mm probe tip at a mid-metacarpal site (see Table 3). The actuator was positioned on the lateral aspect to prevent it from being dislodged when the camel lay down. Data were collected at 10 min intervals with a force rise of 2 N/s.

2.4. Data Analysis

Descriptive data are presented, including range and mean ± SD as appropriate. An unpaired *t*-test on the excursions (ΔT °C = threshold temperature–skin temperature) was used to compare the data collected with wired and wireless systems in cats and horses. Friedman's test was used to compare baseline data with NT after opioid treatment as the data are not normally distributed due to "capping" of the peak TT by the cut-off (GraphPad Prism v8). $p < 0.05$ was regarded as significant.

3. Results

3.1. Thermal System

3.1.1. Evaluation of Remote Control (Table 1)

The IR control was successful in all the species, and there were few incidents where contact was lost. The signal was used either by pointing the transmitter directly at the control unit on the animal or by intentionally reflecting the signal off a reflective surface. The only materials that proved inadequate for reflection were rough high wooden roofing in a barn or horse bedding on the ground. Inclusion of the aerial on the animal and a fixed transmitter position prevented loss of contact through human error in misdirecting the signal and facilitated separation of the tester from the animal. Low ambient temperature did not affect the unit's function, except that more power (i.e., a fully charged battery) was required to heat to threshold temperature from a lower starting point.

Table 1. Skin temperature (°C) (range) and thermal nociceptive threshold (°C) (TT) (mean ± SD (range)) recorded with a remote controlled system during training in cats, dogs, sheep, horses and camels.

Species, Group (site)	Skin Temp (°C)	TT (°C)	Probe Style	Max TT (°C) After Opioid	Cut-Out (c/o) Temp (°C)
Cat					
CDEV *n* = 12 (thorax)	38.6–39.6	43.2 ± 3.0 (41.2–45.8)	old	c/o	55
CNSW *n* = 2 (thorax)	36.2–37.1	45.1 ± 1.8 (43.6–48.1)	new	c/o	55
CWS *n* = 2 (thorax)	36.2–36.4	45.8 ± 0.1 (45.7–45.9)	new	n/a	60
All cats (thorax)		44.7 ± 1.1			
Dog					
DNSW *n* = 2 (thorax)	35.9–36.6	43.5 ± 0.8 (42.7–44.6)	new	c/o	55
DCOL *n* = 6 (thorax)	37.3–38.2	47.9 ± 1.4 (45.9–49.8)	new	n/a	60
DWS (*n* = 3) (leg)	30.4–33.0	42.7 ± 1.4 (38.4–5.3)	new	n/a	60
All dogs (thorax)		45.7 ± 2.5			
Horse					
HBR (*n* = 10) (thorax)	36.3–37.3	54.2 ± 2.0 (51.9–58.4)	new	n/a	55
HBR (*n* = 10) (leg)	31.5–34.0	46.9 ± 3.2 (41.0–51.3)	new	c/o	55
HWS (*n* = 2) (leg)	35.7	54.4	new	n/a	60
(thorax)	34.8	52.7	new	c/o	60
HNSW (*n* = 1) (thorax)	36.2	50.9	new	c/o	60
HNO (*n* = 2) (thorax)	33.7–36.4	50.5 ± 3.4 (45.0–54.4)	new	c/o	60
HHAN (*n* =3) (thorax)	35.8–36.8	50.5 ± 3.2 (47.2–56.0)	old	n/a	60
HPEN (*n* = 4) (thorax)	31.1.–31.7	46.4 ± 1.9 (44.5–48.2)	old	n/a	60
HPEN (*n* = 4) (leg)	21.0–30.5	45.7 ± 1.7 (43.4–47.3)	old	56.2	60
All horses (leg)		49.0 ± 3.8			
All horses (thorax)		50.9 ± 2.4			
Sheep					
SWA (*n* = 3) (leg)	32.3–35.3	51.2 ± 1.9 (48.8–54.1)	new	n/a	60
SZU (*n* = 8) (leg, stifle)	30.5–34.9	53.7 ± 3.0 (48.9–c/o)	old	n/a	70
(leg, metatarsal)	18.0–26.2	55.0 ± 1.6 (53.0–c/o)	old	n/a	70
All sheep (leg)		52.4 ± 2.6			
Camel					
DCSA (*n* = 3) (leg)	24.2–27.2	95.0 ± 14.3 (74.9–106.0)	custom	n/a	130

Probe: old style [8], new style [17]. Cut-out (c/o): temperature set to cease heating automatically when no response is detected. Opioid treatment: butorphanol, methadone, buprenorphine or methadone/sedative combinations as detailed in the text.

3.1.2. Comparison with Wired System

In the four DEV1a cats, mean ± SD baseline with the remote system was 43.6 ± 2.1 °C and with the wired system 42.0 ± 1.7 °C. Excursions (TT − skin temperature) in four cats (DEV1a) were 5.0 ± 1.1, 2.8 ± 0.9, 3.0 ± 0.5, 4.0 ± 2.1 °C (remote) and 4.4 ± 1.3, 3.4 ± 1.3, 5.3 ± 1.3, 6.1 ± 1.0 °C (wired). Excursions measured with both systems were not statistically different. Peak excursions in two cats were >13.1 °C (reached cut-out) 10–105 min after butorphanol, similar to previous data from opioid-treated cats using the original system [23,24].

In group DEV1a cats (n = 4), the skin temperature was always higher when recorded by the remote system compared with the wired (wired 36.8 ± 0.6 and remote 39.4 ± 1.5, p < 0.001). Remote TT was higher than wired in only one cat, and the mean difference was small although significantly different (wired 42.0 ± 1.7 and remote 43.6 ± 2.1, p < 0.01). However, the excursions recorded with the remote system were smaller than with the wired (wired 5.2 ± 1.4 and remote 4.3 ± 2.0, p < 0.05). The TT in all Group DEV1b (n = 4) cats increased after opioid treatment. Peak excursions ranged between 16.1 and 18.2 °C, which is similar to published data (Table 2).

Table 2. Thermal thresholds (TTs) in group DEV1b cats. Peak excursions (skin–TT difference) after treatment with intramuscular (IM) butorphanol 0.4 mg/kg or sub-lingual (SL) buprenorphine 20 µg/kg. Published approximate mean peak delta T data after buprenorphine and butorphanol included for comparison.

	Maximum Excursion °C	
Cat/Reference	*Butorphanol*	*Buprenorphine*
5	18.1	
6	16.4	
7		18.2
8		16.1
Lascelles and Robertson [23]	18	
Robertson et al. [24]		17

The historical data from CDEV1 cats (n = 12) using the wired system [6] and CDEV2 cats (n = 12) using the wireless system [18] showed that mean skin temperature in CDEV1 cats was always 1–2 °C lower than in CDEV2 cats. The TT in both groups increased significantly after buprenorphine treatment, remaining higher than pretreatment from 60 to 240 min in CDEV1 and from 30 to 300 min in CDEV2. The TTs were similar except 1–2 h after treatment, when CDEV2 TTs were higher.

3.1.3. Cats

The cats were undisturbed by the equipment and behaved normally whilst wearing it for several hours. The only impediment in those freely ranging in a room was inability to enter a space only as wide as the cat, when the back pack was caught and slid back. This did not occur in animals loose in a large cage. Cats treated with opioids tended to roll over; this sometimes caused the back pack to slip and require adjustment before a reading could be made. Response at TT was a clear skin flick or turn towards the heated site. Occasionally the cat would jump forwards at TT. Vocalization was rare. Group CNSW baselines ranged from 43.6 to 46.4 °C, reached cut-out (55 °C) 30 min after IM methadone (1 mg/kg) and remained higher than baseline for 3 h. Group CWS cats' baselines were within the same range (see Figure 9a,b).

3.1.4. Dogs

The dogs rapidly became accustomed to the equipment and behaved normally whilst wearing it for several hours. Younger, more active dogs required more acclimatization than calmer animals. They tended to scratch at the back pack with a hindleg when it was first put on. Clear responses included a skin flick, turning to, scratching or biting at the site and jumping forwards. Vocalization was very rare. Baseline TT in group DNSW dogs ranged from 42.7 to 44.6 °C, reached cut- out (55 °C) 10 min after IM methadone (1 mg/kg) and remained higher than baseline for 3 h. In group COL, mean ± SD baseline skin temperature was 37.7 ± 0.3 and TT 47.9 ± 1.4 °C (see Figure 9a,b).

3.1.5. Horses

All the horses accepted the equipment without any reaction. Occasionally it was pushed posteriorly if a horse rubbed on a wall or fence, but repositioning was rarely required. Responses at TT on the leg were a clear leg lift or stamp, or occasionally the horse would nose or even bite at the site. Responses to the thoracic site were more difficult to detect as a skin flick could be obscured by the more bulky testing equipment used in large animals and the horses appeared less responsive at this site. However, a skin flick, a bend of the body away from the testing site or turning to look at the site were also commonly seen. Skin temperature was lower on the leg than the thorax and was more dependent on ambient temperature. The TT was higher on the thorax than on the leg. Group HBR skin temperatures were 36.7 ± 0.4 (thorax) and 32.7 ± 1.0 (leg), and TT 54.8 ± 1.8 (thorax) and 46.9 ± 3.2 °C (leg). Group HPEN skin temperatures were 36.7 ± 0.4 (thorax) and 32.7 ± 1.0 (leg), and TT 54.8 ± 1.8 (thorax) and 46.9 ± 3.2 °C (leg). Leg skin temperature and TT were significantly lower than on the thorax ($p < 0.001$). Skin temperature and TT ranges in the smaller groups (HWS, HNO, HWS) are shown in Table 1. The TT in horses given analgesic treatment (groups HBR, HNSW, HNO, HWS and HPEN) increased above baseline between 10 min and 1–2 h. It reached cut-out (55 or 60 °C) in all except HPEN horses where the highest TT was 53 °C. All TTs had returned to baseline three hours after dosing and were within the same range the following day (see Figure 9a,b).

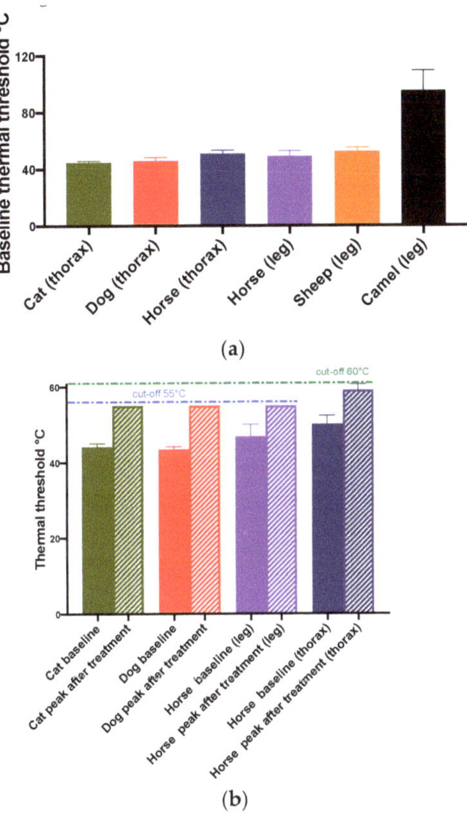

Figure 9. (**a**) Mean ± SD baseline (no treatment) thermal nociceptive thresholds (TTs) (°C) measured on the thorax in cats ($n = 16$), dogs ($n = 8$) and horses ($n = 21$) and on a leg in horses ($n = 15$), sheep ($n = 11$) and camels ($n = 3$). (**b**) Mean ± SD pre treatment and peak post analgesic treatment (see text for detail) TTs (°C) measured on the thorax in cats ($n = 6$), dogs ($n = 2$) and horses ($n = 8$) and on a forelimb in horses ($n = 10$). Thermal threshold reached cut-out (55° or 60 °C as shown) in all cats, dogs and in horses using the leg site. For the horse thoracic site, those reaching cut-out ($n = 4$) were deemed a TT of 60 °C for illustrative purposes.

3.1.6. Sheep

All the sheep accepted wearing the back pack with very little acclimation time. The calm SWA sheep tolerated the testing protocol without any problems, and the response at TT was a clear leg lift or stamp. Skin temperature and TT ranges are shown in Table 1. The young flighty Swiss mountain sheep were much more difficult to test as they appeared to adopt a prey species freeze when agitated or frightened by anything unfamiliar, particularly the presence of people. Many tests went to cut-out with no response detectable. When the tester was hidden outside the room responses were lower, in accordance with values reported in other sheep [25]. This is demonstrated by sheep 5201 with a leg skin temperature of 34.3 °C: with a tester standing in the same room, the temperature consistently reached cut-out with no response. When the tester was obscured from the sheep outside at a window, and outside the pen, TT was 53.7 °C (see Figure 9a).

3.1.7. Camels

The camels willingly accepted the equipment on their backs. Shaving the skin to enable contact with the thermal probe caused more aggravation than wearing the limb bands once they were positioned. The response at threshold was a clear leg lift or stamp. However, the camels tolerated much higher temperatures than other species, and custom probes with a faster heating rate were developed for them. Even after reaching temperatures in excess of 100 °C, there were no skin lesions other than a slight indent under the footprint of the probe. No pain, swelling or abrasion of the skin was seen. Skin temperature in the DCSA group was 26 ± 1.6 and TT 95.0 ± 14.3 °C (see Figure 9a).

3.2. Mechanical System

3.2.1. The System (Table 3)

The unit functioned as intended; the solenoid metered valve resulted in an FRR that did not deviate by more than 0.1 N regardless of the initial reservoir pressure. IR control was successful in both species tested, as for the thermal system. The continuous quiet click of the vent valve during standby and of the pressure valve during the test was ignored by all the subjects, confirming the benefit of the regular opening of the vent valve during standby between tests. If the signal contact was lost for any reason, however, the failsafe system vented the pressure in the actuator and that test was aborted. Incorporating the fixed transmitter and the aerial mounted on the animal's back ensured that this was extremely rare, and there were few incidents where contact was lost. Human MT from five tests was 7.3 ± 0.3 N, consistent with data from previous devices [21], with a coefficient of variation of 3.6%. Cut-out was set at 20–25 N.

Table 3. Mechanical nociceptive thresholds (MTs) (mean ± SD and range) using a limb-mounted actuator under infra-red remote control during training in horses and camels.

Species, Group	MT (N)	Pin Diameter	Max MT after Opioid (N)
Horse			
HPEN ($n = 4$)	10.3 ± 4.6 (3.0–15.0)	3 pin	20.3
HBR ($n = 10$) (leg)	5.6 ± 2.3 (2.7–10.1)	1 mm	n/a
HHAN ($n = 4$)	6.3 ± 2.2 (4.0–9.9)	1 mm	n/a
All horses (1 mm pin)	5.9 ± 2.2		
Camel			
DCSA ($n = 3$)	13.8 ± 2.3 (11.2–15.5)	0.5 mm	n/a

3.2.2. Horses

The horses tolerated the equipment in the same way as the thermal unit, and both could be accommodated on the horse's back at the same time. The response at MT was indistinguishable from the response to thermal stimulation at a similar site, namely a leg lift, stamp or turning to and nosing the actuator. Baseline MT in the HBR, HPEN and HHAN (n = 4) groups was 5.6 ± 2.3 N, 10.3 ± 4.6 N and 6.3 ± 2.2 N, respectively (See Figure 10a). In the HPEN group, MT increased above baseline for 30–60 min after the analgesic treatment. It returned to baseline by 75 min and remained at baseline (7.9 ± 2.5 N) 24 h later ((See Figure 10b). The MT measured simultaneously with the wired system (4.1 ± 1.3 N) in the four HHAN horses was significantly lower (p < 0.05) than when measured by remote control ((See Figure 10c).

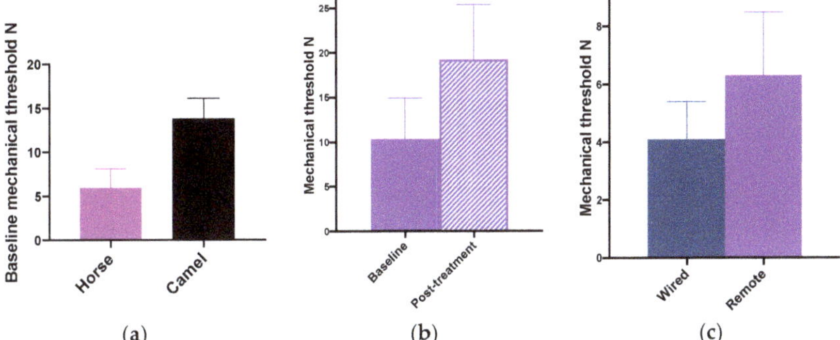

Figure 10. (**a**) Mean ± SD baseline (no treatment) mechanical nociceptive thresholds (MTs) (N) measured on a leg in horses (n = 14) and camels (n = 3). (**b**) Mean ± SD pre treatment and peak post analgesic treatment (see text for detail) MTs (N) measured on a foreleg in four horses using a 3-pin actuator. (**c**) Mean ± SD baseline (no treatment) MTs (N) measured on a foreleg of 4 horses with a wired system (close proximity and contact with the tester) and with the infra-red remote controlled system (horse unrestrained and tester outside the stall). Measurements with each system were made within less than 5 min.

3.2.3. Camels

The camels tolerated the mechanical equipment as easily as the thermal, and both units were worn together for several hours at a time with no effect on normal behaviour. Responses to stimulation were a clear leg lift or stamp. The DCSA group baseline MT was 13.8 ± 2.3 (See Figure 10a).

4. Discussion

This report documents the development and evaluation of remote controlled systems for TT and MT testing in large animals. Remote control by IR of both thermal and mechanical systems proved reliable. The baseline data were consistent, and increased thresholds were detected after analgesic treatment. Although NT testing does not mimic the complexities of true clinical pain, enabling normal behaviour should bring the response to a drug closer to clinical reality than in a restrained animal and therefore provide information that more closely resembles how a drug will behave under clinical conditions. The adaptations required both for power to activate the stimulus and for the control circuits to be placed on the animal were successful. Thermal stimulation depended only on small electrical components producing heat, and the main requirement was sufficient miniaturization for the electronics and sufficient battery power to be carried on the back of an animal as small as the cat. The mechanical system, however, required generation of air pressure to drive the actuator pin, necessitating attachment of a gas reservoir and metered pressure control being fitted to the animal. Wired or handheld MT testing systems driving a pneumatic actuator can make use of manual compression of a syringe [21,22,26]. Both air reservoir (the syringe) and pressure control (e.g., following visual pressure indicators) are under operator control and not fixed to the animal. The system adopted for the remote control system incorporated solenoid metering valves and an air reservoir of at least 60 mL. This proved too large for dogs and cats, but suitable for larger animals.

Most of the criteria for NT testing outlined by Beecher [5] were met. The aspects concerning repeatability of the stimulus, minimal tissue damage, immediate cessation of the stimulus when the animal responds and automatic cut-off if there is no response are no different from wired or handheld devices [8,22]. Features to prevent tissue damage and to ensure stimulus repeatability have been addressed elsewhere [17]; this report concentrates on aspects that allow normal behaviour and a clear escape response at threshold.

All the species tolerated wearing the equipment well. The larger, herbivore animals often required no acclimation to this at all. Young, playful dogs and cats required more "dummy" sessions, but reliable data could be collected in all cases with time and handling appropriate to the species. Complete free roaming was clearly not possible as the IR system would be out of range and the animal must be within sight of the tester for its response to be seen. However, testing was entirely feasible within a room or large kennel for small animals and in a stall without the use of stocks for large animals. Absence of wires and cables was particularly valuable in cats and young, playful dogs.

In contrast to handheld units, remote NT testing limits the stimulus site to body parts where the probe or actuator can be reliably fixed to the animal's body. Mechanical testing in particular is limited as the actuator must react against something to generate the force. A band around a limb provides the necessary configuration, but application to other parts of the body is more difficult. Although fixing the thermal probe on the thorax was both straightforward and suitable for cats and dogs, in the larger animals it was sometimes difficult to detect the end point, in part as the thorax was obscured by the bulkier equipment but also because the response itself was less clear. Independent investigations examined the response to NT at different sites, and both concluded that the thorax was less reliable and NTs were higher [27,28]. Large animals in stalls are more likely to be bothered by flies, resulting in confusing skin twitches over the body. Testing the limb is generally preferable, except when opioid analgesia is employed when an alternative site maybe better. Opioids often produce locomotor stimulation in otherwise pain-free healthy horses and obscure the leg lift or stamp response. In cold weather, a further disadvantage of the limb site for thermal testing was the low skin temperature of the limbs requiring more power to heat to threshold temperature. It is of note that skin temperature and TT in the cats were higher with the remote system than the wired. This was considered likely to be due to the larger bulk of the remote band providing more insulation than the narrow band without circuit boards used for the wired system. The original wired probes also included a substantial aluminium support that conducted heat away, probably contributing to lower indicated skin temperature. However, in spite of the higher temperatures recorded with the remote system, the excursions (TT – skin temperature) were similar. This has led to use of the excursions in many reports [29], at least for statistical analysis, as it removes one source of variation.

The precise NTs recorded are dependent on many factors including the site, the style of the stimulus probe, the environment and the breed or strain of the species under study. This makes results from studies in different laboratories difficult to compare. Remote testing may reduce the variation by allowing normal behaviour but it still remains extremely important that all the conditions are detailed, and strict consistency within one study is essential. A further central source of disparity within a study is individual variation. There is usually a range of skin temperature and both TT and MT even within a group of similar animals [30], and reducing all other sources of variation is necessary to produce useful data. It is important to describe the site, heating rate, probe style and the environment in order to understand the data fully.

It is well recognized that distraction alters the perception of pain [31] and, however caused, may affect recorded NT. Furthermore, the importance of enabling a natural escape response to indicate the NT has been emphasized [32]. Any distraction or, particularly in prey species, anything threatening or frightening will illicit abnormal behaviour, particularly freezing, and may prevent a natural escape response [33]. The potential for the tester to be unseen by the animal subject is of particular benefit in prey species unaccustomed to contact with humans. This was illustrated by the sheep who did not respond when the tester was in the same room. Our own unpublished observations with rabbits support this: there was no response to thermal stimulation unless the tester was hidden, when TTs similar to dogs and cats were recorded. The effect of close proximity of the tester was illustrated even in calm horses who were quite familiar with humans. The MTs in the HHAN horses were significantly lower when tested with the wired compared with the remote controlled system. This is presumably due to an element of anticipation and apprehension with close proximity, when there are visual and perhaps audible clues that the stimulus is coming. This effect is further illustrated in sows [22] where hand-held MT testing was compared with a limb-mounted actuator, albeit still connected by a light pressure line. Baseline MTs were 13–17 N with the handheld device and 18–23 N with the limb fixed actuator, incorporating the same probe tip.

This investigation did not address reproducibility of the NT in the same animals over days, weeks or more. Remote control would in theory probably foster stable baseline thresholds. Learning to anticipate an aversive experience is less likely if the animal's behaviour is unaffected and no cues reinforce anticipation. Dogs undergoing MT testing with a handheld device at several sites at 10–15 s intervals had lower MTs when tested 10–14 days later [34]. It was concluded that the dogs had learned to respond to the same stimulus more quickly. This effect was probably exacerbated by the necessary restraint, the high frequency of the tests and the use of a large probe tip (1 cm diameter), which requires a larger force to produce the same pressure as the smaller probe tips used here. Pain is experienced when the stimulus intensity is sufficient to stimulate nociceptive nerve endings by pressure; the high forces required to generate sufficient pressure with a large probe surface area may squash the dog and be unpleasant even before they produce pain. Remote control does not address all these potential confounders; care with the frequency of repeated stimuli and the forces applied are equally important as lack of restraint and freedom of movement.

In spite of this experience with a handheld device [35], a number of studies using remote controlled NT testing have demonstrated good reproducibility. In dogs, TTs were shown not to change over daily testing for 3 days [36] and in cats over several months [6]. In horses, neither TT nor MT changed significantly over several weeks [27,37–39]. Pastern MT remained stable or even increased with familiarity over a few weeks, even when attached to the operator by a long light pressure line [26]. The ability of remote control to avoid visual cues and allow normal behavior presumably contributes to this.

Neither TT or MT remote control testing were evaluated in cattle, nor MT testing in sheep. In principle, these should both be suitable species for remote controlled systems, being large enough to carry the bigger control boxes on the back. Sheep tolerated the back-mounted remote thermal control system willingly. Actuators applied to limbs have been used in cattle standing in stocks [12,13], and it seems likely that the remote systems would have similar advantages over wired in the same way as in horses. Thermal threshold has been measured successfully in both neonatal and young foals [40].

Camels reacted to MT testing in a similar manner to cattle and horses, and the remote system functioned well in relatively recalcitrant animals once all the equipment was in place [41]. Thermal testing in camels was challenging: the high limb thresholds are presumably a result of adaptation to withstand desert and sandstorm temperatures of over 50 °C.

The IR remote controlled systems have now been used for formal investigations into pain and analgesics in cats [42,43], dogs [29], sheep [25] and horses [37–39,44].

5. Conclusions

Remote controlled thermal nociceptive threshold testing in cats, dogs, sheep, horses and camels and mechanical threshold testing in horses and camels was well tolerated. It allowed free movement and normal behaviour during testing and resulted in data consistent with wired systems. Remote controlled nociceptive threshold testing is a useful refinement for pain research as the animal can behave normally without restraint or proximity of the tester.

Funding: This research received no external funding.

Acknowledgments: Many thanks are due to Michael Dixon who designed and built the systems and to all the colleagues with whom these data were collected and who provided the animals, facilities, enthusiasm as well as moral support and who gave permission for the data to be reported: Tammy Grubb, Andreas Haga, Sabine Kastner, Peter Kronen, Stelio Luna, Khursheed Mama, Adel Mubarak, Jo Murrell, Gabby Musk, Bob Pigott, Louisa Slingsby and Kirsten Wegner.

Conflicts of Interest: Polly Taylor and Michael Dixon are directors of Topcat Metrology Ltd.

References

1. Le Bars, D.; Gozariu, M.; Cadden, S.W. Animal models of nociception. *Pharmacol. Rev.* **2001**, *53*, 597–652. [PubMed]

2. Gugliandolo, E.; D'Amico, R.; Cordaro, M.; Fusco, R.; Siracusa, R.; Crupi, R.; Impellizzeri, D.; Cuzzocrea, S.; Di Paola, R. Effect of PEA-OXA on neuropathic pain and functional recovery after sciatic nerve crush. *J. Neuroinflammation* **2018**, *15*, 264. [CrossRef] [PubMed]

3. Sorge, R.E.; Martin, L.J.; Isbester, K.A.; Sotocinal, S.G.; Rosen, S.; Tuttle, A.H.; Wieskopf, J.S.; Acland, E.L.; Dokova, A.; Kadoura, B.; et al. Olfactory exposure to males, including men, causes stress and related analgesia in rodents. *Nat. Methods* **2014**, *11*, 629–632. [CrossRef] [PubMed]

4. Randall, L.O.; Selitto, J.J. A method for measurement of analgesic activity on inflamed tissue. *Arch. Int. Pharmacodyn. Ther.* **1957**, *111*, 409–419. [PubMed]

5. Beecher, H.K. The measurement of pain; prototype for the quantitative study of subjective responses. *Pharmacol. Rev.* **1957**, *9*, 59–209.

6. Slingsby, L.S.; Taylor, P.M. Thermal antinociception after dexmedetomidine administration in cats: A dose-finding study. *J. Vet. Pharmacol. Ther.* **2008**, *31*, 135–142. [CrossRef]

7. Love, E.J.; Taylor, P.M.; Murrell, J.; Whay, H.R. Effects of acepromazine, butorphanol and buprenorphine on thermal and mechanical nociceptive thresholds in horses. *Equine Vet. J.* **2012**, *44*, 221–225. [CrossRef]

8. Dixon, M.J.; Robertson, S.A.; Taylor, P.M. A thermal threshold testing device for evaluation of analgesics in cats. *Res. Vet. Sci.* **2002**, *72*, 205–210. [CrossRef]

9. Robertson, S.A.; Taylor, P.M.; Lascelles, B.D.; Dixon, M.J. Changes in thermal threshold response in eight cats after administration of buprenorphine, butorphanol and morphine. *Vet. Rec.* **2003**, *153*, 462–465. [CrossRef]

10. Steagall, P.V.; Mantovani, F.B.; Taylor, P.M.; Dixon, M.J.; Luna, S.P. Dose-related antinociceptive effects of intravenous buprenorphine in cats. *Vet. J.* **2009**. [CrossRef]

11. Chambers, J.P.; Waterman, A.E.; Livingston, A. Further development of equipment to measure nociceptive thresholds in large animals. *Vet. Anaesth. Analg.* **1994**, *21*, 66–72. [CrossRef]

12. Whay, H.R.; Webster, A.J.; Waterman-Pearson, A.E. Role of ketoprofen in the modulation of hyperalgesia associated with lameness in dairy cattle. *Vet. Rec.* **2005**, *157*, 729–733. [CrossRef] [PubMed]

13. Kemp, M.H.; Nolan, A.M.; Cripps, P.J.; Fitzpatrick, J.L. Animal-based measurements of the severity of mastitis in dairy cows. *Vet. Rec.* **2008**, *163*, 175–179. [CrossRef] [PubMed]

14. Dixon, M.J.; Taylor, P.M.; Slingsby, L.A.; Waterman-Pearson, A.E. Development of a remote controlled thermal threshold testing system for the cat. In Proceedings of the 9th World Congress of Veterinary Anaesthesia, Santos, Brazil, 12–16 September 2006; p. 168.

15. Dixon, M.J.; Taylor, P.M.; Murrell, J.C.; Love, E.J. Development of a multispecies wireless mechanical threshold testing system. In Proceedings of the Association of Veterinary Anaesthetists Autumn Meeting, Barcelona, Spain, 14–16 October 2008; p. 97.

16. Hoffmann, M.V.; Kastner, S.B.; Kietzmann, M.; Kramer, S. Contact heat thermal threshold testing in beagle dogs: Baseline reproducibility and the effect of acepromazine, levomethadone and fenpipramide. *BMC Vet. Res.* **2012**, *8*, 206. [CrossRef]

17. Dixon, M.J.; Taylor, P.M.; Slingsby, L.C.; Murrell, J.C. Refinement of a thermal threshold probe to prevent burns. *Lab. Anim.* **2016**, *50*, 54–62. [CrossRef] [PubMed]

18. Slingsby, L.S.; Taylor, P.M. Pilot dose response study for intravenous buprenorphine using thermal nociceptive threshold testing in cats. *Vet. Anaesth. Analg. Online Abstr.* **2009**, *36*, 9. [CrossRef]

19. Dixon, M.J.; Taylor, P.M.; Slingsby, L.; Hoffmann, M.V.; Kastner, S.B.; Murrell, J. A small, silent, low friction, linear actuator for mechanical nociceptive testing in veterinary research. *Lab. Anim.* **2010**, *44*, 247–253. [CrossRef]

20. Dixon, M.J.; Taylor, P.M.; Steagall, P.V.; Brondani, J.T.; Luna, S.P. Development of a pressure nociceptive threshold testing device for evaluation of analgesics in cats. *Res. Vet. Sci.* **2007**, *82*, 85–92. [CrossRef]

21. Hoffmann, M.V.; Kastner, S.B.; Kramer, S. Thermal and mechanical threshold testing in Beagle dogs, baseline reproducibility and the influence of sedation and opioid analgesia. In Proceedings of the Association of Veterinary Anaesthetists, Spring Conference, Cambridge, UK, 30–31 March 2010; p. 6.

22. Nalon, E.; Maes, D.; Piepers, S.; Taylor, P.; van Riet, M.M.; Janssens, G.P.; Millet, S.; Tuyttens, F.A. Factors affecting mechanical nociceptive thresholds in healthy sows. *Vet. Anaesth. Analg.* **2016**, *43*, 343–355. [CrossRef]

23. Lascelles, B.D.; Robertson, S.A. Use of thermal threshold response to evaluate the antinociceptive effects of butorphanol in cats. *Am. J. Vet. Res.* **2004**, *65*, 1085–1089. [CrossRef]

24. Robertson, S.A.; Lascelles, B.D.; Taylor, P.M.; Sear, J.W. PK-PD modeling of buprenorphine in cats: Intravenous and oral transmucosal administration. *J. Vet. Pharmacol. Ther.* **2005**, *28*, 453–460. [CrossRef] [PubMed]

25. Musk, G.C.; Murdoch, F.R.; Tuke, J.; Kemp, M.W.; Dixon, M.J.; Taylor, P.M. Thermal and mechanical nociceptive threshold testing in pregnant sheep. *Vet. Anaesth. Analg.* **2014**, *41*, 305–311. [CrossRef] [PubMed]

26. Schambourg, M.; Taylor, P. Mechanical nociceptive thresholds in endurance horses. *Vet. Rec.* **2019**, *185*, 124. [CrossRef]

27. Luna, S.P.; Lopes, C.; Rosa, A.C.; Oliveira, F.A.; Crosignani, N.; Taylor, P.M.; Pantoja, J.C. Validation of mechanical, electrical and thermal nociceptive stimulation methods in horses. *Equine Vet. J.* **2015**, *47*, 609–614. [CrossRef]

28. Poller, C.; Hopster, K.; Rohn, K.; Kastner, S.B. Evaluation of contact heat thermal threshold testing for standardized assessment of cutaneous nociception in horses—Comparison of different locations and environmental conditions. *BMC Vet. Res.* **2013**, *9*, 4. [CrossRef]

29. Niyom, S.; Mama, K.R.; Gustafson, D.L.; Rezende, M.L. Single-and multiple dose pharmacokinetics and multiple dose pharmacodynamics of oral ABT-116 (a TRPV1 antagonist) in dogs. *J. Vet. Pharmacol. Ther.* **2015**, *38*, 336–343. [CrossRef] [PubMed]

30. Sanchis-Mora, S.; Chang, Y.M.; Abeyesinghe, S.; Fisher, A.; Volk, H.A.; Pelligand, L. Development and initial validation of a sensory threshold examination protocol (STEP) for phenotyping canine pain syndromes. *Vet. Anaesth. Analg.* **2017**, *44*, 600–614. [CrossRef]

31. Wiech, K. Deconstructing the sensation of pain: The influence of cognitive processes on pain perception. *Science* **2016**, *354*, 584–587. [CrossRef]

32. Casey, K.L.; Morrow, T.J. Nocifensive responses to cutaneous thermal stimuli in the cat: Stimulus-response profiles, latencies, and afferent activity. *J. Neurophysiol.* **1983**, *50*, 1497–1515. [CrossRef]

33. Leach, M.C.; Allweiler, S.; Richardson, C.; Roughan, J.V.; Narbe, R.; Flecknell, P.A. Behavioural effects of ovariohysterectomy and oral administration of meloxicam in laboratory housed rabbits. *Res. Vet. Sci.* **2009**, *87*, 336–347. [CrossRef]

34. Coleman, K.D.; Schmiedt, C.W.; Kirkby, K.A.; Coleman, A.E.; Robertson, S.A.; Hash, J.; Lascelles, B.D. Learning confounds algometric assessment of mechanical thresholds in normal dogs. *Vet. Surg.* **2014**, *43*, 361–367. [CrossRef]

35. Coleman, D.L.; Slingsby, L.S. Attitudes of veterinary nurses to the assessment of pain and the use of pain scales. *Vet. Rec.* **2007**, *160*, 541–544. [CrossRef]

36. Niyom, S.; Mama, K.; Rezende, M. Repeatability of thermal and mechanical nociceptive threshold testing devices in dogs. In Proceedings of the Association of Veterinary Anaesthetists Spring Meeting, Bari, Italy, 13–16 April 2011; p. 113.

37. Lopes, C.; Luna, S.P.; Rosa, A.C.; Quarterone, C.; Crosignani, N.; Taylor, P.M.; Pantoja, J.C.; Puoli, J.N. Antinociceptive effects of methadone combined with detomidine or acepromazine in horses. *Equine Vet. J.* **2016**, *48*, 613–618. [CrossRef] [PubMed]

38. Gozalo-Marcilla, M.; de Oliveira, A.R.; Fonseca, M.W.; Possebon, F.S.; Pelligand, L.; Taylor, P.M.; Luna, S.P.L. Sedative and antinociceptive effects of different detomidine constant rate infusions, with or without methadone in standing horses. *Equine Vet. J.* **2018**. [CrossRef] [PubMed]

39. Gozalo-Marcilla, M.; Luna, S.P.; Crosignani, N.; Filho, J.N.P.; Possebon, F.S.; Pelligand, L.; Taylor, P.M. Sedative and antinociceptive effects of different combinations of detomidine and methadone in standing horses. *Vet. Anaesth. Analg.* **2017**, *44*, 1116–1127. [CrossRef] [PubMed]

40. McGowan, K.T.; Elfenbein, J.R.; Robertson, S.A.; Sanchez, L.C. Effect of butorphanol on thermal nociceptive threshold in healthy pony foals. *Equine Vet. J.* **2013**, *45*, 503–506. [CrossRef]

41. Almubarak, A. Thermal and mechanical nociceptive threshold testing in one humped camels (Camelus dromedarius). *J. Vet. Sci. Technol.* **2017**, *8*, 58. [CrossRef]

42. Doodnaught, G.M.; Monteiro, B.P.; Benito, J.; Edge, D.; Beaudry, F.; Pelligand, L.; Steagall, P. Pharmacokinetic and pharmacodynamic modelling after subcutaneous, intravenous and buccal administration of a high-concentration formulation of buprenorphine in conscious cats. *PLoS ONE* **2017**, *12*, e0176443. [CrossRef]

43. Slingsby, L.S.; Sear, J.W.; Taylor, P.M.; Murrell, J.C. Effect of intramuscular methadone on pharmacokinetic data and thermal and mechanical nociceptive thresholds in the cat. *J. Feline Med. Surg.* **2016**, *18*, 875–881. [CrossRef]

44. Echelmeyer, J.; Taylor, P.M.; Hopster, K.; Rohn, K.; Delarocque, J.; Kastner, S.B.R. Effect of fentanyl on thermal and mechanical nociceptive thresholds in horses and estimation of anti-nociceptive plasma concentration. *Vet. J.* **2019**, *249*, 82–88. [CrossRef]

Review

The Utility of Grimace Scales for Practical Pain Assessment in Laboratory Animals

Daniel Mota-Rojas [1], Adriana Olmos-Hernández [2], Antonio Verduzco-Mendoza [2], Elein Hernández [3], Julio Martínez-Burnes [4] and Alexandra L. Whittaker [5],*

[1] Neurophysiology, Behavior and Animal Welfare Assessment, DPAA, Universidad Autónoma Metropolitana, Xochimilco Campus, Ciudad de México 04960, CDMX, Mexico; dmota100@yahoo.com.mx

[2] Division of Biotechnology—Bioterio and Experimental Surgery, Instituto Nacional de Rehabilitación-Luis Guillermo Ibarra Ibarra (INR-LGII), Tlalpan 14389, CDMX, Mexico; adrianaolmos05@yahoo.com.mx (A.O.-H.); cnrverduzco@hotmail.com (A.V.-M.)

[3] Department of Clinical Studies and Surgery, Facultad de Estudios Superiores Cuautiltán UNAM, Cuautitlán Izcalli 54714, Estado de México, Mexico; elein_ht@comunidad.unam.mx

[4] Graduate and Research Department, Facultad de Medicina Veterinaria y Zootecnia, Universidad Autónoma de Tamaulipas, Cd Victoria 87000, Tamaulipas, Mexico; jmburnes@docentes.uat.edu.mx

[5] School of Animal and Veterinary Sciences, University of Adelaide, Roseworthy Campus, SA 5116, Australia

* Correspondence: alexandra.whittaker@adelaide.edu.au

Received: 4 September 2020; Accepted: 22 September 2020; Published: 9 October 2020

Simple Summary: Grimace scales for laboratory animals were first reported ten years ago. Yet, despite their promise as pain assessment tools it appears that they have not been implemented widely in animal research establishments for clinical pain assessment. We discuss potential reasons for this based on the knowledge gained to date on their use and suggest avenues for further research, which might improve uptake of their use in laboratory animal medicine.

Abstract: Animals' facial expressions are widely used as a readout for emotion. Scientific interest in the facial expressions of laboratory animals has centered primarily on negative experiences, such as pain, experienced as a result of scientific research procedures. Recent attempts to standardize evaluation of facial expressions associated with pain in laboratory animals has culminated in the development of "grimace scales". The prevention or relief of pain in laboratory animals is a fundamental requirement for in vivo research to satisfy community expectations. However, to date it appears that the grimace scales have not seen widespread implementation as clinical pain assessment techniques in biomedical research. In this review, we discuss some of the barriers to implementation of the scales in clinical laboratory animal medicine, progress made in automation of collection, and suggest avenues for future research.

Keywords: facial expressions; pain; grimace scales; mice; rat; rabbit

---

1. Introduction

Animal welfare is an important societal concern [1,2]. The use of animals in biomedical scientific research is widespread, and globally significant, with approximately 115 million animals used per year [3]. Incontrovertibly, there is an ethical obligation to safeguard welfare of these animals through employing strategies to minimize pain, fear, and distress [4–6], in addition to the promotion of positive welfare states. However, to achieve this, validated methods for identification of animal emotional state are required. Despite significant research attention, ascertaining nature and strength of animal emotion remains a challenging task [7–11].

The study of emotion in laboratory animals has typically focused on aversive states such as pain. This area of study was driven by two perspectives: a scientific and welfare standpoint. The scientific viewpoint, based on the extrinsic value of the animal, relates to the robustness of results acquired from animal models. There is an abundance of data on the impact of pain on a wide range of metabolic, immunologic, and other processes in the body. These alterations introduce variability or confound interpretation of results [12–14]. The welfare viewpoint, considering the intrinsic value of the animal, assumes that pain occurs frequently in animal models and should therefore be avoided or minimized for the benefit of the animal. Notwithstanding, differences between these viewpoints in terms of underlying motivation for study, the requirement for a reliable, practical method for assessment of pain is shared by both.

Recently, evaluation of complex motor responses, such as facial and corporal expression was proposed as a neurobiological readout of mammalian brain neuro-circuitry associated with emotional experience [11,15–17]. The former has received significant research attention, especially in rodents, as a potential assessment method for both positive and negative emotional states [9]. There remains controversy as to the communicative function of facial expressions in rodents, since these species tend to prioritize other senses such as olfaction and touch in communication [8]. However, the finding that in mice, lesions of the insular cortex, modulate facial pain expressions supports the use of facial expression assessment. The insular cortex is associated with human pain perception; hence it is assumed by analogy that facial grimace may represent a negative emotional experience [18]. Furthermore, studies on empathy tends to suggest that rodents are communicating the presence of a painful state to others, to elicit an empathic response [19]. Although not specifically demonstrated, it is feasible that this may be occurring through interpretation of facial expression [8]. Additionally, it was recently shown through the use of machine learning methods that facial expressions in mice may not only indicate direction of effect or valence of emotion (positive or negative), but intensity and persistence [20].

Attempts to standardize evaluation of facial expressions for pain assessment has culminated in the development of the "grimace scales". These were developed originally for mice [18] and were adapted for use in rats [21], rabbits [22,23], sheep [24,25], ferrets [26], cats [27] and horses [28]. Grimace scales are simplified methods for evaluating facial expressions specifically related to pain based on the assessment of action units focusing on the eyes, ears, and cheeks. The utility of the scales was well-established across a range of laboratory animal species and animal model types. However, this evaluation has typically focused on their use via retrospective video recording review, and as a research tool to obtain data relevant to the animal model. There are fewer dedicated studies into the scales as 'bedside' pain assessment tools for rapid evaluation of pain status in laboratory animals in order to implement humane endpoints or provide analgesia. Therefore, the focus of this review is to discuss the practical utility of grimace scales in a range of laboratory animal species, identifying barriers to their use and potential confounders. The focus will be on laboratory animal rodents as the most common species used in biomedical research, but research from other species will be drawn upon. It is anticipated that this review will guide biomedical researchers, animal technicians and ethics committees when implementing pain assessment methods as part of research protocols.

2. History of Facial Expression Scoring for Pain in Laboratory Animals

In recognition of the poor translation of outcomes from animal pre-clinical studies on pain physiology and analgesic development to humans [29,30], there has been a recent focus on development of methods for assessment of the affective pain response using non-evoked (spontaneous) responses [31]. Grimace scales are one such response derived from human facial codification scales [32,33]. The Facial Action Coding System (FACS) systematically catalogs all possible movements of the facial muscles, or combinations of them, such as lowering the eyebrows, tightening and closing the eyelids, wrinkling the nose, and raising the upper lip. Categorization of changes in these muscle movements or so-called "Facial Action Units" (FAU) enables facial recognition and categorization of emotions [16,34]. The finding that facial codification scales could quantify pain in humans with limited or non-existent verbal communication [35], provided the basis for using FAU in the development of grimace scales (GS) for animals (see [36]).

The mouse grimace scale (MGS) was the first to be developed. Langford et al. [18] in 2010 applied a nociceptive abdominal constriction test through administration of acetic acid that allowed the elucidation of facial action units that reliably detected pain. Validation was performed using a variety of traditional preclinical pain assays [18]. Five action units were described: (1) orbital tightening, (2) nose bulge, (3) cheek bulge, (4) ear position and (5) whisker change. A year later, Sotocinal et al. [21] in 2011 published the rat grimace scale (RGS) comprising four action units, due to consolidation of nose and cheek flattening into one unit. Utility of the RGS to detect pain was demonstrated in standard pre-clinical nociceptive tests in addition to following a surgical laparotomy procedure. Furthermore, the RGS was shown to be modified after analgesic administration indicating the specificity to pain [21]. Furthermore, the development of grimace scales in other common laboratory animal species followed, see Table 1.

Table 1. Original studies in which grimace scales were developed for a range of species commonly used as laboratory animals.

Species	Validation Method	Action Units	Study
Mouse Grimace Scale (MGS)	Fourteen commonly used preclinical pain assays.	Five Units: (1) Orbital tightening, (2) Nose bulge, (3) Cheek bulge, (4) Ear position and (5) Whisker change	[18]
Rat Grimace Scale (RGS)	Three pain-eliciting procedures performed. (1) intraplantar administration of Complete Freund's adjuvant (CFA); (2) intra-articular administration of kaolin/carrageenan; and (3) post-operative pain after laparotomy.	Four Units: (1) Orbital tightening, (2) Nose/cheek flattening, (3) Ear changes, (4) Whisker change	[21]
Rabbit Grimace Scale (RbtGS)	Pain caused by ear tattooing, a routine procedure used to identify rabbits. Analgesic test applied in the form of prilocaine/lidocaine (EMLA) local anesthetic	Five Units: (1) Orbital tightening, (2) Cheek flattening, (3) Nose shape, (4) Whisker position, (5) Ear position.	[22]
Sheep Grimace Scales (Sheep Pain Facial Expression Scale—SPFES)	Clinical model based on mastitis and footrot	Five Units: (1) Orbital tightening, (2) Cheek tightness, (3) Ear position, (4) Lip and jaw profile, (5) Nostril and philtrum position	[25]
Ferret (FGS)	Surgery involving the implantation of an intraperitoneal telemetry catheter	Five Units: (1) Orbital tightening, (2) Nose bulging, (3) Cheek bulging, (4) ear changes, (5) Whisker retraction	[26]
Piglets (PGS)	Castration and tail docking. Validated orbital tightening for tail docking but remarked that further validation needed.	Ten used for development, later study [37] modified to three: (1) Ear Position, (2) Cheek Tightening/Nose bulge, (3) Orbital Tightening	[37,38]
Cat (FGS)	Acute pain arising as a result of a variety of clinical conditions	Five units: (1) Ear position, (2) Orbital tightening, (3) Muzzle tension, (4) Whisker change, (5) Head position	[27]
Horse (HGS)	Surgical castration	Six units: (1) Stiffly backward ears, (2) Orbital tightening, (3) Tension above the eye area, (4) Prominent strained chewing muscles, (5) Mouth strained and pronounced chin, (6) Strained nostrils and flattening of the profile	[28]

3. Terminology Around Pain Classification and Assessment

A variety of terms are often used to describe pain and the assessment methods applied to it. Pain is usually classified according to the duration of its effect or its originating source within the body [39,40]. Acute pain arises at the time of injury and is often experienced as different in nature to the alternatively described 'chronic' pain. The latter generally referring to pain experienced over a longer duration, although there appears to be no accepted duration marking the transition from acute to chronic pain [41]. An alternative distinction between the two time-course descriptors was suggested by scientists: that related to functionality. Acute pain is argued to be adaptive, provoking a learned response by the animal to avoid a similar painful insult in the future [39]. Chronic pain on the other hand is said to be maladaptive [42]. However, this latter point is controversial with a variety of studies (see [43] for review) suggesting that pain-related hypervigilance may influence estimation of risk, subsequent behavior, and thus enhance survival.

Pain scales themselves are often described in terms of their validity, reliability, sensitivity [41]. Validity describes the extent to which the scale measures its intended outcome i.e., pain. There are several sub-categories describing validity. The most commonly referred to in the context of grimace scales are face validity and construct validity. Face validity describes what the test appears to be measuring i.e., pain. Construct validity relates to the extent to which the scales measure that specific construct. Therefore, the test needs to be both sensitive and specific to pain [44,45]. In pain studies construct validity is often determined using an applied analgesic test, since this is assumed to reduce pain and thereby reduce grimace scores if the test is truly pain-related [44]. External validity refers to how generalizable the measure is to other settings. In the context of grimace scales this is relevant in taking the scales from research scenarios to the clinical setting. This relates to practicability to perform during the working day, simplicity of the task, as well as the need for equipment and training. To date, this is the area that has received the least attention with regard to grimace scales.

Reliability refers to the scale producing the same result each time it is used both within, and between animals, and time points [46]. In the context of grimace scales, this is determined by the variability resulting in a single observer's measurements (intra-observer variability), the variation between different observers' measurements (inter-observer), and variability between laboratories or research centers [44]. Sensitivity describes the ability of the scale to accurately identify changes in the degree of pain such that subtle changes are recognized [45]. In the context of pain scales this is often indicated when scale changes that occur correlate in direction, and proportion with other measures [45]. It is common in assessment of pain in veterinary species to achieve measurement accuracy in pain scoring by using a smaller number of broad category groups, such as mild, moderate, and severe, rather than expecting sensitivity when small differences in scores are considered. The following will consider how all of these measurement characteristics may influence the clinical applicability of grimace scales for use in biomedical research.

4. Clinical Applicability of Grimace Scales in Biomedical Research

4.1. Development of Real-Time Grimace Scores

There is now an extensive body of literature on the application of grimace scales in a range of animal models used commonly in biomedical research. The majority of this validation work has occurred in rodent models. It is beyond the scope of this review to describe all of the models used but the range includes oncology (see e.g., [47–50]), infectious disease [51], pain models [48,52,53], neurological conditions [33,54,55], genetic conditions [56], and maxillofacial interventions [49,50,57]. However, the vast majority of research to date has performed grimace scoring retrospectively from captured video footage.

Retrospective scoring is likely superior when using grimace scores to inform research outcomes, for example determining efficacy of analgesics or success of model induction. These methods allow for the possibility of replication, by multiple observers where appropriate, with an increased time available for scoring at the researcher's leisure. A cage-side or 'real-time' method on the other hand would ideally provide instant assessment allowing interventions to support welfare, for example by implementing humane endpoints or administering analgesics. Development of the latter is clearly of more interest to ethical review committees and animal carers needing to make rapid clinical decisions. To date there has been substantially less focus on development and validation of real-time methods.

Miller and Leach [58] in 2015 performed the first comprehensive evaluation of a real-time method applied in mice. In this study, both retrospective and real-time scoring were compared. Real-time scoring was performed by observing mice three times over a 10 min period, while animals were being filmed for the retrospective analysis. Grimace scores were calculated by summation of each action unit as described by Langford et al. [18], and totals were then averaged across the observation points. Live scores were always found to be significantly lower than corresponding retrospective video scoring. The authors posed that this could have resulted from the activity levels and changing nature of the face during live scoring. Blinking for instance, resulting in a score of 0 for orbital tightening, will likely be selected at least some of the time as a result of random chance selection of photographs for scoring. In a real-time scenario, the rapid nature of blinking will likely preclude its scoring. Similarly, Chartier et al. [47] in 2020 also found consistently lower scores from live scoring compared to retrospective scoring in a mouse model of colitis-associated colo-rectal cancer. One potential explanation for this trend is that the presence of a human observer influences performance of the facial action units, for example, an increased alertness to the human (predator) could lead to wider eyes and 'pricked' ears, lowering the grimace score. On the contrary, intriguing findings from Sorge et al. [59] demonstrated that not all observers are equal, with no impact of a female observer on scores in rats and mice (obtained retrospectively), but a reduction of scores in the presence of a male [59]. In the first investigation of real-time scoring in rats, Leung et al. [60] in 2016 found that interval observations (15 s of observation) were able to discriminate between control and analgesic–treated groups whereas point observations (conducted several times over a period) showed poor group discrimination. In this study, substantial variability was seen between single observations of either point or interval. Limits of agreement, with a retrospective scoring system were however fairly large with a 0.5 score range either side of the bias meaning there is was a substantial risk of both over or underestimating the score. Furthermore, point scoring became generally unreliable at discriminating groups when done for less than 2 min, assumed to be due to a loss of power due to fewer observations. A later rat study by the same research group [61], investigated the interval method compared to a retrospective method in a colitis model showing the former to be reliable in predicting pain, with scores similar to the standard method.

The implications of these findings for clinical pain assessment are several. Firstly, it needs to be considered that although good discriminant ability was generally found in these studies, results were obtained by statistical combination of multiple scores. In a clinical scenario, an observer is likely to take one score, and not have the means or time to mathematically manipulate the values to arrive at a reliable score. Secondly, the Leung et al. [60] study suggests that variability across the observation period is likely and that at least 2 min of observation is needed. It is unlikely to be practical for a caregiver to spend 2 min per animal performing pain assessment across a study. In this case, some other more general method of distress measurement is likely to be needed to 'triage' animals for secondary grimace assessment. There has been no investigation of the effect of movement to the clear cages, in isolation, as typically occurs in grimace studies as opposed to scoring occurring in the home cage environment. Several factors may influence the grimace scoring between these two scenarios. The novelty of the scoring box may trigger a state of alert influencing grimace scores in a similar vein to that suggested for the presence of a human observer. This novelty may indeed contribute to the variability seen between scores over time since habituation will eventually occur. Alternately, if scoring in the home cage, the presence of cage furniture, a potential more relaxed state of the animal in its familiar environment, or even the influence of circadian rhythms (see later) may all variously influence the action units or ability to see them accurately. A further consideration with real-time scoring is that there may be an inherent observer bias as the animal's overall demeanor, or presence of other pain behaviors such as twitching may be noted leading the observer to err on the side of higher action unit scores when unsure. This is not necessarily an issue per se in a clinical scenario since the goal is to recognize sick animals for further evaluation and treatment. However, these other behaviors may not be unique to pain but represent general sickness behavior that may not be able to be rectified by analgesic administration, and hence inappropriate medication administration may occur. If such biasing were occurring it would be expected that there may be differences in grimace scoring between observers experienced with working with the species in question versus more naïve observers [47].

Notwithstanding, these findings some research groups do appear to have been able to use the MGS or RGS in a point observation, real-time scenario to obtain predicted results. For example, in chemotherapy-induced toxicity models in mice [62], and rats [63] single grimace scores allowed distinguishing between groups and followed the progression of the disease course as expected, after induction of chemotherapy-induced gut toxicity. Alternately, Hsi et al. [64] in 2020 were unable to use point mouse grimace scores to distinguish between groups either supplemented or not with dextrose following bariatric surgery. However, in this experimental design there was no sham group so it is unknown whether the MGS can reliably determine pain in this model [64].

There is clearly a need for further validation of real-time observation methods with a particular focus on one-off observations versus a series of observations, correlation with other established measures of pain assessment, inter-observer variability and home cage versus novel area.

4.2. Impact of Biology and the Environment

4.2.1. Strain and Sex Differences

There is some evidence that features of biology, performance of routine procedures, or aspects of the environment may influence grimace scores. This has implications for setting of intervention scores (see later), and should be a consideration in driving further research or recommendations for application to clinical practice.

Aspects of biology have perhaps been the most researched with regard to their impact on grimace scores. The greatest implication of such changes likely relates to any differences between rodent strains or stocks given the wide range typically used in research. In mice, strain differences in MGS scores in animals not exposed to any painful interventions was demonstrated. Miller and Leach in 2015 [58] found that C3H/He mice showed significantly higher scores than CD-1 and C57BL/6 animals, although the order of effect for the latter two strains was different between males and females. In female BALB/c mice the grimace score was even higher than C3H/He (males were not investigated in this study). Cho et al. [65] in 2019 similarly demonstrated a difference in MGS scores post-craniotomy, with C57BL/6 mice with lower scores than CD-1 animals [65]. However, in pairwise comparisons of the CBA and DBA/2 strains in two further studies, no differences were found [66,67]. It was suggested by some authors that detection of facial features in dark animals may be more difficult [65,68]. Improving the image quality and providing a contrasting background color when recording appear to mitigate the effects [18], hence this may not be a feature of animal pigmentation per se. It should, however, be noted that in the Miller and Leach [58] 2015 study, female C57BL/6 animals were not scored the lowest; that place being taken by the white CD-1 animals. Brown C3H animals also occupied an intermediate position. In a clinical scenario where real-time scoring is likely to take place the issue of poor background contrast on videos is not of concern. However, some investigation of the effects of color on live grimace scoring is warranted since it may be equally as difficult for a human observer to distinguish features such as whiskers against a similar coat color background, especially when trying to observe at a distance so as not to influence the animal's behavior.

Differences between sexes have also been uncovered in research to date on the MGS, but results are complex and suggest there may be strain interactions. For example, Miller and Leach [58] observed no differences in MGS scores between male and female C57BL/6 mice [58]. However in the same study, both CD-1 and C3H/He males had greater scores than their female counterparts [58]. Similarly, male BALB/c mice had higher grimace scores than females [69]. Alternately, Cho et al. [65] found no sex differences in CD-1 mice, although differences in response to analgesic were noted with females appearing to respond to carprofen with a reduction in grimace score more readily than males [65]. In rats, limited studies were carried out into sex differences but no differences were found in the original validation study [21], or in a later study [70]. Unfortunately, it appears that most grimace studies in rats and mice appear to have been conducted in one sex, with a large proportion using male animals, see e.g., [52,71–73]. This bias in study design toward males, coupled with the enhanced understanding of the existence of different pathways and immune-cell types for pain processing between male and female rodents [74], renders extrapolation of findings to female rodents problematic.

4.2.2. Impact of Routine Procedures

It is clear that procedures occurring fairly often as part of vivarium routines may influence responses and should be taken into consideration when considering practical implementation of the grimace scales. For example, several studies evaluated the impact of anesthetics on rodent grimace scales. In general, both inhalational and injectable anesthetics lead to a short-term increase in grimace scores in both rats [73] and mice [66,75,76], although strain differences in the presence of this response were reported [66,75,76]. While this response is generally short-lived, repeated exposures lead to enhanced duration of the increase [68,73]. This is a particular consideration since grimace assessment would typically occur post-operatively to allow rescue analgesia administration and there is suggestion that the score increase may persist for up to a few hours post anesthesia [75,76].

There is a growing body of evidence that non-aversive handling of mice leads to reduced anxiety and improved resilience in the face of accompanying pain [77–79]. Cupping or tunnel handling are proposed as alternatives to the traditional method of picking up by the tail [78]. Perhaps somewhat surprisingly given the reported specificity of the MGS for pain there is some evidence that method of routine handling influences MGS with increased scores in mice handled by the tail compared to those that were tunnel handled [69]. This contradicts the findings of a previous study where no differences between the two methods were reported [67]. This is an area that should be a priority for further investigation for several reasons. Firstly, since non-aversive methods have not been widely incorporated into laboratory animal practice, especially among researchers [80], it is quite likely that mice even within one study will be subject to different handling techniques. Any effect of handling method on grimace score could therefore confound interpretation of grimace scores used to determine research protocol effects on pain. Secondly, while there appears to have been no dedicated study on whether tail handling induces pain, there is suggestion that it is non-painful, yet aversive [78]. If the method is actually non-painful this calls into question the specificity of the MGS for pain, and therefore whether it has construct validity.

Ear tagging or ear notching are routine handling procedures used to permanently identify laboratory animals [81]. These procedures are known to cause acute pain as reflected by alterations in physiological indices such as heart rate and blood pressure [82]. However, the results obtained by Miller and Leach [81] in mice did not reveal any change to MGS scores as a result of ear notching [81]. In a later mouse study, with a factorial study design evaluating handling method with ear tagging or tattooing, MGS increased following ear tagging but tattooing or restraint had no impact on scores [69]. Alternately, Keating et al. [22] in 2012 showed that ear tattooing in rabbits led to increases in rabbit grimace scale scores that were ameliorated by the application of a local topical anaesthetic (lidocaine/prilocaine) [22]. Corticosterone measures in this study suggest that the pain response was short-lived and had resolved by 1-h post-procedure. Given that only three studies, performed in different species, evaluated these common procedures, it would be unwise to draw firm conclusions. However, the lack of grimace score increase in the Miller and Leach [81] study does imply that the scale may not be sensitive to pain of a mild and short-lived nature either intrinsically, or as a result of practical features whereby the pain is missed due to the scoring process required. Conversely, this finding provides some evidence that routine procedures may have minimal effect on grimace scales, thus reducing potential confounding when using the scales for humane endpoint implementation. When reconciling the difference in findings between this [81], and the later study, Roughan and Sevenoaks [69] in 2019 speculated that ear tagging may be perceived as more painful than notching due to the prolonged irritation by the tag [69].

4.2.3. Environmental Impacts

If the grimace scales are to be used as a practical tool they need to be repeatable across time and conditions, and not subject to extraneous influences. This requirement also relates to their face validity as reliable indicators of pain. In common with the other factors that may influence the scales, there has been limited research in this sphere.

Miller and Leach in 2015 [58] performed a comprehensive evaluation of some of the factors that might be predicted to influence grimace responses. One factor that may have an impact is the circadian cycle and whether differences in score occur across the day. For both live scoring and retrospective scoring, there were largely no differences seen between scores dependent on whether scoring took place in the morning, lunchtime or at the end of the day. There were some exceptions to this with BALB/c mice showing a greater live MGS score at noon compared to am and C57BL/6 mice showing higher retrospective MGS scores in both afternoon time points in comparison to the morning. It should be noted that in this, as in the majority of the studies examining grimace scores, animals were scored during the light phase of the circadian cycle when they would be expected to be inactive. There is some evidence that grimace scores may not be comparable between dark and light conditions with the finding that MGS was higher in the dark than in bright light in CD1 mice treated with a peptide believed to induce pain and migraine symptoms [83]. Analysis of the action units showed that the transition to light caused a significant decrease in orbital tightening and nose bulge [83]. Given that this finding was also observed in vehicle controls it appears unrelated to the migraine symptoms, and may be an aspect of normal biology needing consideration. Alternately, Matsumiya et al. 2012 found no difference in baseline MGS scores between morning and evening but did find that in operated animals scores were higher in the dark cycle, implying that pain was greater in the mice active phase [84]. However, in consideration of the use of the scales as a practical tool the reality is that most scoring will occur during the working day, in the light cycle, and therefore the findings of Miller and Leach in 2015 [58] provide confidence that time of scoring should not influence the score. A further important outcome from Miller and Leach [58] was that there was no effect of repeatedly being placed in the photography boxes on grimace score i.e., a habituation effect over the three occasions used [58]. The later study by Jirkof et al. [85] in 2020 supports this finding. This provides assurance that longitudinal monitoring post-procedure could occur throughout the day without the need to account for time of day or habituation to the box. However, as discussed earlier the need to remove animals to a separate box does impede the practical application of the test. Further study should consider time of day effects in the non-stimulated home environment.

There is further evidence of an impact of the external environment on grimace scores. Sorge et al. [59] compared grimace responses of mice and rats recorded after a painful insult, in the presence of a male compared to a female. Significant decreases in grimace response were recorded compared to the situation with no observer in the room. Females did not induce such a change. The findings therefore suggest that olfactory cues from human males lead to a physiological stress response, and associated stress-induced analgesia.

There is also some evidence of inter-laboratory variation in the outputs obtained from behavioral testing, to include MGS scores. In a multicenter study, Jirkof et al. [85] in 2020 demonstrated some quantitative differences in scores, although they were qualitatively comparable (direction of effect). However, variability between research centers in the MGS, especially when presented as a median score, was less pronounced than in burrowing behavior readouts [84]. This inter-lab variability was recognized across the spectrum of preclinical research pursuits, arising as a result of environmental variables leading to stress [86]. While this issue may be a concern when considering basic-to-clinical translation and reproducibility, it is less likely to be of concern for clinical application of the grimace scales. As a clinical tool, provided good inter and intra-observer, and thus intra-site agreement is obtained, grimace scores may be relied on locally for welfare determination subject to some of the other caveats discussed in this paper.

4.3. Validity

If grimace scales are to be implemented as a routine clinical assessment tool in biomedical research facilities, there needs to be a clear understanding of whether they are specific to pain, and can reliably measure pain in the models being used. This is important because it influences the animal caretaker's decision as regard to treatment options, for example, whether analgesics will be effective in mitigating clinical signs. It can be seen from the above discussion that there are a range of external factors that affect grimace scores, speaking to their validity as a pain assessment tool; anesthetics are a prime example. Setting aside the lack of study into their application in a real-time scenario, which influences their generalizability, another key concern is whether they are valid for all pain types. Results of the original Langford et al. [18] in 2010 study suggested that the technique was only applicable for acute pain states [18], since changes were not recorded after the application of traditional models of chronic pain, such as chronic constriction injury (CCI). However, there have now been a range of studies, largely performed in mice, which suggest that the grimace scale may be applicable for pain that is chronic or neuropathic in nature, or of a non-surgical origin (see [36] for detailed discussion).

The study findings of Akintola et al. [52] in 2017 contradict the previous results of Langford et al. 2010 with both RGS and MGS increasing after application of the CCI model in these species. Pain arising from cancer has also been shown to cause an elevation of the MGS, for example in colo-rectal cancer [49] and in a metastatic breast cancer model [49,50]. The MGS was successfully used in models expected to produce pain of a neuropathic nature, for example in headache and migraine [55,87] and craniotomy [65]. There is also suggestion that pain of a visceral nature elevates scores based on studies evaluating colonic nociception [88], pelvic pain [89], colitis [61], and alimentary mucositis [62,63]. Hereditary sickle cell disease frequently leads to painful episodes in human patients. Cold treatment of transgenic sickle mice led to increased grimace scores, which were alleviated using a known analgesic agent. Furthermore, body changes of decreased length and increased back curvature were also correlated with the change in grimace scores [56]. These findings lend support to the proposition that the grimace scales have good construct validity for non-acute pain.

Despite these findings, results from other studies implies that further evaluation of the grimace techniques are necessary to ascertain validity. For example, in contradiction to later work [62,63] demonstrating elevations in scores in rats and mice with mucositis, Whittaker et al. [90] in 2015 found no change in grimace scores in a rat model, albeit using retrospective rather than real-time scoring. However, this study did find increases in frequency of established behavioral indicators of pain such as back arching and twitching [90]. Alternately, Leung et al. [61] in a rat DSS- colitis model found grimace score increases in the absence of an increase in composite behavioral score.

Other studies also raise questions of whether the grimace scales are truly unique to pain. Caecal ligation and puncture models are commonly used to study sepsis [91]. While sepsis is undoubtedly a painful condition based on human reports [92], there is also an overwhelming cytokine response causing sickness behavior. Studies to date on this model [51,93] have not teased apart the possible contribution of this sickness response to the facial expression changes. There is a study that lends support to this idea; the work of Yamamoto et al. in 2016 [94], while not employing the published rat grimace scale, provides evidence that nausea influences the eye action unit. Toxin administration, which might also be expected to cause dual symptoms of pain and sickness, similarly elevated the MGS [95]. Furthermore, analgesic administration was not always successful in reducing the scores implying an alternate cause of the facial action unit response. Finally, head injury may alter the animal's ability to influence the facial action units via neural mechanisms and render grimace scores unreliable [44].

4.4. Automation of Techniques

One of the main current barriers to widespread clinical application of the grimace scales is the lack of understanding as to their validity and reliability when used for live scoring. However, as illustrated, there is now a wealth of literature on the validity and application of retrospective techniques using video or photo footage. In a clinical scenario these methods have limited application due to the time taken to extract the images, perform the scoring and potentially combine scores using statistical methods. However, there was some investigation of a range of technologies which minimize the time taken for various aspects of this process. At the simplest level use of freeware video to JPG converter software can reduce the time associated with manual searching and capture of images from recorded video footage by automating the capture process [48]. However, this still requires manual viewing of the selected images to obtain unobstructed head shots. Sotocinal et al. [21] in 2011 developed Rodent Face Finder® which is able to detect rodent eyes and ears to generate stills of rodent faces. This software was used in a range of studies measuring grimace scores in both rats and mice see e.g., [44,52,84,96]. Recently, another research group generated an algorithm to generate repeatable, non-observer biased, standardized and randomized pictures in one step. The authors suggest that their system offers benefits in scoring animals with dark fur and allowing several animals to be filmed and generate images simultaneously [97]. They further went on to show that the system was robust across several facilities potentially minimizing issues around inter-laboratory variability as discussed previously [98].

This process of semi-automation makes grimace scoring somewhat more applicable to a clinical environment but the time taken to manually score images is still likely to be a barrier to implementation. In recent years, there has been some progress on further automation of facial expression recognition using machine learning techniques. Deep learning methods allow classification and predictions on the data without previous feature design [99]. Tuttle et al. 2018 [99] first trained a neural network using human scored mouse images. Their system was highly accurate (94% agreement with human scores) for a binary (pain versus no pain) output, with scores correlating highly with human-assigned scores. Other groups have similarly demonstrated the promise of deep learning methods for use with the MGS when based on binary outputs [100,101]. Progress has also been made toward automating a facial pain expression system in sheep using techniques used in human facial recognition [102,103].

These automation methods are in their infancy and no doubt there will be further development of these techniques over the next few years. A key issue currently is that they lack sensitivity-being only able to distinguish a painful from a non-painful state. This renders their current use for welfare assessment and endpoint implementation limited. However, given the success and practical implementation of machine learning methods in recognition of human facial expression, it is likely to be only a matter of time before a similar level of sensitivity of scoring will be possible in animal–focused methods [104].

5. Practical Considerations

The above discussion highlights some areas in need for future research particularly in regard to practical usage of the grimace scales in laboratory animal medicine. A key issue is what to do with the data when it is acquired, and what it means for the animal. In research use of the grimace scales, statistically significant differences in grimace scores in comparison with controls are typically reported. However, in a clinical scenario, a mass of data or control animals' results may not be available to make this comparison on the spot. Moreover, statistical significance may not always equate with clinical significance. There needs to be ascertainment of the level of grimace score at which pain is actually occurring, since the evidence suggests that grimace scores in healthy animals are rarely zero [58]. Some attempts were made to address this issue with the development of intervention thresholds. Scores that are above this level signify that the animal is in pain, and consideration should be given to providing rescue analgesia [105]. These thresholds would need to be derived based on the method of combining individual action unit scores used, for example in the MGS summation of scores leads to a maximum of 10, whereas averaging leads to a maximum of 2. Oliver et al. [105] in 2014 determined for rats that 0.67/2 was a suitable intervention threshold. An intervention threshold has also been suggested for sheep (above 5/10) [25], and cats (0.39/1) [27]. It was considered by the authors in the sheep study that false positives for pain were unlikely above this cut-off score, although it was acknowledged that due to low test sensitivity some animals scoring below five may have a painful condition [25]. Since individuals experience pain differently, and there are associated sex differences in both pain experience and response to analgesics, work is needed to tailor intervention thresholds considering these factors. Additionally, consideration should be given to the fluctuating nature of pain [106], rendering regular monitoring, scoring and comparison with previous scores critical [44].

Monitoring staff need to consider tailoring of analgesic regimes due to animals potentially being in more pain in their active phase (see e.g., [84]), which may fall outside of staffed hours.

Given, the lack of established intervention thresholds perhaps the best current advice would be to use a holistic approach in pain assessment and consider grimace scores alongside other measures of well-being such as standard clinical scoring, and where possible look for trends in score progression within the same animal to guide decision-making. Animal carers also need to consider the potential impacts of inter and intra-observer variability on scoring which may be significant when statistical methods on group data are not used to smooth out variability. A prudent approach, where possible, would be to use the same scorer in a clinical case. This concern also brings up the issue of training of scorers which has received minimal research attention. Some studies implied that minimal training, such as the provision of online instructions, is all that is necessary to achieve consistent results between expert and novice scorers [69,107]. However, another study has shown that more in-depth training, using practice scoring associated with structured opportunities for discussion, enhanced scoring ability [108].

6. Conclusions and Future Directions

Despite 10 years of investigation, widespread uptake of grimace scoring in biomedical research has not occurred. The grimace scales offer enormous potential for clinical use in biomedical research. They are simple, require no equipment and were shown through research study to have good construct validity for most conditions. However, the methodology used in research on grimace scales is unlikely to lend to practical implementation due to its time intensive and retrospective nature. To date, few studies have investigated the validity of grimace scales in scenarios requiring on the spot pain assessment and clinical decision-making. Key areas for focus are on grimace score validity in animals housed in home cages, the reliability of using a limited number of real-time observation points, the impact of observers on scores, and the need for observer training. This is an area in urgent need for future research to realize the potential value of grimace scales.

One area that has received attention is the automation of scales using machine learning and algorithmic methods. This is a welcome development and will enhance the practical potential of grimace scales. It is hoped that in future years, grimace scale scoring may just be one of several outcome measures acquired routinely through facility-automated systems. This scenario is most likely to address the practical issues inherent when dealing with large numbers of animals, going some way toward addressing public concern around ethical decision-making in biomedical research.

Author Contributions: Conceptualization, D.M.-R. and A.L.W.; investigation, D.M.-R., A.L.W., A.O.-H., A.V.-M., E.H., J.M.-B.; writing—original draft preparation, D.M.-R., A.L.W., A.O.-H., A.V.-M., E.H., J.M.-B.; writing—review and editing, D.M.-R. and A.L.W.; project administration, D.M.-R. and A.L.W. All authors have read and agreed to the published version of the manuscript.

Funding: This research received no external funding. A.W. is supported by an Australian Government, NHMRC Peter Doherty Biomedical Research Fellowship (APP1140072).

Conflicts of Interest: The authors declare that they have no conflict of interest.

References

1. Mota-Rojas, D.; Velarde, A.; Maris-Huertas, S.; Cajiao, M.N. (Eds.) *Animal Welfare, A Global Vision in Ibero-America*, 3rd ed.; Elsevier: Barcelona, Spain, 2016; pp. 1–516.
2. Lewejohann, L.; Schwabe, K.; Häger, C.; Jirkof, P. Impulse for animal welfare outside the experiment. *Lab. Anim.* **2020**, *54*, 150–158. [CrossRef]
3. Taylor, K.; Gordon, N.; Langley, G.; Higgins, W. Estimates for worldwide laboratory animal use in 2005. *Altern. Lab. Anim.* **2008**, *36*, 327–342. [CrossRef]
4. Mota-Rojas, D.; Orihuela, A.; Martínez-Burnes, J.; Gómez, J.; Mora-Medina, P.; Alavez, B.; Ramírez, L.; González-Lozano, M. Neurological modulation of facial expressions in pigs and implications for production. *J. Anim. Behav. Biometeorol.* **2020**, *8*, 232–243. [CrossRef]

5.   Mota-Rojas, D.; Olmos-Hernández, A.; Verduzco-Mendoza, A.; Lecona-Butrón, H.; Martínez-Burnes, J.; Mora-Medina, P.; Gómez-Prado, J.; Orihuela, A. Infrared thermal imaging associated with pain in laboratory animals. *Exp. Anim.* **2020**, *70*, 20-0052. [CrossRef] [PubMed]

6.   Baumans, V. Science-based assessment of animal welfare: Laboratory animals. *Rev. Sci. Tech. OIE* **2005**, *24*, 503–513. [CrossRef]

7.   Lezama-García, K.; Orihuela, A.; Olmos-Hernández, A.; Reyes-Long, S.; Mota-Rojas, D. Facial expressions and emotions in domestic animals. *CAB Rev. Perspect. Agric. Vet. Sci. Nutr. Nat. Resour.* **2019**, *14*, 1–12. [CrossRef]

8.   Finlayson, K.; Lampe, J.; Hintze, S.; Würbel, H.; Melotti, L. Facial indicators of positive emotions in rats. *PLoS ONE* **2016**, *11*, e0166446. [CrossRef] [PubMed]

9.   Whittaker, A. The role of behavioural assessment in determining 'positive' affective states in animals. *CAB Rev. Perspect. Agric. Vet. Sci. Nutr. Nat. Resour.* **2019**, *14*, 1–13. [CrossRef]

10.   Boissy, A.; Manteuffel, G.; Jensen, M.B.; Moe, R.O.; Spruijt, B.; Keeling, L.J.; Winckler, C.; Forkman, B.; Dimitrov, I.; Langbein, J.; et al. Assessment of positive emotions in animals to improve their welfare. *Physiol. Behav.* **2007**, *92*, 375–397. [CrossRef]

11.   Panksepp, J. Affective consciousness: Core emotional feelings in animals and humans. *Conscious. Cogn.* **2005**, *14*, 30–80. [CrossRef]

12.   Carbone, L.; Austin, J. Pain and laboratory animals: Publication practices for better data reproducibility and better animal welfare. *PLoS ONE* **2016**, *11*, e0155001. [CrossRef] [PubMed]

13.   Peterson, N.C.; Nunamaker, E.A.; Turner, P.V. To treat or not to treat: The effects of pain on experimental parameters. *Comp. Med.* **2017**, *67*, 469–482. [PubMed]

14.   Zurlo, J.; Hutchinson, E. Refinement. *ALTEX* **2014**, *31*, 4–10. [CrossRef] [PubMed]

15.   Bennett, V.; Gourkow, N.; Mills, D. Facial correlates of emotional behaviour in the domestic cat (Felis catus). *Behav. Process.* **2017**, *141*, 342–350. [CrossRef]

16.   Ekman, P. Are there basic emotions? *Psychol. Rev.* **1992**, *99*, 550–553. [CrossRef]

17.   Mota-Rojas, D.; Orihuela, A.; Strappini-Asteggiano, A.; Cajiao-Pachón, M.N.; Agüera-Buendía, E.; Mora-Medina, P.; Ghezzi, M.; Alonso-Spilsbury, M. Teaching animal welfare in veterinary schools in Latin America. *Int. J. Vet. Sci. Med.* **2018**, *6*, 131–140. [CrossRef]

18.   Langford, D.J.; Bailey, A.L.; Chanda, M.L.; Clarke, S.E.; Drummond, T.E.; Echols, S.; Glick, S.; Ingrao, J.; Klassen-Ross, T.; LaCroix-Fralish, M.L.; et al. Coding of facial expressions of pain in the laboratory mouse. *Nat. Methods* **2010**, *7*, 447–449. [CrossRef]

19.   Langford, D.J. Social modulation of pain as evidence for empathy in mice. *Science* **2006**, *312*, 1967–1970. [CrossRef]

20.   Dolensek, N.; Gehrlach, D.A.; Klein, A.S.; Gogolla, N. Facial expressions of emotion states and their neuronal correlates in mice. *Science* **2020**, *368*, 89–94. [CrossRef]

21.   Sotocinal, S.G.; Sorge, R.E.; Zaloum, A.; Tuttle, A.H.; Martin, L.J.; Wieskopf, J.S.; Mapplebeck, J.C.S.; Wei, P.; Zhan, S.; Zhang, S.; et al. The Rat Grimace Scale: A partially automated method for quantifying pain in the laboratory rat via facial expressions. *Mol. Pain* **2011**, *7*, 55. [CrossRef]

22.   Keating, S.C.J.; Thomas, A.A.; Flecknell, P.A.; Leach, M.C. Evaluation of EMLA cream for preventing pain during tattooing of rabbits: Changes in physiological, behavioural and facial expression responses. *PLoS ONE* **2012**, *7*, e44437. [CrossRef] [PubMed]

23.   Hampshire, V.; Robertson, S. Using the facial grimace scale to evaluate rabbit wellness in post-procedural monitoring. *Lab. Anim.* **2015**, *44*, 259–260. [CrossRef] [PubMed]

24.   Häger, C.; Biernot, S.; Buettner, M.; Glage, S.; Keubler, L.M.; Held, N.; Bleich, E.M.; Otto, K.; Müller, C.W.; Decker, S.; et al. The Sheep Grimace Scale as an indicator of post-operative distress and pain in laboratory sheep. *PLoS ONE* **2017**, *12*, e0175839. [CrossRef] [PubMed]

25.   McLennan, K.M.; Rebelo, C.J.; Corke, M.J.; Holmes, M.A.; Leach, M.C.; Constantino-Casas, F. Development of a facial expression scale using footrot and mastitis as models of pain in sheep. *Appl. Anim. Behav. Sci.* **2016**, *176*, 19–26. [CrossRef]

26.   Reijgwart, M.L.; Schoemaker, N.J.; Pascuzzo, R.; Leach, M.C.; Stodel, M.; De Nies, L.; Hendriksen, C.F.M.; Van Der Meer, M.; Vinke, C.M.; Van Zeeland, Y.R.A. The composition and initial evaluation of a grimace scale in ferrets after surgical implantation of a telemetry probe. *PLoS ONE* **2017**, *12*, e0187986. [CrossRef]

27. Evangelista, M.C.; Watanabe, R.; Leung, V.S.Y.; Monteiro, B.P.; O'Toole, E.; Pang, D.S.J.; Steagall, P.V. Facial expressions of pain in cats: The development and validation of a Feline Grimace Scale. *Sci. Rep.* **2019**, *9*, 1–11. [CrossRef]

28. Costa, E.D.; Minero, M.; Lebelt, D.; Stucke, D.; Canali, E.; Leach, M.C. Development of the horse grimace scale (HGS) as a pain assessment tool in horses undergoing routine castration. *PLoS ONE* **2014**, *9*, e92281. [CrossRef]

29. Apkarian, A.V.; Hashmi, J.A.; Baliki, M.N. Pain and the brain: Specificity and plasticity of the brain in clinical chronic pain. *Pain* **2011**, *152*, S49–S64. [CrossRef]

30. Blackburn-Munro, G. Pain-like behaviours in animals—How human are they? *Trends Pharmacol. Sci.* **2004**, *25*, 299–305. [CrossRef]

31. Nagakura, Y. The need for fundamental reforms in the pain research field to develop innovative drugs. *Expert Opin. Drug Discov.* **2016**, *12*, 39–46. [CrossRef]

32. Nagakura, Y.; Miwa, M.; Yoshida, M.; Miura, R.; Tanei, S.; Tsuji, M.; Takeda, H. Spontaneous pain-associated facial expression and efficacy of clinically used drugs in the reserpine-induced rat model of fibromyalgia. *Eur. J. Pharmacol.* **2019**, *864*, 172716. [CrossRef] [PubMed]

33. Serizawa, K.; Tomizawa-Shinohara, H.; Yasuno, H.; Yogo, K.; Matsumoto, Y. Anti-IL-6 receptor antibody inhibits spontaneous pain at the pre-onset of experimental autoimmune encephalomyelitis in Mice. *Front. Neurol.* **2019**, *10*, 341. [CrossRef] [PubMed]

34. LeResche, L. Facial expression in pain: A study of candid photographs. *J. Nonverbal Behav.* **1982**, *7*, 46–56. [CrossRef]

35. Williams, A.C.D.C. Facial expression of pain: An evolutionary account. *Behav. Brain Sci.* **2002**, *25*, 439–455. [CrossRef]

36. Mogil, J.S.; Pang, D.S.; Dutra, G.G.S.; Chambers, C.T. The development and use of facial grimace scales for pain measurement in animals. *Neurosci. Biobehav. Rev.* **2020**, *116*, 480–493. [CrossRef]

37. Viscardi, A.V.; Hunniford, M.; Lawlis, P.; Leach, M.; Turner, P.V. Development of a piglet grimace scale to evaluate piglet pain using facial expressions following castration and tail docking: A pilot study. *Front. Vet. Sci.* **2017**, *4*, 51. [CrossRef]

38. Di Giminiani, P.; Brierley, V.L.; Scollo, A.; Gottardo, F.; Malcolm, E.M.; Edwards, S.A.; Leach, M.C. The assessment of facial expressions in piglets undergoing tail docking and castration: Toward the development of the piglet grimace scale. *Front. Vet. Sci.* **2016**, *3*, 100. [CrossRef]

39. Bateson, P. Assessment of pain in animals. *Anim. Behav.* **1991**, *42*, 827–839. [CrossRef]

40. Whittaker, A.L.; Howarth, G.S. Use of spontaneous behaviour measures to assess pain in laboratory rats and mice: How are we progressing? *Appl. Anim. Behav. Sci.* **2014**, *151*, 1–12. [CrossRef]

41. Rutherford, K. Assessing pain in animals. *Anim. Welf.* **2002**, *11*, 31–53.

42. De C Williams, A.C. Persistence of pain in humans and other mammals. *Philos. Trans. R. Soc. B Biol. Sci.* **2019**, *374*, 20190276. [CrossRef] [PubMed]

43. Walters, E.T.; De C Williams, A.C. Evolution of mechanisms and behaviour important for pain. *Philos. Trans. R. Soc. B Biol. Sci.* **2019**, *374*, 20190275. [CrossRef]

44. McLennan, K.M.; Miller, A.L.; Costa, E.D.; Stucke, D.; Corke, M.J.; Broom, D.M.; Leach, M.C. Conceptual and methodological issues relating to pain assessment in mammals: The development and utilisation of pain facial expression scales. *Appl. Anim. Behav. Sci.* **2019**, *217*, 1–15. [CrossRef]

45. Bendinger, T.; Plunkett, N. Measurement in pain medicine. *BJA Educ.* **2016**, *16*, 310–315. [CrossRef]

46. Good, M.; Stiller, C.; Zauszniewski, J.A.; Anderson, G.C.; Stanton-Hicks, M.; Grass, J.A. Sensation and distress of pain scales: Reliability, validity, and sensitivity. *J. Nurs. Meas.* **2001**, *9*, 219–238. [CrossRef]

47. Chartier, L.C.; Hebart, M.L.; Howarth, G.S.; Whittaker, A.L.; Mashtoub, S. Affective state determination in a mouse model of colitis-associated colorectal cancer. *PLoS ONE* **2020**, *15*, e0228413. [CrossRef]

48. George, R.P.; Howarth, G.S.; Whittaker, A.L. Use of the rat grimace scale to evaluate visceral pain in a model of chemotherapy-induced mucositis. *Animals* **2019**, *9*, 678. [CrossRef] [PubMed]

49. De Almeida, A.S.; Rigo, F.K.; De Prá, S.D.T.; Milioli, A.M.; Dalenogare, D.P.; Pereira, G.C.; Ritter, C.D.S.; Peres, D.S.; Antoniazzi, C.T.D.; Stein, C.; et al. Characterization of cancer-induced nociception in a murine model of breast carcinoma. *Cell. Mol. Neurobiol.* **2019**, *39*, 605–617. [CrossRef]

50. De Almeida, A.S.; Rigo, F.K.; De Prá, S.D.T.; Milioli, A.M.; Pereira, G.C.; Lückemeyer, D.D.; Antoniazzi, C.T.; Kudsi, S.Q.; Araújo, D.; Oliveira, S.M.; et al. Role of transient receptor potential ankyrin 1 (TRPA1) on nociception caused by a murine model of breast carcinoma. *Pharmacol. Res.* **2020**, *152*, 104576. [CrossRef]

51. Mai, S.H.C.; Sharma, N.; Kwong, A.C.; Dwivedi, D.J.; Khan, M.; Grin, P.; Fox-Robichaud, A.E.; Liaw, P.C. Body temperature and mouse scoring systems as surrogate markers of death in cecal ligation and puncture sepsis. *Intensiv. Care Med. Exp.* **2018**, *6*, 20. [CrossRef]

52. Akintola, T.; Raver, C.; Studlack, P.; Uddin, O.; Masri, R.; Keller, A. The grimace scale reliably assesses chronic pain in a rodent model of trigeminal neuropathic pain. *Neurobiol. Pain* **2017**, *2*, 13–17. [CrossRef] [PubMed]

53. Akintola, T.; Tricou, C.; Raver, C.; Castro, A.; Colloca, L.; Keller, A. In search of a rodent model of placebo analgesia in chronic orofacial neuropathic pain. *Neurobiol. Pain* **2019**, *6*, 100033. [CrossRef] [PubMed]

54. Duffy, S.S.; Perera, C.J.; Makker, P.G.S.; Lees, J.G.; Carrive, P.; Moalem-Taylor, G. Peripheral and central neuroinflammatory changes and pain behaviors in an animal model of multiple sclerosis. *Front. Immunol.* **2016**, *7*, 369. [CrossRef] [PubMed]

55. Hassler, S.N.; Ahmad, F.B.; Burgos-Vega, C.C.; Boitano, S.; Vágner, J.; Price, T.J.; Dussor, G. Protease activated receptor 2 (PAR2) activation causes migraine-like pain behaviors in mice. *Cephalalgia* **2018**, *39*, 111–122. [CrossRef]

56. Mittal, A.; Gupta, M.; Lamarre, Y.; Jahagirdar, B.; Gupta, K. Quantification of pain in sickle mice using facial expressions and body measurements. *Blood Cells Mol. Dis.* **2016**, *57*, 58–66. [CrossRef]

57. Gao, M.; Long, H.; Ma, W.; Liao, L.; Yang, X.; Zhou, Y.; Shan, D.; Huang, R.; Jian, F.; Wang, Y.; et al. The role of periodontal ASIC3 in orofacial pain induced by experimental tooth movement in rats. *Eur. J. Orthod.* **2015**, *38*, 577–583. [CrossRef]

58. Miller, A.L.; Leach, M.C. The mouse grimace scale: A clinically useful tool? *PLoS ONE* **2015**, *10*, e0136000. [CrossRef]

59. Sorge, R.E.; Martin, L.J.; Isbester, K.A.; Sotocinal, S.G.; Rosen, S.; Tuttle, A.H.; Wieskopf, J.S.; Acland, E.L.; Dokova, A.; Kadoura, B.; et al. Olfactory exposure to males, including men, causes stress and related analgesia in rodents. *Nat. Methods* **2014**, *11*, 629–632. [CrossRef]

60. Leung, V.; Zhang, E.; Pang, D.S. Real-time application of the Rat Grimace Scale as a welfare refinement in laboratory rats. *Sci. Rep.* **2016**, *6*, 31667. [CrossRef]

61. Leung, V.S.; Benoit-Biancamano, M.O.; Pang, D.S. Performance of behavioral assays: The Rat Grimace Scale, burrowing activity and a composite behavior score to identify visceral pain in an acute and chronic colitis model. *PAIN Rep.* **2019**, *4*, e718. [CrossRef]

62. Wardill, H.R.; Gibson, R.J.; Van Sebille, Y.Z.; Secombe, K.R.; Coller, J.K.; White, I.A.; Manavis, J.; Hutchinson, M.R.; Staikopoulos, V.; Logan, R.; et al. Irinotecan-induced gastrointestinal dysfunction and pain are mediated by common TLR4-dependent mechanisms. *Mol. Cancer Ther.* **2016**, *15*, 1376–1386. [CrossRef] [PubMed]

63. Gibson, R.J.; Coller, J.K.; Wardill, H.R.; Hutchinson, M.R.; Smid, S.; Bowen, J.M. Chemotherapy-induced gut toxicity and pain: Involvement of TLRs. *Support. Care Cancer* **2015**, *24*, 2251–2258. [CrossRef] [PubMed]

64. Hsi, Z.Y.; Stewart, L.A.; Lloyd, K.C.K.; Grimsrud, K.N. Hypoglycemia after bariatric surgery in mice and optimal dosage and efficacy of glucose supplementation. *Comp. Med.* **2020**, *70*, 111–118. [CrossRef] [PubMed]

65. Cho, C.; Michalidis, V.; Lecker, I.; Collymore, C.; Hanwell, D.; Loka, M.; Danesh, M.; Pham, C.; Urban, P.; Bonin, R.P.; et al. Evaluating analgesic efficacy and administration route following craniotomy in mice using the grimace scale. *Sci. Rep.* **2019**, *9*, 359. [CrossRef] [PubMed]

66. Miller, A.L.; Kitson, G.; Skalkoyannis, B.; Leach, M. The effect of isoflurane anaesthesia and buprenorphine on the mouse grimace scale and behaviour in CBA and DBA/2 mice. *Appl. Anim. Behav. Sci.* **2015**, *172*, 58–62. [CrossRef] [PubMed]

67. Miller, A.L.; Leach, M.C. The effect of handling method on the mouse grimace scale in two strains of laboratory mice. *Lab. Anim.* **2016**, *50*, 305–307. [CrossRef]

68. Costa, E.D.; Pascuzzo, R.; Leach, M.C.; Dai, F.; Lebelt, D.; Vantini, S.; Minero, M. Can grimace scales estimate the pain status in horses and mice? A statistical approach to identify a classifier. *PLoS ONE* **2018**, *13*, e0200339. [CrossRef]

69. Roughan, J.V.; Sevenoaks, T. Welfare and scientific considerations of tattooing and ear tagging for mouse identification. *J. Am. Assoc. Lab. Anim. Sci.* **2019**, *58*, 142–153. [CrossRef]

70. Waite, M.E.; Tomkovich, A.; Quinn, T.L.; Schumann, A.P.; Dewberry, L.S.; Totsch, S.K.; Sorge, R.E. Efficacy of common analgesics for postsurgical pain in rats. *J. Am. Assoc. Lab. Anim. Sci.* **2015**, *54*, 420–425.

71. Wang, S.; Kim, M.; Ali, Z.; Ong, K.; Pae, E.K.; Chung, M.K. Trpv1 and trpv1-expressing nociceptors mediate orofacial pain behaviors in a mouse model of orthodontic tooth movement. *Front Physiol* **2019**, *10*, 1207. [CrossRef]

72. Zhu, Y.; Wang, S.; Long, H.; Zhu, J.; Jian, F.; Ye, N.; Lai, W. Effect of static magnetic field on pain level and expression of P2X3 receptors in the trigeminal ganglion in mice following experimental tooth movement. *Bioelectromagnetics* **2016**, *38*, 22–30. [CrossRef] [PubMed]

73. Miller, A.L.; Golledge, H.D.R.; Leach, M.C. The influence of isoflurane anaesthesia on the rat grimace scale. *PLoS ONE* **2016**, *11*, e0166652. [CrossRef] [PubMed]

74. Sorge, R.E.; Mapplebeck, J.C.S.; Rosen, S.; Beggs, S.; Taves, S.; Alexander, J.K.; Martin, L.J.; Austin, J.-S.; Sotocinal, S.G.; Chen, D.; et al. Different immune cells mediate mechanical pain hypersensitivity in male and female mice. *Nat. Neurosci.* **2015**, *18*, 1081–1083. [CrossRef] [PubMed]

75. Hohlbaum, K.; Bert, B.; Dietze, S.; Palme, R.; Fink, H.; Thöne-Reineke, C. Severity classification of repeated isoflurane anesthesia in C57BL/6JRj mice—Assessing the degree of distress. *PLoS ONE* **2017**, *12*, e0179588. [CrossRef]

76. Hohlbaum, K.; Bert, B.; Dietze, S.; Palme, R.; Fink, H.; Thöne-Reineke, C. Impact of repeated anesthesia with ketamine and xylazine on the well-being of C57BL/6JRj mice. *PLoS ONE* **2018**, *13*, e0203559. [CrossRef]

77. Gouveia, K.; Hurst, J.L. Optimising reliability of mouse performance in behavioural testing: The major role of non-aversive handling. *Sci. Rep.* **2017**, *7*, 44999. [CrossRef]

78. Hurst, J.L.; West, R.S. Taming anxiety in laboratory mice. *Nat. Methods* **2010**, *7*, 825–826. [CrossRef]

79. Gouveia, K.; Hurst, J.L. Reducing mouse anxiety during handling: Effect of experience with handling tunnels. *PLoS ONE* **2013**, *8*, e66401. [CrossRef]

80. Henderson, L.J.; Smulders, T.V.; Roughan, J.V. Identifying obstacles preventing the uptake of tunnel handling methods for laboratory mice: An international thematic survey. *PLoS ONE* **2020**, *15*, e0231454. [CrossRef]

81. Miller, A.L.; Leach, M. Using the mouse grimace scale to assess pain associated with routine ear notching and the effect of analgesia in laboratory mice. *Lab. Anim.* **2014**, *49*, 117–120. [CrossRef]

82. Kasanen, I.H.E.; Voipio, H.-M.; Leskinen, H.; Luodonpää, M.; Nevalainen, T.O. Comparison of ear tattoo, ear notching and microtattoo in rats undergoing cardiovascular telemetry. *Lab. Anim.* **2011**, *45*, 154–159. [CrossRef] [PubMed]

83. Rea, B.J.; Wattiez, A.-S.; Waite, J.S.; Castonguay, W.C.; Schmidt, C.M.; Fairbanks, A.M.; Robertson, B.R.; Brown, C.J.; Mason, B.N.; Moldovan-Loomis, M.-C.; et al. Peripherally administered calcitonin gene–related peptide induces spontaneous pain in mice. *Pain* **2018**, *159*, 2306–2317. [CrossRef] [PubMed]

84. Matsumiya, L.C.; Sorge, R.E.; Sotocinal, S.G.; Tabaka, J.M.; Wieskopf, J.S.; Zaloum, A.; King, O.D.; Mogil, J.S. Using the mouse grimace scale to reevaluate the efficacy of postoperative analgesics in laboratory mice. *J. Am. Assoc. Lab. Anim. Sci.* **2012**, *51*, 42–49. [PubMed]

85. Jirkof, P.; Abdelrahman, A.; Bleich, A.; Durst, M.; Keubler, L.M.; Potschka, H.; Struve, B.; Talbot, S.R.; Vollmar, B.; Zechner, D.; et al. A safe bet? Inter-laboratory variability in behaviour-based severity assessment. *Lab. Anim.* **2019**, *54*, 73–82. [CrossRef]

86. Mogil, J.S. Laboratory environmental factors and pain behavior: The relevance of unknown unknowns to reproducibility and translation. *Lab. Anim.* **2017**, *46*, 136–141. [CrossRef]

87. Burgos-Vega, C.C.; Quigley, L.D.; Dos Santos, G.T.; Yan, F.; Asiedu, M.; Jacobs, B.; Motina, M.; Safdar, N.; Yousuf, H.; Avona, A.; et al. Non-invasive dural stimulation in mice: A novel preclinical model of migraine. *Cephalalgia* **2018**, *39*, 123–134. [CrossRef]

88. Hassan, A.M.; Jain, P.; Mayerhofer, R.; Fröhlich, E.E.; Farzi, A.; Reichmann, F.; Herzog, H.; Holzer, P. Visceral hyperalgesia caused by peptide YY deletion and Y2 receptor antagonism. *Sci. Rep.* **2017**, *7*, 40968. [CrossRef]

89. Bu, X.; Liu, Y.; Lu, Q.; Jin, Z. Effects of "Danzhi Decoction" on chronic pelvic pain, hemodynamics, and proinflammatory factors in the murine model of sequelae of pelvic inflammatory disease. *Evid. Based Complement. Altern. Med.* **2015**, *2015*, 1–12. [CrossRef]

90. Whittaker, A.L.; Leach, M.C.; Preston, F.L.; Lymn, K.A.; Howarth, G.S. Effects of acute chemotherapy-induced mucositis on spontaneous behaviour and the grimace scale in laboratory rats. *Lab. Anim.* **2015**, *50*, 108–118. [CrossRef]

91. Toscano, M.G.; Ganea, I.; Gamero, A.M. Cecal ligation puncture procedure. *J. Vis. Exp.* **2011**. [CrossRef]

92. Nguyen, H.B.; Rivers, E.P.; Abrahamian, F.M.; Moran, G.J.; Abraham, E.; Trzeciak, S.; Huang, D.T.; Osborn, T.M.; Stevens, D.; Talan, D.A. Severe sepsis and septic shock: Review of the literature and emergency department management guidelines. *Ann. Emerg. Med.* **2006**, *48*, 54. [CrossRef] [PubMed]

93. Dwivedi, D.J.; Grin, P.; Khan, M.; Prat, A.; Zhou, J.; Fox-Robichaud, A.E.; Seidah, N.G.; Liaw, P.C. Differential expression of PCSK9 modulates infection, inflammation, and coagulation in a murine model of sepsis. *Shock* **2016**, *46*, 672–680. [CrossRef] [PubMed]

94. Yamamoto, K.; Tatsutani, S.; Ishida, T. Detection of nausea-like response in rats by monitoring facial expression. *Front. Pharmacol.* **2017**, *7*, 100. [CrossRef]

95. Herrera, C.; Bolton, F.; Arias, A.; Harrison, R.A.; Gutiérrez, J.M. Analgesic effect of morphine and tramadol in standard toxicity assays in mice injected with venom of the snake Bothrops asper. *Toxicon* **2018**, *154*, 35–41. [CrossRef] [PubMed]

96. Wong, S.M.; Tan, S.J.X.; Koh, J.; Zainul, M.; Phang, G.S.S.; Toh, A.; Babu, K.R.; Chooi, K.F. The Rat Face Finder and Improved Assessment of Visceral Pain. In Proceedings of the 9th SALAS Annual Regional Conference—Neuroscience: A New Frontier, New York, NY, USA, 4–6 December 2013.

97. Ernst, L.; Kopaczka, M.; Schulz, M.; Talbot, S.R.; Zieglowski, L.; Meyer, M.; Bruch, S.; Merhof, D.; Tolba, R.H. Improvement of the Mouse Grimace Scale set-up for implementing a semi-automated Mouse Grimace Scale scoring (Part 1). *Lab. Anim.* **2019**, *54*, 83–91. [CrossRef] [PubMed]

98. Ernst, L.; Kopaczka, M.; Schulz, M.; Talbot, S.R.; Struve, B.; Häger, C.; Bleich, A.; Durst, M.; Jirkof, P.; Arras, M.; et al. Semi-automated generation of pictures for the Mouse Grimace Scale: A multi-laboratory analysis (Part 2). *Lab. Anim.* **2019**, *54*, 92–98. [CrossRef] [PubMed]

99. Tuttle, A.H.; Molinaro, M.J.; Jethwa, J.F.; Sotocinal, S.G.; Prieto, J.C.; Styner, M.A.; Mogil, J.S.; Zylka, M.J. A deep neural network to assess spontaneous pain from mouse facial expressions. *Mol. Pain* **2018**, *14*, 1744806918763658. [CrossRef]

100. Andresen, N.; Wöllhaf, M.; Hohlbaum, K.; Lewejohann, L.; Hellwich, O.; Thöne-Reineke, C.; Belik, V. Towards a fully automated surveillance of well-being status in laboratory mice using deep learning: Starting with facial expression analysis. *PLoS ONE* **2020**, *15*, e0228059. [CrossRef]

101. Eral, M.; Aktas, C.C.; Kocak, E.E.; Dalkara, T.; Halici, U. Assessment of pain in mouse facial images. In Proceedings of the 2016 20th National Biomedical Engineering Meeting (BIYOMUT), Izmir, Turkey, 3–5 November 2016; pp. 1–4. [CrossRef]

102. Mahmoud, M.; Lu, Y.; Hou, X.; McLennan, K.; Robinson, P. Estimation of Pain in Sheep Using Computer Vision. In *Handbook of Pain and Palliative Care*; Springer: Berlin/Heidelberg, Germany, 2018; pp. 145–157.

103. McLennan, K.M.; Mahmoud, M. Development of an automated pain facial expression detection system for sheep (ovis aries). *Animals* **2019**, *9*, 196. [CrossRef]

104. Bartlett, M.; Littlewort, G.; Frank, M.; Lainscsek, C.; Fasel, I.; Movellan, J. Recognizing Facial Expression: Machine Learning and Application to Spontaneous Behavior. In Proceedings of the 2005 IEEE Computer Society Conference on Computer Vision and Pattern Recognition (CVPR'05)—Workshops, San Diego, CA, USA, 21–23 September 2005; Volume 2, pp. 568–573. [CrossRef]

105. Oliver, V.; De Rantere, D.; Ritchie, R.; Chisholm, J.; Hecker, K.G.; Pang, D.S. Psychometric assessment of the rat grimace scale and development of an analgesic intervention score. *PLoS ONE* **2014**, *9*, e97882. [CrossRef]

106. Baliki, M.N.; Chialvo, D.R.; Geha, P.Y.; Levy, R.M.; Harden, R.N.; Parrish, T.B.; Apkarian, A.V. Chronic pain and the emotional brain: Specific brain activity associated with spontaneous fluctuations of intensity of chronic back pain. *J. Neurosci.* **2006**, *26*, 12165–12173. [CrossRef] [PubMed]

107. Roughan, J.V.; Bertrand, H.G.; Isles, H.M. Meloxicam prevents COX-2-mediated post-surgical inflammation but not pain following laparotomy in mice. *Eur. J. Pain* **2015**, *20*, 231–240. [CrossRef] [PubMed]
108. Zhang, E.Q.; Leung, V.S.; Pang, D.S. Influence of rater training on inter- and intrarater reliability when using the rat grimace scale. *J. Am. Assoc. Lab. Anim. Sci.* **2019**, *58*, 178–183. [CrossRef] [PubMed]

Review

Grimace Scores: Tools to Support the Identification of Pain in Mammals Used in Research

Shari Cohen [1,*] **and Thierry Beths** [2]

1   Office of Research, Ethics, and Integrity, University of Melbourne, Melbourne 3010, Australia
2   Anesthesia and Veterinary Clinical Sciences, University of Melbourne, Melbourne 3010, Australia;
    thierry.beths@unimelb.edu.au
*   Correspondence: sharicohen.vet@gmail.com; Tel.: +61-0429-725-806

Received: 16 August 2020; Accepted: 16 September 2020; Published: 23 September 2020

Simple Summary: The ability to identify and assess pain is paramount in animal research to address the 'refinement' principle of the 3Rs (Reduction, Refinement, Replacement), satisfy public acceptability of animal use in research and address ethical and legal obligations. Many physiological, behavioural and physical pain assessments are commonly used, but all have their limitations. Grimace scales are a promising adjunctive behavioural pain assessment technique in some mammalian species used in research. This paper reviews the extant literature studying pain assessment techniques in general, and grimace scales specifically, in animal research. The results indicate that the grimace scale technique is simple and able to be used spontaneously at the 'cage side', is non-invasive in its application, highly repeatable, reliable between interobserver and intraobserver applications and easy to train and use. The use of grimace scales should be more frequently considered as an important parameter of interest in research and animal wellbeing. Further research into the use of grimace scales is required to develop scales for a wider range of animal species, increase applicability in studies specifically related to pain assessment and for further validation of the technique.

Abstract: The 3Rs, Replacement, Reduction and Refinement, is a framework to ensure the ethical and justified use of animals in research. The implementation of refinements is required to alleviate and minimise the pain and suffering of animals in research. Public acceptability of animal use in research is contingent on satisfying ethical and legal obligations to provide pain relief along with humane endpoints. To fulfil this obligation, staff, researchers, veterinarians, and technicians must rapidly, accurately, efficiently and consistently identify, assess and act on signs of pain. This ability is paramount to uphold animal welfare, prevent undue suffering and mitigate possible negative impacts on research. Identification of pain may be based on indicators such as physiological, behavioural, or physical ones. Each has been used to develop different pain scoring systems with potential benefits and limitations in identifying and assessing pain. Grimace scores are a promising adjunctive behavioural technique in some mammalian species to identify and assess pain in research animals. The use of this method can be beneficial to animal welfare and research outcomes by identifying animals that may require alleviation of pain or humane intervention. This paper highlights the benefits, caveats, and potential applications of grimace scales.

Keywords: grimace scores; pain; laboratory animals; pain assessment

---

1. Introduction

The 3Rs, Replacement, Reduction and Refinement, is a fundamental framework used internationally to ensure the ethical and justified use of animals in research [1]. The implementation of refinements is required to alleviate and minimise the pain and suffering of animals used in research. Public acceptability of animal use in research is contingent on satisfying the ethical and legal obligations to provide appropriate pain relief along with humane endpoints for potentially painful procedures. To fulfil this obligation, staff, researchers, veterinarians, and technicians must rapidly, accurately, efficiently and consistently identify and assess signs of pain in their target species, and act accordingly. The ability to identify and assess pain and suffering is paramount to animal welfare in research to prevent undue suffering and any possible consequent negative impact on research outcomes.

Identification of pain may be based on several indicators such as physiological, behavioural, or physical ones. Each of these has been used to develop different pain scoring systems, with potential benefits and limitations in identifying and assessing pain. Grimace scores are a promising adjunctive behavioural technique in some mammalian species to facilitate in identifying and assessing pain in research animals. The use of this method can be beneficial to animal welfare and research outcomes by identifying animals that may require alleviation of pain or humane intervention. A discussion of the benefits of grimace scales, including their potential applications, is included in this paper.

1.1. Eligibility Criteria

The inclusion criteria were: publication in English; assessments of pain in research animals and livestock; mammalian grimaces scales in animals; studies key to the development of grimace scales; studies that used facial units as an indicator of pain assessments in animals.

1.2. Search Strategy

The search strategy aimed to only find articles published in English or translated into English. There was no restriction on the date of publication. Articles were searched for between June–July 2020. Keywords used to search all databases and references sources were animal grimace score, animal grimace scale, animal pain assessment, animal pain indicators, animal pain face, animal pain scales and the NC3Rs website. All papers were retrieved and downloaded into Endnote with X8.0.1 with any duplicates removed. What is pain and why does it matter?

The International Association for the Study of Pain defines pain as: 'An unpleasant sensory and emotional experience associated with, or resembling that associated with, actual or potential tissue damage' [2]. Pain can be further categorised as acute, visceral or chronic. Acute pain serves an evolutionary and adaptive function to signal and avoid potential or actual damage to tissues. This type of pain may result from an injury or surgical wound [3–5]. Visceral pain is due to the activation of stretch or pressure receptors in visceral organs. Unalleviated or poorly treated acute pain can progress to chronic pain. The latter is the result of neuroplastic changes occurring within the nervous system, rendering the body more sensitive to pain and can even create sensations of pain without any external stimuli [3–7]. It is important to be able to manage and assess all types of pain in research mammals and avoid the inadvertent development of chronic pain. While the types of pain may manifest differently, research staff must be able to assess and alleviate pain to maintain optimal animal wellbeing (mental and physical).

There are moral, legal and ethical obligations that require those working with animals to manage pain [5,7–13]. The recognition, assessment and treatment of pain is an essential aspect in public support and acceptability in the use of animals for research [7,12,13]. Using the precautionary principle, animal ethics committees and research staff must acknowledge the potential for pain [5,10,12,14]. They must also consider the experimental and animal welfare consequences of pain and take steps to ensure pain is adequately managed during procedures [3–5,15]. Regulatory frameworks often apply, as a precaution, the anthropomorphic principle, by which any procedure causing, or expected to cause, pain in humans, may produce pain in animals. The European Union, United Kingdom and Australian regulations operate on this principle and require the alleviation of pain for research animals. Exceptions to the alleviation of pain may be granted for studies that measure pain and/or distress. However, even in these exceptional cases, there is always a maximum threshold level of pain before intervention is required [5,10,12,14–17]. Although management of pain and associated humane interventions will vary due to the nature of the experimental outcomes, researchers are required to intervene with predetermined criteria to alleviate pain or, if necessary, humanely euthanise animals. For practical purposes, animal users can only fulfil this obligation for humane intervention if they are able to identify pain rapidly, consistently and accurately in their target species.

Unalleviated pain results in alterations in animal behaviour, physiology, and physical states [18–20]. These changes can be identified in various ways through behavioural observation, biochemistry, haematology, endocrinology and physical alterations in locomotion or posture [4,21]. In addition to the suffering of the animal, these changes may impact experimental outcomes and become a confounder by increasing experimental variability and producing negative affective states. Conversely, positive emotional states of animals are linked to less experimental variability and more robust experimental results [22,23]. The full spectrum of potential confounders to unmitigated pain is not entirely understood; however, the literature supports that there are experimental and animal welfare benefits in identifying and subsequently treating pain and alleviating negative effects on animals used for research [3,17,19,21–24].

1.3. Pain Faces

Humans are known to display a series of facial expressions linked to the experience of pain [25,26]. Those so-called 'pain faces' are used in human medicine to detect, as well as assess, pain in non-verbal humans (i.e., infants) [7,25,26]. These pain faces can be used to develop grimace scales and capitalise on the human propensity to focus on the facial area [27,28]. The conservation of these pain faces is also present in many non-human mammalian species [26,29–34] and are a naturally useful method in the identification and assessment of pain. However, as with any technique, Grimace scales have benefits and limitations. These are important to acknowledge and take into considerations before their use.

1.4. Pain Assessment Requirements

There are a series of important considerations when determining if a method is an appropriate test in identifying and/or assessing pain. The testing method must reliably produce the same result independent of the observer and the number of times an animal is observed. These are, respectively, known as intraobserver and interobserver agreement. It should also be consistent between testing timepoints and observers [7,35,36]. An ideal method should be easy to train and not require specialist knowledge or equipment [7,33,36–38].

A suitable test must demonstrate validity by accurately determining or reflecting the presence or absence of pain [7,35,36,39]. To determine the validity of a pain assessment technique, we should test the animals before the painful stimulus, after the introduction of the painful stimulus and once pain relief has been provided. The test should demonstrate an absence of pain before the painful stimulus, an increase in pain at the introduction of the painful stimulus, and a subsequent reduction in observed pain on the delivery of an appropriate analgesic [7,36]. Ideally, the test should be able to demonstrate a dose–responsive curve to pain based on the administration of appropriate analgesia [40–44].

The specificity and sensitivity of a test are also crucial to ensure animals are correctly identified when pain or welfare concerns arise. If the specificity is too low, there is a risk of pain being incorrectly identified, potentially leading to unnecessary interventions such as pain relief or humane euthanasia [3,7,15,36]. Alternatively, if the sensitivity is too low, experimental animals may reach their threshold for intervention while being inaccurately identified as not painful, therefore remaining in pain possibly even beyond their humane endpoint. An appropriate method would demonstrate both high sensitivity and specificity to ensure correct assessment and correct management of arising pain or welfare issues [3,7,15,36].

Cage or pen-side pain identification techniques should rely on spontaneous rather than retrospective indicators of pain. It ensures humane intervention can be applied promptly with animals not left in distress for any extended length of time [9,45,46]. The assessment of pain should preferably be a non-invasive method, to avoid the risk of eliciting a pseudoanalgesic stress response, inhibiting the ability of the observer to detect pain accurately [47–50]. Techniques such as assessing the quality of nest-building in mice [3,15,18,51–53] or degree of burrowing in rodents are non-invasive, observatory, proxy measures to wellbeing and potentially pain [3,15,18,54,55].

1.5. Confounders to Pain Identification

Some caveats must be maintained when selecting a pain assessment technique. Many pain assessment indicators may be ambiguous. The choice of a pain identification tool or methodology must be specific to the species and validated for the procedures or experimental work being performed [56–60]. It well accepted that not all animals demonstrate the same signs of pain, even for a similar nociceptive stimulus [36,61]. Many research animals are prey animals and as such, are prone to hide signs of pain or demonstrate a freeze response, rendering pain assessments challenging [21,46,50,62–66]. The types of procedures or experiment performed should not obscure the ability of the technique to detect pain [39,56–58,67]. Pain identification should be consistent across the species regardless of sex, strain or breed; however, differences in pain thresholds between sexes or strains may exist [67,68]. Additionally, some natural behaviours (i.e., flehmen response, aggression) [3,4,32,69–72] or physiological indicators (i.e., cortisol, heart rate) [3–5,7,15,17,21,66] may be equivocal and require differentiation. Whenever possible, the choice of technique should accurately identify an animal in pain, independent of the procedure or behaviour performed, species, affective or physiological state, sex, strain or breed.

1.6. Non-Grimace Scale Pain Assessment

The individual expression, magnitude and experience of pain can vary between animals [67,68]. There are known difficulties in measuring the magnitude of a particular animal's pain or distress which can make the absolute measurement or degree of pain challenging to assess [4,21,62,68,73–75].

Before the development and use of grimace scales, a variety of indicators have been used in an attempt to identify and assess pain. These can be grouped into behavioural, physiological and physical indicators (Table 1). Typically, behavioural indicators have the benefit of being non-invasive, observational, requiring limited equipment and offering an opportunity to capture signs of pain in species or individuals that may hide signs of pain (i.e., prey) [3,21,50–55,66]. Many behavioural assessment techniques take time (>5 min), require extensive training, are more retrospective than spontaneous, and may be non-specific proxy indicators to pain [4,18,76–78]. Physiological indicators (neuroendocrine or sympathetic nervous system) are often non-specific markers related to stress or distress [3,5,21,79,80]. They do have the benefit of being relatively quantifiable but often require specialised equipment, are retrospective, usually require animal handling or restraint with the potential for a confounding pseudoanalgesic effect [47–49] or are a non-specific stress response [3,4,8,81]. Physical indicators such as changes in posture, locomotion and production yields, have been correlated with the presence of pain in animals [3,8,15,17,21,64,66,82–85]. However, physical indicators are just as often non-specific indicators of non-painful animal wellbeing or environmental factors [3,5,8,15,17,21,64,66,82,84].

Ideally, a pain assessment technique should ensure accurate pain identification and minimal opportunity for the confounding of experimental outcomes due to experimental procedures, sex, breed/strain, or species. At present, there is not a single non-invasive, low-cost behavioural, physical or physiological pain assessment technique that is spontaneous, pain-specific, easy to train and quick to use (Table 1) [5,15,16,21,38,46,53,78,86–89]. With the exception of some behavioural ethograms [63,78,90,91], other methods are unable to give a reliable dose-dependent response to pain. While many pain identification methods have their use and benefits, their use in the cage or pen-side management of animal pain and/or in the timely and appropriate application of humane intervention is limited.

Thus, a myriad of techniques has been developed in an attempt to assess and capture the various expressions of pain in animals. These tools usually revolve around three dimensions: behavioural, physiological, and physical [5,73]. Table 1 categorises and reviews some commonly used assessments [3,4,7,8,15,21,80] in terms of their dimension, ability to be timely (spontaneous), non-invasiveness, spontaneous, easiness to train, and low-cost with minimal or no equipment requirements.

Table 1. Behavioural, physiological, and physical pain assessment techniques.

Category	Assessment	Non-Invasive	Easy Training	High Cost	Special Equipment	Time Required >5 min	Spontaneous	Publications
Behaviour	Ethogram	Y	N	N	N	Y	Y	[4,5,7,15,17,18,21,52,63,66,78,86,87,91–93]
	Nesting	Y	Y	N	N	Y	N	[3,15,18,51,52]
	Burrowing	Y	N	N	N	Y	N	[3,15,18,54,55]
	Vocalisation	Y	Y	Y	Y/N *	Y	Y	[4,5,8,17,21,45,66,68,81,88,93,94]
	Grooming	Y	Y	N	Y	Y	N	[3–5,8,15,17,18,21,52,66]
	Real-time Grimace Score	Y	Y	N	N	N	Y	[37,54,55,70,85,92,95,96]
Physiological	Heart Rate or Respiratory Rate	N	Y	N	Y	N	Y	[3–5,15,17,21,66,93]
	Biochemical marker	N	N	Y	Y	Y	N	[3,4,15,17,21,93]
Physical	Weight loss or failure to gain weight	Y	N	N	Y	N	N	[3,5,8,15,17,21,64,66]
	Reduction in production **	Y	N	N	Y/N ***	Y	N	[21,66,84]
	Lameness	Y	Y/N ****	Y	N	N	Y	[3,5,15,17,21,37,66,82,85,93]
	Postural change	Y	N	Y	N	N	Y	[3–5,9,15,17,58,83,93]

* Ultrasonic vocalization monitoring requires special equipment; ** e.g., Egg or Milk; *** Eggs can easily be counted; **** Obvious signs of lameness are easy to train but more subtle lameness may be more difficult.

1.7. Grimace Scales in Animals

Grimace scales are proving to be a useful methodology for the identification of pain in research that meets most of the prerequisites for identifying and assessing pain in research animals. A range of research species-specific grimace scores has been developed (Table 2) and used in a wide range of experimental studies and research settings (Table 3). The initial methodology in the mapping of pain and the development of a facial action coding system (FACS) was developed in humans [97,98]. A FACS is an anatomical classification system used to map facial movements and facial muscles areas involved in facial contraction and relaxation. Photographs and videos scored by blinded observers serve as the base of facial mapping for FACS. FACSs offers the ability to code and identify expressions of pain via the individual components of facial expressions known as facial action units (FAUs) [99]. FAUs consistent with the expression of pain can then be used to develop a pain face or 'grimace' [99]. Regions of the face that have been found to change during the expression of pain include the eye, nose, cheek, mouth, ear and whiskers [8,81,100]. The position or carriage of the head is also found to change in some species as well [33,64,68,85,101,102]. The FAUs related to the expression of a grimace face in mammalian animals used in research are included in Table 4. From this known 'grimace face', the severity of the pain experienced can be objectively scored from images and/or film of animals in a known naturally (i.e., lameness, mastitis) [37,82,85] or experimentally induced (i.e., plantar incision [41,44,103] state of pain.

Table 2 summarises many of the available studies that demonstrate a successful use of grimace scales in research animals. The table outlines which species-specific grimace scales have been validated, shown to be pain-specific, demonstrated a dose-dependent relationship, used in real-time and were easy to use. The different pain states to which they are applicable is also listed. In all but one species (guinea pigs) [63,78,90], observers were found to correctly, reliably and objectively identify pain in animals when using facial expressions or facial action units.

Table 2. Summary of species-specific available grimace scores.

Species	Validated *	Specific To Pain	Dose-Dependent Relationship	Real-Time	Easy To Train	Acute Pain	Chronic Pain	Visceral Pain	Publications
Cattle	Y	Y	Y	Y	Y	Y	Y	N/R	[7,8,64,68,85]
Equine	Y	Y	Y	Y	Y	Y	N/R	Y	[33,35,37,39,87,92,102,104,105]
Feline	Y	Y	N/R	Y	Y	Y	N/R	N/R	[101]
Ferret	N/R	Y	N/R	Y	Y	Y	N/R	N/R	[106]
Guinea Pig	N/R	N/R	N/R	N/R	N/R	N/R	N/R	N/R	[63,78,90]
Lamb	Y	Y	N/R	Y	Y	Y	N/R	N/R	[107]
Mouse	Y	Y	Y	Y	Y	Y	Y	Y	[3,35,40,57,67,76,77,83,100,108–113]
Pig	Y	Y	N/R	Y	Y	Y	N/R	N/R	[114]
Piglet	Y	Y	Y	Y	Y	Y	N/R	N/R	[91,115–117]
Rabbit	Y	Y	Y	Y	Y	Y	N/R	N/R	[24,28,45,96]
Rat	Y	Y	Y	Y	Y	Y	Y	Y	[41–44,55,60,94,95,103,113,118–130]
Sheep	Y	Y	Y	Y	Y	Y	Y	N/R	[70,82]

* Validated by reduction in grimace sccres on receipt of pain relief and/or corroborated with other pain behaviours or testing; N/R denotes lack of publications available.

Table 3. Grimaces Scales Facial Action Units by Species.

Facial Action Unit or Indicator	Species
Orbital Tightening and/or Change in Orbital Area	Cattle [68,85] Equine [33,102] Feline [101] Ferret [106] Lamb [107] Pig [114] Piglet [116] Mouse [100] Rabbit [45] Rat [41] Sheep [70,82]
Cheek Tightening or Flattening	Cattle [68,85] Equine [33] Lamb [107] Pig [114] Piglet [116] Sheep [70,82] Rabbit [45] Rat [41]
Cheek Bulge	Ferret [106] Mouse [100]
Nose Bulge	Ferret [106] Mouse [100] Pig [114] Piglet [116] Rabbit [45]
Nose Flattening	Equine [33] Lamb [107] Rat [41]
Lowered Head Carriage	Equine [33,102] Cattle [68,85] Feline [101] Sheep [70]
Lip curling	Equine [92] Sheep [70]
Abnormal Nostril or Philtrum shape	Cattle [68,85] Equine [33,102] Lamb [107] Rabbit [45] Sheep [70]
Eye Rolling	Cattle [68]
Ear Position	Cattle [68] Equine [33,102] Feline [101] Ferret [106] Lamb [107] Mouse [100] Pig [114] Piglet [116] Rabbits [45] Sheep [70,82]
Whisker Position	Feline [101] Ferret [106] Mouse [100] Rabbit [45] Rat [41]
Abnormal Lip or mouth shape	Equine [33,102] Feline [101] Lamb [107] Sheep [70]
Open Mouth +/− Tongue Extruded	Cattle [68]

Table 4. Grimace Scales by Experimental Study or Pain Type.

Pain or Study Type	Species
Visceral	Cattle [85] Equine [92] Mouse [100] Rat [44,55,121,123]
Chronic	Mouse [108] Rat [55,120,128,129]
Acute	Equine [33,35,37,102,105] Cattle Gleerup [68,85] Ferret [106] Lamb [107] Mouse [35,40,76,77,100,109–111] Pig [114] Piglet [91,115–117] Rabbit [28,45] Rat [41,43,44,55,60,95,103,118,120,122–132] Sheep [70,82]
Neuropathic	Mouse [108] Rats [95,126]
Soft Tissues Surgery	Equine [33,35] Ferret [106] Lamb [107] Mouse [40,76,77,111] Pig [114] Piglet [115–117] Rabbit [28] Rat [41,44,60,125,127,130,132]
Orthopaedic Surgery	Mouse [109,110] Sheep [70]
Surgical, Mechanical, Branding, or Hypersensitivity Injury	Equine [102] Cattle [68] Mouse [100,112] Rabbit [45] Rat [41,103]
Dental	Equine [104,105] Mouse [112] Rat [43,124,128,129,131]
Stifle injury	Mouse [100] Rat [120]
Intraplantar CFA	Rat [41,103,118]
Intracerebral Haemorrhage	Rat [42]
Head and Ocular Pain	Mouse [109] Equine [105]
Footrot	Sheep [82]
Mastitis	Cattle [85] Sheep [82]
Lameness	Cattle [85] Sheep [44,82,118,130,132]
Sickle Anaemia	Mouse [83]
Cold hypersensitivity	Mouse [83]
Myocardial Infarction	Mouse [110]
Laminitis	Equine [37]
Cystitis	Mouse [100]
Sepsis	Rat [121]

Control animals (negative or positive) were also included throughout this process and a simple species-specific grimace scale was developed [25,33,41,107]. The scoring system most commonly used in grimace scales is a three-point scale to determine if a specific FAU is not (score = 0), moderately (score = 1), or obviously present (score = 2) [41,45,100]. The scale must then demonstrate a dose-dependent change in pain scores on the delivery of analgesia [7]. Further research is typically performed to ensure the applicability of the grimace scale across multiple pain scenarios or environments, sex, strain/breed, age, as well as type and length of painful stimuli [7]. The scoring system can be used three ways. Firstly, it can determine either the absence or presence of pain. Secondly, it can offer some distinction between the intensity of pain via the summation of total scores. A change by two or more points is considered to be a legitimate alteration in pain intensity [133]. Thirdly, a threshold score can be set to offer guidance to research staff as to when to intervene to provide pain alleviation or humane euthanasia of research animals. The process of developing a grimace scale is time intensive but once developed and validated is relatively easy to train research staff to use [7,38,70,85,101].

2. Advantages and Uses

Grimace scales have been applied across numerous research models, species and environmental contexts [41,128] (Table 4). They are a technique that can also be used to detect pain in existing pain research models as well as analgesic drug studies [40–42,45,60,77,109,110,128–130]. Grimace scales offer the ability to detect and assess the severity of pain, determine the potential benefit of any analgesic intervention and assist in identifying humane interventions. The technique is of practical value as it can be used at the cage or pen-side level as a spontaneous indicator of pain [39,41,55,75,92]. As a methodology, it has the added benefit of being easy to teach to a range of observers including research staff, clinical veterinarians, animal scientists and undergraduate and graduate students [38,41,55,75,129]. Overall, the grimace scale methodology appears to be acceptably conserved and validated across a number of mammalian species and range of experiments. It is likely this technique has the capacity to be applied across an even greater range of mammalian species and experimental settings (Tables 3 and 4). However, a careful systematic assessment will always be required to ensure applicability, accuracy and validity.

Grimace scale facial expressions are proving to be a useful [81] complement to existing tools in the assessment of animal wellbeing. The scores generated from the grimace scale should be used in conjunction with the context in which the animal is scored, its history, the procedure performed and the general parameters for wellbeing and signalment (sex, strain, species). When used appropriately, it is an excellent method to identify pain and as an adjunct to maintaining animal wellbeing in research studies [3,64,70,82,85,87]. Using this technique has the potential to improve pain detection in research animals and enable observers (i.e., research staff) a better opportunity to provide analgesia, humane euthanasia or identify animals requiring reassessment. The use of these grimace scales can be a vital tool to enable mitigation of the experience of pain in animals and refine animal welfare outcomes [41,60,66,75,76,82,100,114,128]. Unlike other types of pain assessment, grimace scales are spontaneous and usable in real-time [7,45,55,76,87,91,92,101]. They can also be matched and corroborated against other known indicators of pain or painful diseases including, but not limited to, lameness [37,64,82,85], cortisol [70], behavioural ethograms [81,85,91,92], acute laminitis [37], mastitis and foot rot [82]. A future area for development and benefit is the use of software automation in the development and scoring of facial expressions. The use of scoring software along with the installation of video cameras into enclosures may be able to enhance and hasten the development of grimaces, offer highly accurate grimace scores for animals in pain but also allow the remote monitoring and scoring of affected animals [41,59,134,135].

Another benefit of the system is simplicity, as it enables staff to distinguish a painful face from a non-painful one. Using a three-point scale is thought to be very useful in the reduction in subjectivity and offers observers greater clarity, confidence and support as to when to administer pain relief or humane intervention [7,75,81]. Reduction in grimace scores has been shown to occur on the application of pain relief [33,35,41,45,82,85,100,102] in a dose-dependent manner [40,41,128]. Therefore, grimace scales have the potential to assess both the presence and severity of pain. The use of grimace scales can alert research staff to animal discomfort, which may require additional monitoring, assessment or analgesia.

Grimace scales are a non-invasive method in the detection of pain [7,81,100]. Many of the animals utilised in research are known 'prey' species with a high degree of stoicism and evolutionary adaptation to minimise expressions of pain or poor welfare states [50,63–65,74]. Consequently, an ideal pain identification and assessment should be non-invasive and should reduce the possibility for these prey animals to minimise their expression of pain or for the potential of stress-induced analgesia [47–50].

Both experienced and inexperienced observers can identify pain with a significant intraobserver and interobserver agreement [41,57,60,66,76,82,100,114]. A potential benefit of using grimace scales to identify and assess pain in animals is that extensive animal experience is not required. Observers varied in their background experience to research and animal work and their training in pain assessment techniques. The observers ranged from students (undergraduate and postgraduate), veterinarians, animal care professionals, and early to late-career researchers [33,38,70,75,101,106,114,116]. Another favourable outcome when using grimace scales is that a natural empathy or innate understanding of animal behaviour is not necessary nor is a belief in the ability of an animal to experience pain. Through the use of a grimace scale, pain identification and assessment can be more objective (for or against the presence of pain). It also requires research staff to formally record a score and monitor animals for signs of pain, which can offer a more precise framework to determine when humane intervention or pain relief is needed [36].

The apparent usefulness of grimace scales could be related to several factors. One of which is that it capitalises on the innate human tendency to focus on the facial area when observing an animal [28]. Interestingly, many FAUs (orbital tightening, ear position and cheek area) appear to be conserved across mammalian species [33,41,45,82,100] (Table 3) and may be tapping into an evolutionary conservation repertoire of known FAUs. It may help explain how even the single identification of a few potentially evolutionarily conserved FAUs can still be useful in detecting pain [35]. It is supported via statistical modelling which has identified the FAUs most strongly correlated with a pain face, thereby offering the potential to isolate which FAUs are critical for use in a grimace scale (i.e., statistically significant) and which ones may detract from the scale (i.e., equines have four and mice have two critical FAUs) [35]. It may explain why grimace scales are one of the few techniques proven to be robust across several different mammalian species when compared to other pain assessment techniques [34]. However, by using only the minimum number required of FAUs to score pain, the ability to determine appropriate intervention thresholds and assess pain intensity may be reduced.

The use of FACS and subsequent combinations of FAUs appears to be an excellent method to identify changes in facial features, which are consistent with the experience of pain in animals [33,40,41,76,82,85,92,101,104,106,114,130]. The grimace scale method seems to meet many of the requirements for an ideal pain identification technique. It is known to be a reliable and validated method of assessing pain in many of the commonly used research animals [33,35,40,41,45,70,82,85,100,106,114].

3. Limitations

Similar to any tool, grimace scales have their caveats and limitations. The creation of pain and grimace scales takes considerable time to develop [3,41,61,62,99]. FAUs can be species-specific with each FAU requiring validation and ideally, statistical modelling and weighting, to determine its significance in the system [8]. This means the number of FAUs can vary amongst species with mice having five FAUs [100], rats four [41], sheep three to five [70,82], lambs five [107], equines six [37], ferrets three [106], cattle three to four [68,85], rabbits five [45] and pigs and piglets three [91,114]. While some FAUs movements appear to be tightly conserved (i.e., orbital tightening), others vary amongst species. These variations can be contradictory between species and may be due to age of the animal [107,115] and/or musculature of the face. The nose and philtrum areas tend to be areas with greater variation amongst mammals [37,82]. For example, rats and rabbits [41,45] will flatten their nose when in pain while mice and ferrets bulge their noses [100,106]. Therefore, each species requires the development of its own precise facial or grimace scoring system. Currently there are several commonly used mammalian research species that either do not have developed grimace scales or are yet to be fully developed. These include hamsters, dogs, guinea pigs, and non-human primates. Further work is needed to develop and determine the validity of grimace scales in these species.

Pain expression and threshold levels can also vary slightly amongst breed or strain [67,68]. Baseline grimace scores need to be taken for every cohort daily for approximately three days before the initiation of an experiment or potentially painful stimuli to minimise these variations [75]. False positives are known to occur in a small range of scenarios such as sedation/anaesthesia, sleeping status [41,56,75,100], or during bouts of aggression [32]. Therefore, grimace scales should not be used during those times. Additionally, it is important to note that facial variations may occur between individuals. As a result, absolute scores may be less important than a change in the score by two points or more (i.e., trends) [133], and a more 'trends-based' approach could be more useful. There are also times when the grimace scales can result in a false positive with animals not demonstrating a pain face during a known painful procedure. For example, ear clipping in mice did not demonstrate any changes in grimace scores [57] and neither did experimentally induced gastrointestinal mucositis in rats [58]. There is discussion around the differences found in the length of time of post-painful stimuli that an animal may display a pain face and hence a grimace score. Early peer-reviewed publications questioned the ability of grimace scales to be useful for more than 24 h after a painful event [59,100]. More current studies have demonstrated that pain can be identified in animals via grimace scales for more than 24 h and more than 14 days after a painful stimulus [55,82,108,120,128,129]. From the recent literature and available publications, it is clear this technique has applications beyond its initial use.

The history of the animal, the species, breed/strain, environmental context, procedures performed, and general parameters of wellbeing must be considered when using grimace scores [7,64,75,81]. There is still research to be conducted to explore the use of grimace scales. Currently, not every grimace scale has been fully validated (ferret, piglet, lamb) and additional species may yet benefit from their development (goats or other small mammals). Preliminary work does suggest that guinea pigs do not appear to be good candidates for facial pain scales. These studies used behaviourial ethograms which included elements of commonly and strongly conserved facial expressions (i.e., orbital tightening) and did not find any significant correlation of these expressions as indicators of pain [63,78,90]. It may be that grimace scales are not appropriate for these species or the FAUs associated with pain are different to other mammalian species. Many scales have only been used in specific settings or studies and need further work to determine if they are affected by common agricultural or animal procedures such as restraint in lambs [107] or piglets [115]. There is still variability in the available literature as to the length of time a grimace score can be detected in some species and studies [58,128] as well as its applicability of use [56] which should be further explored. While grimace scales have been developed and validated for several mammalian species, it is known there are species-specific variations in the expression of pain faces (guinea pigs) which may determine the development of a grimace to be unsuitable or require a different approach.

4. Application and Summary

In an ideal situation, a single pain identification technique would be sufficient across all species and scenarios; however, this currently does not exist and may never exist, given pain is an individual multifactorial experience [61]. Nonetheless, the growing body of literature is demonstrating that, overall, pain faces in mammalian species are often expressed and can be identified during most procedures, pain types, and contexts. Most of the variation found when using grimace scores to identify and assess pain is in the strength of association, the magnitude of certainty and the consistency of grimace score expression. Even with these variations, the use of grimace scales appears to be good at detecting pain in mammals [41,45,64,70,85,87,91,100–102,106,114]. However, if studies could be more standardised in their approach and the use of grimace scales, this may be beneficial in reducing minor confounding elements (i.e., handling, conspecifics) or in identifying areas of improvement. Future studies and the day to day practical and experimental application of this technique would benefit from having a formally validated and consistent training program, complete with video and photographic materials. A standard training program would be useful for grimace score users and has been useful for other pain scoring systems [38,46,86]. Part of the development, training and implementation of grimaces could be enhanced by the use of various technologies such as automated or semiautomated software for scale development and scoring via video surveillance [41,59,134,135]. These nascent technologies are often unfeasible due to cost, infrastructure constraints and a lack of development but in the future their use may play a greater role in grimace scoring systems.

The identification and mitigation of pain fulfil an essential and required aspect of refinement when working with animals in research. As of yet, no single indicator or technique is considered sufficient in the identification and assessment of pain. Several peer-reviewed publications have advocated multiple measures of animal welfare, and pain should be employed to mitigate the potential negative effects of pain on animal welfare and research outcomes [3,4,9,15,21,61,64]. Using a combination of relevant retrospective and spontaneous techniques applied on a case by case basis can maximise the opportunity to detect and assess pain in research animals. It minimises the chance for pain to be undetected and maximises the opportunity to preserve animal welfare and research outcomes. While there are known limitations, grimace scales to at least identify potential indicators of pain are useful tools [60]. The use of grimace scales with other parameters of pain and/or animals wellbeing is likely to increase the ability of research staff to identify and assess pain in mammals and offer appropriate humane interventions. At this time grimace scales are a potentially promising and important pain identification tool; however, further work should be performed in a consistent manner to validate existing work as well as explore new applications to other species, conditions and experimental studies.

To achieve good animal welfare and research outcomes and meet legal and ethical obligations, it is paramount to utilise a consistent and accurate pain identification method. The use of a grimace score can assist in fulfilling these obligations by identifying pain and allowing a timely intervention via analgesia or humane endpoints. Grimace scales are thus proving to be a valuable tool with a myriad of applications. Their use can offer improvements in animal welfare and more robust animal research outcomes [9,64]. While grimace scales are not without limitations, there is a growing body of literature and evidence to suggest they can be a significantly useful adjuncts in the detection and assessment of pain in a variety of species and research studies [7,35,41,66,76,100]. When used correctly by trained individuals along with an animal's history and basic wellbeing criteria, grimace scales can be a practical, accurate and easy method to identify pain in research animals to provide refinements in experimental animal welfare and outcomes [38,61,75]. Future applications of their use could focus on different types of experimental studies, new species, neonates, standardisation in training protocols, and correlation of multiple observations over time.

5. Conclusions

While there are some identified limitations, grimace scales appear to be a valid tool for pain assessment in many mammalian animals, and have many benefits compared to non-grimace pain assessment techniques. Due to the simplicity of spontaneous use, non-invasive application, repeatability of results, interobserver and intraobserver reliability and ease of training, the use of grimace scales should be more frequently considered as an important parameter of interest in research and animal wellbeing. In addition, this technique has the capacity to satisfy the requirement for refinement in accordance with the 3Rs. Additional research into the use of grimace scales is required for other species, pain-related or other specific studies, and further validation.

Author Contributions: Conceptualization, S.C. and T.B.; writing—original draft preparation, S.C.; writing—review and editing, S.C. and T.B.; visualization, S.C.; supervision, T.B. All authors have read and agreed to the published version of the manuscript.

Funding: This research received no external funding.

Acknowledgments: Special thank you to Manuel Christie, Natalie Roadknight, Kat Littlewood, and Natarsha Williams for their assistance in the preparation of this manuscript.

Conflicts of Interest: The authors declare no conflict of interest.

References

1. Russell, W.M.S. *The Principles of Humane Experimental Technique*; Methuen: London, UK, 1959.
2. Raja, S.N.; Carr, D.B.; Cohen, M.; Finnerup, N.B.; Flor, H.; Gibson, S.; Keefe, F.J.; Mogil, J.S.; Ringkamp, M.; Sluka, K.A.; et al. The revised international association for the study of pain definition of pain: Concepts, challenges, and compromises. *Pain* **2020**, *161*, 1976–1982. [CrossRef]
3. Turner, P.V.; Pang, D.S.; Lofgren, J.L. A review of pain assessment methods in laboratory rodents. *Comp. Med.* **2019**, *69*, 451–467. [CrossRef]
4. Carstens, E.; Moberg, G.P. Recognizing pain and distress in laboratory animals. *ILAR J.* **2000**, *41*, 62–71. [CrossRef]
5. National Research Council Committee; Alleviation of Pain in Laboratory Animals. The national academies collection: Reports funded by national institutes of health. In *Recognition and Alleviation of Pain in Laboratory Animals*; National Academies Press (US): Washington, DC, USA, 2010; National Academy of Sciences: Washington, DC, USA, 2009.
6. Larson, C.M.; Wilcox, G.L.; Fairbanks, C.A. The study of pain in rats and mice. *Comp. Med.* **2019**, *69*, 555–570. [CrossRef]
7. McLennan, K.M.; Miller, A.L.; Dalla Costa, E.; Stucke, D.; Corke, M.J.; Broom, D.M.; Leach, M.C. Conceptual and methodological issues relating to pain assessment in mammals: The development and utilisation of pain facial expression scales. *Appl. Anim. Behav. Sci.* **2019**, *217*, 1–15. [CrossRef]
8. McLennan, K.J.A. Why pain is still a welfare issue for farm animals, and how facial expression could be the answer. *Agriculture* **2018**, *8*, 127. [CrossRef]
9. Guatteo, R.; Levionnois, O.; Fournier, D.; Guémené, D.; Latouche, K.; Leterrier, C.; Mormède, P.; Prunier, A.; Servière, J.; Terlouw, C.; et al. Minimising pain in farm animals: The 3s approach—'suppress, substitute, soothe'. *Anim. Int. J. Anim. Biosci.* **2012**, *6*, 1261–1274. [CrossRef]
10. Parliament, E.; Council, E. Directive 2010/63/eu on the protection of animals used for scientific purposes. *EU Off. J.* **2010**, *V276*.
11. Olsson, I.A.S.; Silva, S.P.D.; Townend, D.; Sandøe, P. Protecting animals and enabling research in the european union: An overview of development and implementation of directive 2010/63/eu. *ILAR J.* **2017**, *57*, 347–357. [CrossRef]
12. National, Health and Medical Research Council. *Australian Code of Practice for the Care and Use of Animals for Scientific Purposes/National Health and Medical Research Council*; National Health and Medical Research Council: Canberra, Australia, 2004.
13. NC3R. The 3rs. Available online: https://www.nc3rs.org.uk/the-3rs (accessed on 15 July 2020).
14. Jennings, M.; Berdoy, M.; Hawkins, P.; Kerton, A.; Law, B.; Reed, B.; Sinnett-Smith, P.; Smith, D.; Farmer, A.M.; Jennings, M. *Guiding Principles on Good Practice for Ethical Review Processes*; RSPCA: Wales, UK; LASA: Pittsburgh, PA, USA, 2010.
15. Kohn, D.F.; Martin, T.E.; Foley, P.L.; Morris, T.H.; Swindle, M.M.; Vogler, G.A.; Wixson, S.K. Public statement: Guidelines for the assessment and management of pain in rodents and rabbits. *J. Am. Assoc. Lab. Anim. Sci. JAALAS* **2007**, *46*, 97–108.
16. National Research Council Committee for the Update of the Guide for the, Care and Use of Laboratory Animals. The national academies collection: Reports funded by national institutes of health. In *Guide for the Care and Use of Laboratory Animals*; National Academies Press (US): Washington, DC, USA, 2011; National Academy of Sciences: Washington, DC, USA, 2011.
17. Hawkins, P.; Morton, D.B.; Burman, O.; Dennison, N.; Honess, P.; Jennings, M.; Lane, S.; Middleton, V.; Roughan, J.V.; Wells, S.; et al. A guide to defining and implementing protocols for the welfare assessment of laboratory animals: Eleventh report of the bvaawf/frame/rspca/ufaw joint working group on refinement. *Lab. Anim.* **2011**, *45*, 1–13. [CrossRef]
18. Flecknell, P. Rodent analgesia: Assessment and therapeutics. *Vet. J.* **2017**, *232*, 70–77. [CrossRef]
19. Peterson, N.C.; Nunamaker, E.A.; Turner, P.V. To treat or not to treat: The effects of pain on experimental parameters. *Comp. Med.* **2017**, *67*, 469–482.
20. Magalhães Sant'Ana, M.; Sandøe, P.; Olsson, A. Painful dilemmas: The ethics of animal-based pain research. *Anim. Welf.* **2009**, *18*, 49–63.

21. Prunier, A.; Mounier, L.; Le Neindre, P.; Leterrier, C.; Mormède, P.; Paulmier, V.; Prunet, P.; Terlouw, C.; Guatteo, R. Identifying and monitoring pain in farm animals: A review. *Anim. Int. J. Anim. Biosci.* **2013**, *7*, 998–1010. [CrossRef] [PubMed]

22. Würbel, H. Ideal homes? Housing effects on rodent brain and behaviour. *Trends Neurosci.* **2001**, *24*, 207–211. [CrossRef]

23. Poole, T. Happy animals make good science. *Lab. Anim.* **1997**, *31*, 116–124. [CrossRef]

24. Graham, D.M.; Hampshire, V. Methods for measuring pain in laboratory animals. *Lab. Anim.* **2016**, *45*, 99–101. [CrossRef]

25. Prkachin, K.M. *Assessing Pain by Facial Expression: Facial Expression as Nexus*; Pulsus Group: Oakville, ON, Canada, 2009; Volume 14, pp. 53–58.

26. Williams, A.C. Facial expression of pain: An evolutionary account. *Behav. Brain Sci.* **2002**, *25*, 439–455, discussion 455–488. [CrossRef]

27. Deyo, K.S.; Prkachin, K.M.; Mercer, S.R. Development of sensitivity to facial expression of pain. *Pain* **2004**, *107*, 16–21. [CrossRef]

28. Leach, M.C.; Coulter, C.A.; Richardson, C.A.; Flecknell, P.A. Are we looking in the wrong place? Implications for behavioural-based pain assessment in rabbits (oryctolagus cuniculi) and beyond? *PLoS ONE* **2011**, *6*, e13347. [CrossRef] [PubMed]

29. Darwin, C. *The Expression of the Emotions in Man and Animals*; John Murray: London, UK, 1872; p. 374.

30. Waller, B.M.; Micheletta, J. Facial expression in nonhuman animals. *Emot. Rev.* **2013**, *5*, 54–59. [CrossRef]

31. Diogo, R.; Wood, B.; Diogo, R.; Wood, B. Origin and evolution of primate and human muscles, anatomical variations and anomalies, and evolutionary developmental biology. In *Evolutionary Developmental Anthropology*; Boughner, J., Rolian, C., Eds.; John Wiley & Sons: Hoboken, NJ, USA, 2015; pp. 167–174.

32. Defensor, E.B.; Corley, M.J.; Blanchard, R.J.; Blanchard, D.C. Facial expressions of mice in aggressive and fearful contexts. *Physiol. Behav.* **2012**, *107*, 680–685. [CrossRef] [PubMed]

33. Dalla Costa, E.; Minero, M.; Lebelt, D.; Stucke, D.; Canali, E.; Leach, M.C. Development of the horse grimace scale (hgs) as a pain assessment tool in horses undergoing routine castration. *PLoS ONE* **2014**, *9*, e92281. [CrossRef] [PubMed]

34. Chambers, C.T.; Mogil, J.S. Ontogeny and phylogeny of facial expression of pain. *Pain* **2015**, *156*, 798–799. [CrossRef]

35. Dalla Costa, E.; Pascuzzo, R.; Leach, M.C.; Dai, F.; Lebelt, D.; Vantini, S.; Minero, M. Can grimace scales estimate the pain status in horses and mice? A statistical approach to identify a classifier. *PLoS ONE* **2018**, *13*, e0200339. [CrossRef]

36. Reid, J.; Scott, M.; Nolan, A.; Wiseman-Orr, L. Pain assessment in animals. *Practice* **2013**, *35*, 51. [CrossRef]

37. Dalla Costa, E.; Stucke, D.; Dai, F.; Minero, M.; Leach, M.C.; Lebelt, D. Using the horse grimace scale (hgs) to assess pain associated with acute laminitis in horses (equus caballus). *Animals* **2016**, *6*, 47. [CrossRef]

38. Zhang, E.Q.; Leung, V.S.Y.; Pang, D.S.J. Influence of rater training on inter- and intrarater reliability when using the rat grimace scale. *J. Am. Assoc. Lab. Anim. Sci.* **2019**, *58*, 178–183. [CrossRef]

39. Dalla Costa, E.; Bracci, D.; Dai, F.; Lebelt, D.; Minero, M. Do different emotional states affect the horse grimace scale score? A pilot study. *J. Equine Vet. Sci.* **2017**, *54*, 114–117. [CrossRef]

40. Matsumiya, L.C.; Sorge, R.E.; Sotocinal, S.G.; Tabaka, J.M.; Wieskopf, J.S.; Zaloum, A.; King, O.D.; Mogil, J.S. Using the mouse grimace scale to reevaluate the efficacy of postoperative analgesics in laboratory mice. *J. Am. Assoc. Lab. Anim. Sci. JAALAS* **2012**, *51*, 42–49. [PubMed]

41. Sotocinal, S.G.; Sorge, R.E.; Zaloum, A.; Tuttle, A.H.; Martin, L.J.; Wieskopf, J.S.; Mapplebeck, J.C.S.; Wei, P.; Zhan, S.; Zhang, S.; et al. The rat grimace scale: A partially automated method for quantifying pain in the laboratory rat via facial expressions. *Mol. Pain* **2011**, *7*, 55. [PubMed]

42. Saine, L.; Hélie, P.; Vachon, P. Effects of fentanyl on pain and motor behaviors following a collagenase-induced intracerebral hemorrhage in rats. *J. Pain Res.* **2016**, *9*, 1039–1048. [CrossRef]

43. Long, H.; Liao, L.; Gao, M.; Ma, W.; Zhou, Y.; Jian, F.; Wang, Y.; Lai, W. Periodontal cgrp contributes to orofacial pain following experimental tooth movement in rats. *Neuropeptides* **2015**, *52*, 31–37. [CrossRef] [PubMed]

44. Kawano, T.; Eguchi, S.; Iwata, H.; Yamanaka, D.; Tateiwa, H.; Locatelli, F.M.; Yokoyama, M. Effects and underlying mechanisms of endotoxemia on post-incisional pain in rats. *Life Sci.* **2016**, *148*, 145–153. [CrossRef] [PubMed]

45. Keating, S.C.J.; Thomas, A.A.; Flecknell, P.A.; Leach, M.C. Evaluation of emla cream for preventing pain during tattooing of rabbits: Changes in physiological, behavioural and facial expression responses. *PLoS ONE* **2012**, *7*, e44437. [CrossRef]

46. Leach, M.; Allweiler, S.; Richardson, C.; Roughan, J.; Narbe, R.; Flecknell, P. Behavioural effects of ovariohysterectomy and oral administration of meloxicam in laboratory housed rabbits. *Res. Vet. Sci.* **2009**, *87*, 336–347. [CrossRef]

47. Amit, Z.; Galina, Z.H. Stress induced analgesia plays an adaptive role in the organization of behavioral responding. *Brain Res. Bull.* **1988**, *21*, 955–958. [CrossRef]

48. Jacobson, R. Stress-induced analgesia. Edited by M. D. Tricklebank and G. Curzon. Chichester: John wiley. 1984. Pp. 194. *Br. J. Psychiatry* **1985**, *146*, 676–677. [CrossRef]

49. Watkins, L.R.; Mayer, D.J. Organization of endogenous opiate and nonopiate pain control systems. *Science* **1982**, *216*, 1185–1192. [CrossRef]

50. Stasiak, K.L.; Maul, D.; French, E.; Hellyer, P.W.; VandeWoude, S. Species-specific assessment of pain in laboratory animals. *Contemp. Top. Lab. Anim. Sci.* **2003**, *42*, 13–20. [PubMed]

51. Jirkof, P.; Fleischmann, T.; Cesarovic, N.; Rettich, A.; Vogel, J.; Arras, M. Assessment of postsurgical distress and pain in laboratory mice by nest complexity scoring. *Lab. Anim.* **2013**, *47*, 153–161. [CrossRef] [PubMed]

52. Oliver, V.L.; Thurston, S.E.; Lofgren, J.L. Using cageside measures to evaluate analgesic efficacy in mice (mus musculus) after surgery. *J. Am. Assoc. Lab. Anim. Sci. JAALAS* **2018**, *57*, 186–201.

53. Rock, M.L.; Karas, A.Z.; Rodriguez, K.B.; Gallo, M.S.; Pritchett-Corning, K.; Karas, R.H.; Aronovitz, M.; Gaskill, B.N. The time-to-integrate-to-nest test as an indicator of wellbeing in laboratory mice. *J. Am. Assoc. Lab. Anim. Sci. JAALAS* **2014**, *53*, 24–28. [PubMed]

54. Deuis, J.R.; Dvorakova, L.S.; Vetter, I. Methods used to evaluate pain behaviors in rodents. *Front. Mol. Neurosci.* **2017**, *10*, 284. [CrossRef]

55. Leung, V.S.Y.; Benoit-Biancamano, M.O.; Pang, D.S.J. Performance of behavioral assays: The rat grimace scale, burrowing activity and a composite behavior score to identify visceral pain in an acute and chronic colitis model. *Pain Rep.* **2019**, *4*, e718. [CrossRef]

56. Miller, A.L.; Golledge, H.D.; Leach, M.C. The influence of isoflurane anaesthesia on the rat grimace scale. *PLoS ONE* **2016**, *11*, e0166652. [CrossRef]

57. Miller, A.L.; Leach, M.C. Using the mouse grimace scale to assess pain associated with routine ear notching and the effect of analgesia in laboratory mice. *Lab. Anim.* **2015**, *49*, 117–120. [CrossRef]

58. Whittaker, A.L.; Leach, M.C.; Preston, F.L.; Lymn, K.A.; Howarth, G.S. Effects of acute chemotherapy-induced mucositis on spontaneous behaviour and the grimace scale in laboratory rats. *Lab. Anim.* **2016**, *50*, 108–118. [CrossRef]

59. Whittaker, A.L.; Howarth, G.S. Use of spontaneous behaviour measures to assess pain in laboratory rats and mice: How are we progressing? *Appl. Anim. Behav. Sci.* **2014**, *151*, 1–12. [CrossRef]

60. Oliver, V.; De Rantere, D.; Ritchie, R.; Chisholm, J.; Hecker, K.G.; Pang, D.S.J. Psychometric assessment of the rat grimace scale and development of an analgesic intervention score. *PLoS ONE* **2014**, *9*, e97882. [CrossRef] [PubMed]

61. Sneddon, L.U.; Elwood, R.W.; Adamo, S.A.; Leach, M.C.J.A.B. Defining and assessing animal pain. *Anim. Behav.* **2014**, *97*, 201–212. [CrossRef]

62. Rutherford, K.M.D. Assessing pain in animals. *Anim. Welf.* **2002**, *11*, 31–53.

63. Dunbar, M.L.; David, E.M.; Aline, M.R.; Lofgren, J.L. Validation of a behavioral ethogram for assessing postoperative pain in guinea pigs (cavia porcellus). *J. Am. Assoc. Lab. Anim. Sci. JAALAS* **2016**, *55*, 29–34.

64. Gleerup, K. Identifying Pain Behaviors in Dairy Cattle. *WCDS Adv. Dairy Technol.* **2017**, *55*, 231–239.

65. Koolhaas, J.M.; Korte, S.M.; De Boer, S.F.; Van Der Vegt, B.J.; Van Reenen, C.G.; Hopster, H.; De Jong, I.C.; Ruis, M.A.; Blokhuis, H.J. Coping styles in animals: Current status in behavior and stress-physiology. *Neurosci. Biobehav. Rev.* **1999**, *23*, 925–935. [CrossRef]

66. Goldberg, M.E. Pain recognition and scales for livestock patients. *J. Dairy Vet. Anim. Res.* **2018**, *7*, 236–239. [CrossRef]

67. Miller, A.L.; Leach, M.C. The effect of handling method on the mouse grimace scale in two strains of laboratory mice. *Lab. Anim.* **2016**, *50*, 305–307. [CrossRef]

68. Müller, B.R.; Soriano, V.S.; Bellio, J.C.B.; Molento, C.F.M. Facial expression of pain in nellore and crossbred beef cattle. *J. Vet. Behav.* **2019**, *34*, 60–65. [CrossRef]

69. Roelvink, M.E.; Goossens, L.; Kalsbeek, H.C.; Wensing, T. Analgesic and spasmolytic effects of dipyrone, hyoscine-n-butylbromide and a combination of the two in ponies. *Vet. Rec.* **1991**, *129*, 378–380. [CrossRef]

70. Häger, C.; Biernot, S.; Buettner, M.; Glage, S.; Keubler, L.M.; Held, N.; Bleich, E.M.; Otto, K.; Müller, C.W.; Decker, S.; et al. The sheep grimace scale as an indicator of post-operative distress and pain in laboratory sheep. *PLoS ONE* **2017**, *12*, e0175839. [CrossRef] [PubMed]

71. Stahlbaum, C.C.; Houpt, K.A.J.P. The role of the flehmen response in the behavioral repertoire of the stallion. *Physiol. Behav.* **1989**, *45*, 1207–1214. [CrossRef]

72. Pritchett, L.; Ulibarri, C.; Roberts, M.; Schneider, R.; Sellon, D. Identification of potential physiological and behavioral indicators of postoperative pain in horses after exploratory celiotomy for colic. *Appl. Anim. Behav. Sci. Appl. Anim. Behav. Sci.* **2003**, *80*, 31–43. [CrossRef]

73. Mogil, J.S.; Crager, S.E. What should we be measuring in behavioral studies of chronic pain in animals? *Pain* **2004**, *112*, 12–15. [CrossRef]

74. Stafford, K. Recognition and assessment of pain in ruminants. *Pain Manag. Vet. Pract.* **2013**, *8*, 349–357.

75. Miller, A.L.; Leach, M.C. The mouse grimace scale: A clinically useful tool? *PLoS ONE* **2015**, *10*, e0136000. [CrossRef]

76. Leach, M.C.; Klaus, K.; Miller, A.L.; Scotto di Perrotolo, M.; Sotocinal, S.G.; Flecknell, P.A. The assessment of post-vasectomy pain in mice using behaviour and the mouse grimace scale. *PLoS ONE* **2012**, *7*, e35656. [CrossRef]

77. Roughan, J.V.; Bertrand, H.G.M.J.; Isles, H.M. Meloxicam prevents cox-2-mediated post-surgical inflammation but not pain following laparotomy in mice. *Eur. J. Pain* **2016**, *20*, 231–240. [CrossRef]

78. Oliver, V.L.; Athavale, S.; Simon, K.E.; Kendall, L.V.; Nemzek, J.A.; Lofgren, J.L. Evaluation of pain assessment techniques and analgesia efficacy in a female guinea pig (cavia porcellus) model of surgical pain. *J. Am. Assoc. Lab. Anim. Sci. JAALAS* **2017**, *56*, 425–435.

79. Earley, B.; Buckham-Sporer, K.; Gupta, S.; Pang, W.Y.; Ting, S. Biologic response of animals to husbandry stress with implications for biomedical models. *Open Access Anim. Physiol.* **2010**, *2*, 25–42. [CrossRef]

80. Nicol, C. The biology of animal stress: Basic principles and implications for animal welfare: G.P. Moberg, J.A. Mench. (eds.), cab international, wallingford, uk, 2000, 377 pp., uk£ 55.00, us$ 100.00, isbn 0-85199-359-1 (hard cover). *Appl. Anim. Behav. Sci.* **2001**, *72*, 375–378. [CrossRef]

81. Descovich, K.A.; Wathan, J.; Leach, M.C.; Buchanan-Smith, H.M.; Flecknell, P.; Farningham, D.; Vick, S.J. Facial expression: An under-utilised tool for the assessment of welfare in mammals. *Altex* **2017**, *34*, 409–429. [CrossRef] [PubMed]

82. McLennan, K.M.; Rebelo, C.J.B.; Corke, M.J.; Holmes, M.A.; Leach, M.C.; Constantino-Casas, F. Development of a facial expression scale using footrot and mastitis as models of pain in sheep. *Appl. Anim. Behav. Sci.* **2016**, *176*, 19–26. [CrossRef]

83. Mittal, A.; Gupta, M.; Lamarre, Y.; Jahagirdar, B.; Gupta, K. Quantification of pain in sickle mice using facial expressions and body measurements. *Blood Cells Mol. Dis.* **2016**, *57*, 58–66. [CrossRef]

84. Fourichon, C.; Seegers, H.; Bareille, N.; Beaudeau, F. Effects of disease on milk production in the dairy cow: A review. *Prev. Vet. Med.* **1999**, *41*, 1–35. [CrossRef]

85. Gleerup, K.B.; Andersen, P.H.; Munksgaard, L.; Forkman, B. Pain evaluation in dairy cattle. *Appl. Anim. Behav. Sci.* **2015**, *171*, 25–32. [CrossRef]

86. Minero, M.; Dalla Costa, E.; Dai, F.; Murray, L.; Canali, E.; Wemelsfelder, F. Use of qualitative behaviour assessment as an indicator of welfare in donkeys. *Appl. Anim. Behav. Sci.* **2015**, *174*, 147–153. [CrossRef]

87. Van Loon, J.P.A.M.; Van Dierendonck, M.C. Objective pain assessment in horses (2014–2018). *Vet. J.* **2018**, *242*, 1–7. [CrossRef]

88. Gigliuto, C.; De Gregori, M.; Malafoglia, V.; Raffaeli, W.; Compagnone, C.; Visai, L.; Petrini, P.; Avanzini, M.A.; Muscoli, C.; Viganò, J.; et al. Pain assessment in animal models: Do we need further studies? *J. Pain Res.* **2014**, *7*, 227–236.

89. Molony, V.; Kent, J. Assessment of acute pain in farm animals using behavioral and physiological measurements. *J. Anim. Sci.* **1997**, *75*, 266–272. [CrossRef]

90. Ellen, Y.; Flecknell, P.; Leach, M. Evaluation of using behavioural changes to assess post-operative pain in the guinea pig (cavia porcellus). *PLoS ONE* **2016**, *11*, e0161941. [CrossRef] [PubMed]

91. Viscardi, A.V.; Turner, P.V. Use of meloxicam or ketoprofen for piglet pain control following surgical castration. *Front. Vet. Sci.* **2018**, *5*, 299. [CrossRef] [PubMed]

92. Van Loon, J.P.; Van Dierendonck, M.C. Monitoring acute equine visceral pain with the equine utrecht university scale for composite pain assessment (equus-compass) and the equine utrecht university scale for facial assessment of pain (equus-fap): A scale-construction study. *Vet. J.* **2015**, *206*, 356–364. [CrossRef] [PubMed]

93. Ison, S.H.; Clutton, R.E.; Di Giminiani, P.; Rutherford, K.M.D. A review of pain assessment in pigs. *Front. Vet. Sci.* **2016**, *3*, 108. [CrossRef] [PubMed]

94. Finlayson, K.; Lampe, J.F.; Hintze, S.; Würbel, H.; Melotti, L. Facial indicators of positive emotions in rats. *PLoS ONE* **2016**, *11*, e0166446. [CrossRef]

95. Schneider, L.E.; Henley, K.Y.; Turner, O.A.; Pat, B.; Niedzielko, T.L.; Floyd, C.L. Application of the rat grimace scale as a marker of supraspinal pain sensation after cervical spinal cord injury. *J. Neurotrauma.* **2017**, *34*, 2982–2993. [CrossRef]

96. Hampshire, V.; Robertson, S. Using the facial grimace scale to evaluate rabbit wellness in post-procedural monitoring. *Lab. Anim.* **2015**, *44*, 259–260. [CrossRef]

97. Ekman, P.; Friesen, W.V. Measuring facial movement. *Environ. Psychol. Nonverbal Behav.* **1976**, *1*, 56–75. [CrossRef]

98. Ekman, P.; Friesen, W.V. The repertoire of nonverbal behavior: Categories, origins, usage, and coding. *Nonverbal Commun. Interact. Gesture* **1969**, *1*, 57–106. [CrossRef]

99. Ekman, P.A.R.E. *What the Face Reveals: Basic and Applied Studies of Spontaneous Expression Using the Facial Action Coding System (Facs)*, 2nd ed.; Oxford University Press: New York, NY, USA, 2005; p. 639.

100. Langford, D.J.; Bailey, A.L.; Chanda, M.L.; Clarke, S.E.; Drummond, T.E.; Echols, S.; Glick, S.; Ingrao, J.; Klassen-Ross, T.; LaCroix-Fralish, M.L.; et al. Coding of facial expressions of pain in the laboratory mouse. *Nat. Methods* **2010**, *7*, 447–449. [CrossRef]

101. Evangelista, M.C.; Watanabe, R.; Leung, V.S.Y.; Monteiro, B.P.; O'Toole, E.; Pang, D.S.J.; Steagall, P.V. Facial expressions of pain in cats: The development and validation of a feline grimace scale. *Sci. Rep.* **2019**, *9*, 19128. [CrossRef] [PubMed]

102. Gleerup, K.B.; Forkman, B.; Lindegaard, C.; Andersen, P.H. An equine pain face. *Vet. Anaesth. Analg.* **2015**, *42*, 103–114. [CrossRef] [PubMed]

103. De Rantere, D.; Schuster, C.J.; Reimer, J.N.; Pang, D.S. The relationship between the rat grimace scale and mechanical hypersensitivity testing in three experimental pain models. *Eur. J. Pain* **2016**, *20*, 417–426. [CrossRef]

104. Marcantonio Coneglian, M.; Duarte Borges, T.; Weber, S.H.; Godoi Bertagnon, H.; Michelotto, P.V. Use of the horse grimace scale to identify and quantify pain due to dental disorders in horses. *Appl. Anim. Behav. Sci.* **2020**, *225*, 104970. [CrossRef]

105. Van Loon, J.; Dierendonck, M. Monitoring equine head-related pain with the equine utrecht university scale for facial assessment of pain (equus-fap). *Vet. J.* **2017**, *220*, 88–90. [CrossRef]

106. Reijgwart, M.L.; Schoemaker, N.J.; Pascuzzo, R.; Leach, M.C.; Stodel, M.; de Nies, L.; Hendriksen, C.F.M.; van der Meer, M.; Vinke, C.M.; van Zeeland, Y.R.A. The composition and initial evaluation of a grimace scale in ferrets after surgical implantation of a telemetry probe. *PLoS ONE* **2017**, *12*, e0187986. [CrossRef]

107. Guesgen, M.J.; Beausoleil, N.J.; Leach, M.; Minot, E.O.; Stewart, M.; Stafford, K.J. Coding and quantification of a facial expression for pain in lambs. *Behav. Process.* **2016**, *132*, 49–56. [CrossRef]

108. Akintola, T.; Raver, C.; Studlack, P.; Uddin, O.; Masri, R.; Keller, A. The grimace scale reliably assesses chronic pain in a rodent model of trigeminal neuropathic pain. *Neurobiol. Pain* **2017**, *2*, 13–17. [CrossRef]

109. Cho, C.; Michailidis, V.; Lecker, I.; Collymore, C.; Hanwell, D.; Loka, M.; Danesh, M.; Pham, C.; Urban, P.; Bonin, R.P.; et al. Evaluating analgesic efficacy and administration route following craniotomy in mice using the grimace scale. *Sci. Rep.* **2019**, *9*, 359. [CrossRef]

110. Faller, K.M.E.; McAndrew, D.J.; Schneider, J.E.; Lygate, C.A. Refinement of analgesia following thoracotomy and experimental myocardial infarction using the mouse grimace scale. *Exp. Physiol.* **2015**, *100*, 164–172. [CrossRef]

111. Miller, A.L.; Kitson, G.L.; Skalkoyannis, B.; Flecknell, P.A.; Leach, M.C. Using the mouse grimace scale and behaviour to assess pain in cba mice following vasectomy. *Appl. Anim. Behav. Sci.* **2016**, *181*, 160–165. [CrossRef] [PubMed]

112. Rossi, H.L.; See, L.P.; Foster, W.; Pitake, S.; Gibbs, J.; Schmidt, B.; Mitchell, C.H.; Abdus-Saboor, I. Evoked and spontaneous pain assessment during tooth pulp injury. *Sci. Rep.* **2020**, *10*, 2759. [CrossRef] [PubMed]

113. Wood, J.N.; Mogil, J.S. *The Measurement of Pain in the Laboratory Rodent*; Oxford University Press: Oxford, UK, 2019.

114. Vullo, C.; Barbieri, S.; Catone, G.; Graïc, J.-M.; Magaletti, M.; Di Rosa, A.; Motta, A.; Tremolada, C.; Canali, E.; Dalla Costa, E. Is the piglet grimace scale (pgs) a useful welfare indicator to assess pain after cryptorchidectomy in growing pigs? *Animals* **2020**, *10*, 412. [CrossRef]

115. Di Giminiani, P.; Brierley, V.L.M.H.; Scollo, A.; Gottardo, F.; Malcolm, E.M.; Edwards, S.A.; Leach, M.C. The assessment of facial expressions in piglets undergoing tail docking and castration: Toward the development of the piglet grimace scale. *Front. Vet. Sci.* **2016**, *3*, 100. [CrossRef]

116. Viscardi, A.V.; Hunniford, M.; Lawlis, P.; Leach, M.; Turner, P.V. Development of a piglet grimace scale to evaluate piglet pain using facial expressions following castration and tail docking: A pilot study. *Front. Vet. Sci.* **2017**, *4*, 51. [CrossRef]

117. Viscardi, A.V.; Turner, P.V.J.B.V.R. Efficacy of buprenorphine for management of surgical castration pain in piglets. *BMC Vet. Res.* **2018**, *14*, 318. [CrossRef] [PubMed]

118. Asgar, J.; Zhang, Y.; Saloman, J.L.; Wang, S.; Chung, M.K.; Ro, J.Y. The role of trpa1 in muscle pain and mechanical hypersensitivity under inflammatory conditions in rats. *Neuroscience* **2015**, *310*, 206–215. [CrossRef]

119. Chi, H.; Kawano, T.; Tamura, T.; Iwata, H.; Takahashi, Y.; Eguchi, S.; Yamazaki, F.; Kumagai, N.; Yokoyama, M. Postoperative pain impairs subsequent performance on a spatial memory task via effects on n-methyl-d-aspartate receptor in aged rats. *Life Sci.* **2013**, *93*, 986–993. [CrossRef]

120. Iqbal, S.M.; Leonard, C.; Regmi, S.C.; De Rantere, D.; Tailor, P.; Ren, G.; Ishida, H.; Hsu, C.; Abubacker, S.; Pang, D.S.; et al. Lubricin/proteoglycan 4 binds to and regulates the activity of toll-like receptors in vitro. *Sci. Rep.* **2016**, *6*, 18910. [CrossRef]

121. Jeger, V.; Arrigo, M.; Hildenbrand, F.F.; Müller, D.; Jirkof, P.; Hauffe, T.; Seifert, B.; Arras, M.; Spahn, D.R.; Bettex, D.; et al. Improving animal welfare using continuous nalbuphine infusion in a long-term rat model of sepsis. *Intensive. Care Med. Exp.* **2017**, *5*, 23. [CrossRef]

122. Kawano, T.; Takahashi, T.; Iwata, H.; Morikawa, A.; Imori, S.; Waki, S.; Tamura, T.; Yamazaki, F.; Eguchi, S.; Kumagai, N.; et al. Effects of ketoprofen for prevention of postoperative cognitive dysfunction in aged rats. *J. Anesth.* **2014**, *28*, 932–936. [CrossRef] [PubMed]

123. Khoo, S.Y.; Lay, B.P.P.; Joya, J.; McNally, G.P. Local anaesthetic refinement of pentobarbital euthanasia reduces abdominal writhing without affecting immunohistochemical endpoints in rats. *Lab. Anim.* **2018**, *52*, 152–162. [CrossRef] [PubMed]

124. Gao, M.; Long, H.; Ma, W.; Liao, L.; Yang, X.; Zhou, Y.; Shan, D.; Huang, R.; Jian, F.; Wang, Y.; et al. The role of periodontal asic3 in orofacial pain induced by experimental tooth movement in rats. *Eur. J. Orthod.* **2016**, *38*, 577–583. [CrossRef]

125. Gao, Z.; Cui, F.; Cao, X.; Wang, D.; Li, X.; Li, T. Local infiltration of the surgical wounds with levobupivacaine, dexibuprofen, and norepinephrine to reduce postoperative pain: A randomized, vehicle-controlled, and preclinical study. *Biomed. Pharm.* **2017**, *92*, 459–467. [CrossRef]

126. Philips, B.H.; Weisshaar, C.L.; Winkelstein, B.A. Use of the rat grimace scale to evaluate neuropathic pain in a model of cervical radiculopathy. *Comp. Med.* **2017**, *67*, 34–42. [PubMed]

127. Préfontaine, L.; Hélie, P.; Vachon, P. Postoperative pain in sprague dawley rats after liver biopsy by laparotomy versus laparoscopy. *Lab. Anim.* **2015**, *44*, 174–178. [CrossRef] [PubMed]

128. Sperry, M.M.; Yu, Y.-H.; Welch, R.L.; Granquist, E.J.; Winkelstein, B.A. Grading facial expression is a sensitive means to detect grimace differences in orofacial pain in a rat model. *Sci. Rep.* **2018**, *8*, 13894. [CrossRef]

129. Yu, Y.-H.; Sperry, M.; Winkelstein, B.; Granquist, E. Using the Rat Grimace Scale to Detect Orofacial Pain in Mechanically-Induced Temporomandibular Joint Pain in Rats. Master's Thesis, University of Pennsylvania Scholarly Commons, Philadelphia, PA, USA, 2018.

130. Waite, M.E.; Tomkovich, A.; Quinn, T.L.; Schumann, A.P.; Dewberry, L.S.; Totsch, S.K.; Sorge, R.E. Efficacy of common analgesics for postsurgical pain in rats. *J. Am. Assoc. Lab. Anim. Sci. JAALAS* **2015**, *54*, 420–425.

131. Liao, L.; Long, H.; Zhang, L.; Chen, H.; Zhou, Y.; Ye, N.; Lai, W. Evaluation of pain in rats through facial expression following experimental tooth movement. *Eur. J. Oral. Sci.* **2014**, *122*, 121–124. [CrossRef]

132. Fujita, M.; Fukuda, T.; Sato, Y.; Takasusuki, T.; Tanaka, M. Allopregnanolone suppresses mechanical allodynia and internalization of neurokinin-1 receptors at the spinal dorsal horn in a rat postoperative pain model. *Korean J. Pain* **2018**, *31*, 10–15. [CrossRef]

133. Farrar, J.T.; Young, J.P., Jr.; LaMoreaux, L.; Werth, J.L.; Poole, R.M. Clinical importance of changes in chronic pain intensity measured on an 11-point numerical pain rating scale. *Pain* **2001**, *94*, 149–158. [CrossRef]
134. Andresen, N.; Wöllhaf, M.; Hohlbaum, K.; Lewejohann, L.; Hellwich, O.; Thöne-Reineke, C.; Belik, V. Towards a fully automated surveillance of well-being status in laboratory mice using deep learning: Starting with facial expression analysis. *PLoS ONE* **2020**, *15*, e0228059. [CrossRef] [PubMed]
135. Ernst, L.; Kopaczka, M.; Schulz, M.; Talbot, S.R.; Zieglowski, L.; Meyer, M.; Bruch, S.; Merhof, D.; Tolba, R.H. Improvement of the mouse grimace scale set-up for implementing a semi-automated mouse grimace scale scoring (part 1). *Lab. Anim.* **2020**, *54*, 83–91. [CrossRef] [PubMed]

Article

Fentanyl Plasma Concentrations after Application of a Transdermal Patch in Three Different Locations to Refine Postoperative Pain Management in Rabbits

Valentina Mirschberger [1], Christian von Deimling [1], Anja Heider [2], Claudia Spadavecchia [3], Helene Rohrbach [3] and Stephan Zeiter [1,*]

[1] AO Research Institute Davos, 7270 Davos Platz, Switzerland; valentina_riehl@web.de (V.M.); deimling.ch@gmail.com (C.v.D.)
[2] Swiss Institute of Allergy and Asthma Research (SIAF), University Zurich, 7265 Davos Wolfgang, Switzerland; anja.heider@siaf.uzh.ch
[3] Vetsuisse Faculty, Department for Veterinary Medicine, University of Bern, 3012 Bern, Switzerland; claudia.spadavecchia@vetsuisse.unibe.ch (C.S.); helene.rohrbach@vetsuisse.unibe.ch (H.R.)
* Correspondence: stephan.zeiter@aofoundation.org; Tel.: +41-81-414-2311

Received: 31 July 2020; Accepted: 25 September 2020; Published: 1 October 2020

Simple Summary: Fentanyl patches offer "stress free" postoperative pain management in rabbits. It has been shown that fentanyl uptake is dependent on exogenous and endogenous factors of the area where the patch is applied. The purpose of the study was to investigate three different locations (neck, inner and outer surface of the ear) to obtain reliable fentanyl plasma concentrations above those previously shown to be analgesic. The fentanyl plasma concentration was measured at several time points after patch application. In addition, the practicability of the proposed methods was evaluated. The group with application on the neck had the fastest uptake and equal to or over the analgesic plasma concentration for up to 72 h. The outer surface of the ear had slightly slower uptake and shorter analgesic duration whereas fentanyl uptake at the inner surface of the ear was insufficient to provide plasma analgesic concentration. The preparation of the neck proved to be the most laborious because of the thin and dense fur and the removal of the patch resulted in erythema. In conclusion, depending on how long potent analgesia is required, either the neck or the outer surface of the ear are suitable for patch application enabling "stress free" and reliable postoperative analgesia in rabbits.

Abstract: Transdermal patches allow a noninvasive and "stress free" analgesia in rabbits. As fentanyl uptake is dependent on exogenous and endogenous factors of the area where the patch is applied, this study investigated three different locations (neck, inner and outer surfaces of the ear) for fentanyl patch application to provide adequate and reliable fentanyl plasma concentrations above those previously shown to be analgesic. Fentanyl plasma concentration was measured at different time points (3, 6, 9, 12, 18, 24, 36, 48, 72, 96, 120 h) and rabbits were assessed for their general conditions and treatment-related side effects. Practicability of the proposed methods was evaluated. Following patch application on the neck, fentanyl plasma concentrations equal to or above the analgesic value were measured in all rabbits between 6 and 72 h. Comparable concentrations were reached between 9 and 48 h in all animals for the outer ear surface. However, for the inner ear surface, analgesic concentrations were not reached, even if practicability was considered the best for this location. Preparation of the neck skin was judged as the most cumbersome due to the clipping of the dense fur and patch removal resulted in erythema. In summary, the application of the fentanyl patch on the neck and outer ear surface allowed the reach of reliable plasma concentrations above the analgesic threshold in rabbits. When applied on the neck, fentanyl patches provided the longest duration of analgesic plasma concentrations, whereas patch application and removal were easier on the outer ear surface.

Keywords: transdermal fentanyl patch; rabbit; postoperative analgesia; refinement

1. Introduction

Rabbits are commonly used as animal models in biomedical research, e.g., orthopedic studies concerning bone healing, cartilage repair and infection studies [1–3]. In these experimental studies, the animals often undergo invasive surgeries which require systemic analgesia for several days. Looking into the literature, the administration of systemic analgesia in invasive experimental studies involving rabbits increased significantly from 16% (1995–997) to 50% (2005–2007) [4,5]. Additionally, despite improvement in the last 20 years it is reported that pain management is still inadequate for pet rabbits [6]. Nevertheless, optimization of postoperative analgesia protocols is still highly needed as studies investigating effective analgesia for surgical interventions have been rarely reported for this species [7]. Among opioids, the most commonly used drug was buprenorphine as the first-line therapy over this period [4].

Buprenorphine, a partial μ-receptor agonist opioid, is commonly used for perioperative pain control in laboratory animal species [4]. Despite the long duration of action, multiple injections are required daily to guarantee a continuous analgesic coverage [8]. As rabbits are highly susceptible to stress and anxiety [9,10], repeated handling and medical interventions should be kept to a minimum to improve animal welfare. As an alternative, titrating analgesic drugs to effect would carry the risk of letting pain go undertreated, as pain recognition in a prey species like the rabbit represents a true challenge [10,11], even though the rabbit grimace scale is a promising tool to assess postoperative pain [12].

An opportunity to ensure a noninvasive and "stress free" analgesia is through the use of transdermal patches which provide a continuous release of medications into the bloodstream over a considerable period. Fentanyl, a potent m-agonist synthetic opioid characterized by a short half time, would not be adequate for postoperative analgesia in rabbits as an injectable formulation, but becomes interesting as a transdermal patch. Fentanyl patches were developed by the Alza Corporation and came on the market in 1991. These patches provide a sustained and constant drug release into the systemic circulation for at least 72 h, and are commonly prescribed in human medicine to treat chronic cancer pain [13]. In 1995, the first animal study investigating these opioid releasing patches in dogs was published [14]. Later, their off-label use in the veterinary medicine, especially in the experimental fields for postoperative pain therapy, became popular due to their properties as a noninvasive long-term delivery form. In a study from Foley et al., the use of fentanyl patches placed on the neck of rabbits was investigated for the first time and considered to be efficacious [15]. In other species such as dogs, cats, sheep, horses and minipigs, the application of the analgesic patches has further been investigated [14,16–23]. It was shown that exogenous and endogenous factors such as the amount of subcutaneous fat, skin integrity, structure and arrangement of hair follicles, possible dermal depots, body core temperature, skin thickness, first-pass cutaneous biotransformation as well as the environmental temperature caused strong intra- and inter-individual differences in the uptake of fentanyl into the bloodstream. Therefore, the prediction of fentanyl plasma concentrations is difficult [5,8,14,18,24]. Based on this described variability of drug absorption depending on the application locations, it was hypothesized that there is a variation in the uptake of fentanyl into the blood when different application locations are used [24]. The goal of this study was to investigate different locations for fentanyl patch application (neck, inner and outer surfaces of the ear) in rabbits. We assumed that the application on the inner or outer surface of the ear would provide a more reliable absorption of fentanyl with less variability due to a more homogenous anatomy and easier site preparation.

2. Materials and Methods

2.1. Study Design

In this experimental, prospective study, the New Zealand White rabbits (NZW) were randomly and equally assigned to three groups, differing in the location of patch application: outer surface of the ear (group 1, n = 6), inner surface of the ear (group 2, n = 6) and neck between the scapulas (group 3, n = 6). In all groups, fentanyl patches were covered with a tape (Leukoplast, BSN Medical GmbH, Hamburg, Germany) to ensure a continuous contact throughout the study period.

Venous blood samples withdrawn at different time points (Figure 1) were centrifuged and frozen at −80 °C until analysis of fentanyl plasma concentration. Additionally, treatment-related side effects were recorded, and practicability of the proposed methods was evaluated to determine the most adequate location for patch application in rabbits. No surgical procedures were performed as only the drug's absorption curve was investigated in this trial.

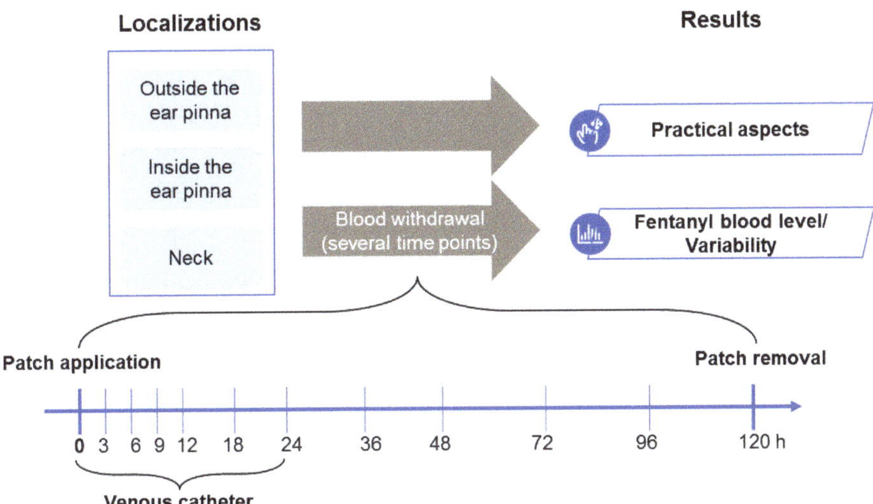

Figure 1. Study design. Fentanyl plasma concentrations higher than 0.5 ng/mL were considered to be analgesic based on previous reports [13,19]. Outcome parameters were time spent above threshold concentration, variability of fentanyl plasma concentrations and practicability of patch application.

2.2. Animals

Eighteen specific pathogen free (according to FELASA guidelines), female New Zealand White rabbits, weighing 3.05 to 3.80 kg with an age of 23 to 24 weeks at the start of the study were purchased from Charles River Laboratories (Sulzfeld, Germany). The study was performed in an AAALAC (Association for Assessment and Accreditation of Laboratory Animal Care) approved institution according to the Swiss animal welfare regulations and approved by the ethical committee of the canton Graubünden, Switzerland (No. 2016_40).

Upon delivery, the animals were given at least two weeks to acclimatize to the housing conditions (17 +/− 2 °C, > 30% humidity, 10 to 20 air changes per hour 12:12 h light-dark cycle). Animals were group housed with a maximum of 15 animals per group, on straw and fed with hay, supplemental food (3140, Kaninchen & Meerschweinchen, Haltung, Standard, Kliba NAFAG, Switzerland) as well as some fresh carrots daily. Animals had free access to water. For the 5–day study duration, rabbits were single housed in stainless steel cages with ground floor area of 5600 cm². The cages are 62 cm high and there is an increase area with an area of 1971 cm², underneath which the rabbits can hide. The cages were littered down with straw. Feeding was done the same manner as before, providing a piece of wood for gnawing and offering the hay in a fodder rack as additional environmental enrichment. Before inclusion into the study, the rabbits underwent a general clinical examination by a veterinarian. During the experimental period, the animals' welfare was regularly evaluated at least twice a day by a veterinarian. After the end of the study, the rabbits were reintegrated in their initial group.

2.3. Materials

In this study, a 12 μg/h transdermal fentanyl patch (Fentanyl-Mepha Matrixpfl, Mepha Pharma AG, Switzerland) was used. Due to the recommendation of a dosage of 2 μg/kg/h in sheep and up to 5 μg/kg/h in small animals (dogs, cats) described in several former studies, the 12 μg/h patch size was chosen [5,7,22]. The mean administered dose rate of fentanyl was 3.60 μg/kg/h (Dmax = 3.93 μg/kg/h, Dmin = 3.16 μg/kg/h).

2.4. Fentanyl Patch Application Process

Rabbits were sedated with Medetomidin (200 μg/kg) and Midazolam (0.5 mg/kg) both mixed in a syringe and given intramuscularly approximately 15 min before starting skin preparation. Thereafter, the rabbits were placed in sternal recumbency, eye ointment (Vitamin A Blache Augensalbe, Bausch & Lomb Swiss AG, Switzerland) was applied and oxygen (flow rate 1 L/min) was provided via face mask. The required site was carefully clipped using two different clippers depending on the location. For the location on the ear, the clipper Isis (GT420 Aesculap, Germany; cutting length 0.5 mm) was utilized. For the neck, another mechanical fur clipper (model GH703/10 (Favorita II GT104), Aesculap, Germany; cutting length 0.1 mm) was used in the interscapular region due to the increased density of hair at this site. During clipping extreme precautions were taken not to traumatize the application site to avoid influencing fentanyl absorption.

Afterwards, skin was degreased with swabs (Mesoft, Moelnlycke, Sweden) soaked in alcohol (Softasept, B. Braun Vet Care GmbH, Germany) to ensure a good contact of the patch with the skin. While waiting for 5 min to air dry, the ear not used for patch application was clipped and prepared aseptically with alcohol for the placement of a 22 G catheter (Vasofix Safety, B.Braun Meisungen AG, Germany) in the marginal ear vein. The catheter, inserted to facilitate blood withdrawal, was wrapped with tape (Durapore, 3 M (Schweiz) GmbH, Switzerland) and a roll of gauze (Mesoft) to hold it in place. After each blood withdrawal, the catheter was flushed with approximately 0.5 mL of heparinized 0.9% NaCl (B.Braun Meisungen AG, Germany) and a mandrin (Mandrin Vasofix, B.Braun Meisungen AG, Germany) was inserted. On the dry application site, the sticky side of the patch was then pressed on the skin with the palm of the hand and kept there for 1 min to ensure a good patch-skin contact. Finally, the patch was fixed with tape (Leukoplast) to prevent the loosening of the patch detachment during animal handling.

2.5. Sample Collection and Plasma Fentanyl Analysis via ELISA

Venous blood sampling was performed immediately prior to the application of the patch and 3, 6, 9, 12, 18, 24, 36, 48, 72, 96, 120 h thereafter. In a few cases arterial blood was sampled, if no venous sample could be obtained. At each sampling time point, approximately 1 mL of blood was collected in a 1.3 mL EDTA-covered tube (1.3 mL K3E, Sarstedt, AG & Co, Nümbrecht, Germany). A venous catheter was always used to withdraw blood samples. For the first 24 h the catheter was left in the marginal ear vein. Thereafter, a new catheter was inserted for each blood sampling. EDTA-tubes were then centrifuged for 15 min, at 23 °C and 1000 rcf (Centrifuge 5810R, Vaudaux-Eppendorf AG, Switzerland) within 2 h of collection.

Subsequently, the obtained plasma was transferred into Eppendorf tubes (Vaudaux-Eppendorf AG, Switzerland) and stored at −80 °C until analysis. Plasma fentanyl concentrations were measured by using a commercially available human enzyme-linked immunoabsorbent assay (ELISA). Fentanyl Kits (Adnova, Fentanyl (Human) ELISA Kit, Taiwan) were used according to the manufacturer's instructions. Additionally, serial dilutions of fentanyl standard (Fentanyl Sintetica 0.5 mg/10 mL) in fentanyl negative rabbit plasma were measured on the ELISA-plate, as a standard curve. The absorbance was read at 450 nm using a Mithras microplate reader (Berthold Technologies, Bad Wildbad, Germany). Each sample was tested in duplicates.

2.6. Scoring and Practicability

During the study, the rabbits were scored twice a day. The scoring system included observation of food and water intake, assessment of the general conditions, body weight and rectal temperature as well as defecation and coprophagy. Additionally, practicability was assessed based on the following criteria: ease of preparation, quality of patch adhesiveness, ease of daily checks, occurrence of undesired patch detachment before the end of the study, ease of patch removal and skin condition after patch removal. Any findings in respect to these criteria were noted and at the end a subjective three level scoring system (positive, neutral and negative assessment) comparing the three groups was applied by one veterinarian.

2.7. Statistical Analysis

For all locations descriptive statistics was used to describe fentanyl uptake over time. The goal of this study was not to find significant difference between groups. As we are aiming towards a reliable postoperative analgesia for all animals, the main outcome was the time all animals of one group had a fentanyl plasma concentration above the threshold of 0.5 ng/mL.

3. Results

One rabbit was excluded from the study and subsequently replaced due to CNS symptoms occurring 18 h after patch application. Necropsy revealed a thromboembolism as the most likely reason for the symptoms, caused by repeated blood withdrawal.

3.1. Plasma Fentanyl Concentrations

In group 1 (outer surface of the ear; Figure 2) plasma concentrations considered to be analgesic (0.5 ng/mL) was reached in 3 out of 6 rabbits 3 h after patch application, while all the animals in this group were above threshold at 9 h. In group 3 (neck; Figure 3), 2 out of 6 animals reached threshold at 3 h, while all 6 animals at 6 h (mean plasma concentration of 1.26 ng/mL). In contrast, only 1 out of 6 rabbits in group 2 (inner surface of the ear; Figure 4) reached the threshold at 18 h, while the others stayed below the threshold for the whole duration of the study.

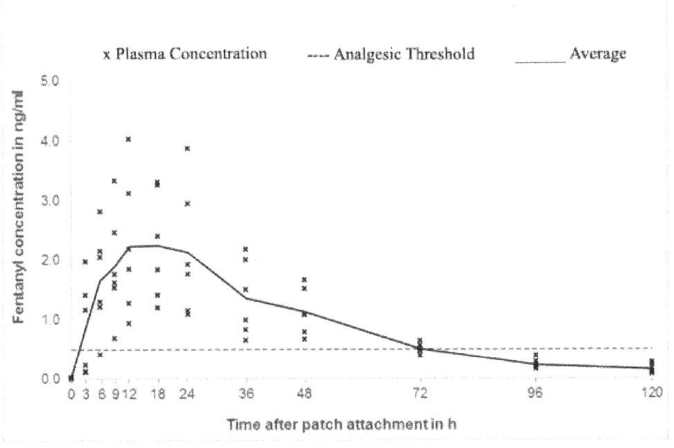

Figure 2. Fentanyl plasma concentrations group 1 (outer surface of the ear).

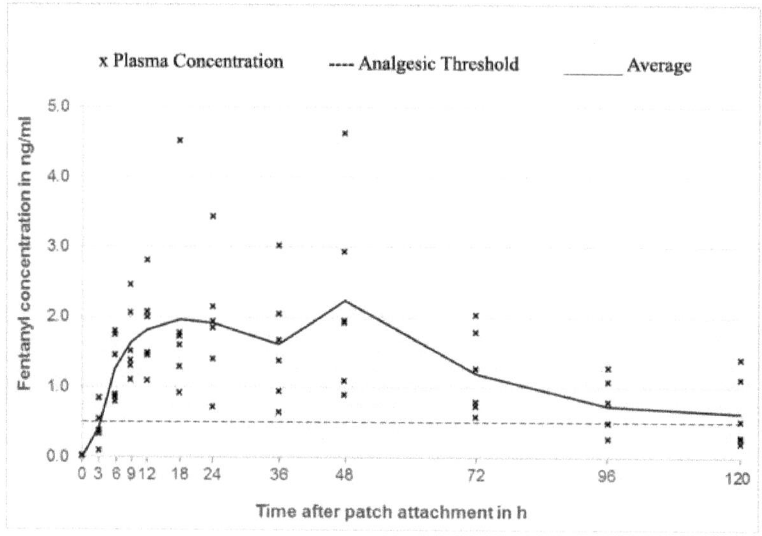

Figure 3. Fentanyl plasma concentrations group 3 (neck).

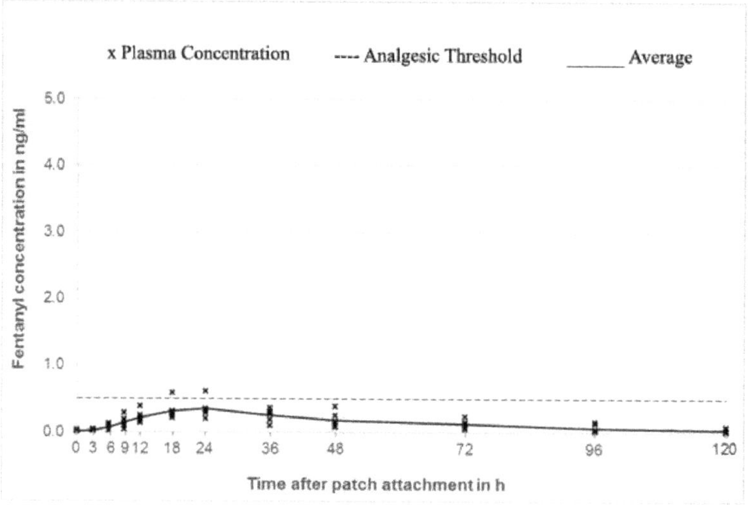

Figure 4. Fentanyl plasma concentrations group 2 (inner surface of the ear).

Mean steady state concentrations in group 1 and 3 were similar (2.24 and 2.23 ng/mL) but the timing was different. Group 1 reached the peak concentration at 18 h, whereas group 3 at 48 h. Every animal of group 1 reached concentrations above 0.5 ng/mL between 9 and 48 h, while animals of group 3 between 6 and 72 h after patch application (Figure 5).

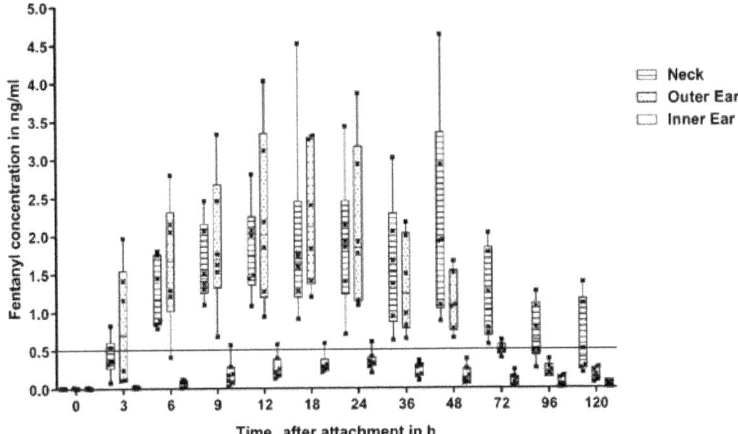

Figure 5. Fentanyl plasma concentrations of all groups.

3.2. Practicability

Findings of the practicability assessment are summarized in Table 1.

Table 1. Assessment of the practicability for the different locations. For each criteria, a subjective three level scoring system (positive, neutral and negative assessment) comparing the three groups was applied.

Aspect	Group 1 (Outside Ear)	Group 2 (Inside Ear)	Group 3 (Neck)
Ease of preparation	Neutral	Positive	Negative
Quality of patch adhesiveness	Positive	Positive	Positive
Ease of daily checks	Positive	Neutral	Negative
Occurrence of undesired patch detachment	Positive	Negative	Neutral
Ease of patch removal	Negative	Positive	Negative
Skin condition after patch removal	Negative	Positive	Negative

Ease of preparation: For patch application on the outer surface of the ear, group 1 needed a moderate effort to clip and disinfect the skin. Compared to the outer surface of the ear, the inner surface (group 2) needed minimal effort to clip due to almost no hair growth. However, it took longer to degrease the skin. Group 3, with the patch located on the neck, showed the highest density of hair growth combined with very thin hair. As a result, even with another more powerful clipper, it took longer to clip the fur in this location. Additionally, fighting wounds were discovered in this location while clipping, which made it hard to place the patch in exactly the same position on each rabbit avoiding these skin lesions.

Quality of patch adhesiveness was comparable in all groups—no difference could be noticed.

Ease of daily checks: Animals of group 1 were the easiest to check, as the outer surface of the ear was promptly visualized. Tape loosening could be seen without manipulation. In contrast, in group 2 loosening of the tape and patch underneath could only be noticed while handling the rabbit. Group 3 differed from the other groups as a check of the patch was only feasible while holding the rabbit, as the patch on the neck resulted covered by the surrounding fur.

Occurrence of undesired patch detachment: Patch loosening in terms of detaching edges occurred starting 48 h after patch application particularly in group 2 and 3 but also in one animal of group 1, 72 h after patch application. Signs of mild manipulation (scratching) of the tape by the rabbits could be observed in all groups suggesting some irritation engendered by the presence of the tape/ patch. While in group 1 manipulation lead to partial patch loosening in just one animal, detaching of the patches occurred in 5 of 6 animals of group 2 and 3 of 6 in group 3. Obvious and strong manipulation of the tape covering the patch was visible in 2 of 18 animals namely in group 2 and 3.

Ease of patch removal: Patch removal 120 h after application was the easiest and most stress-free for the animals in group 2 due to greasy skin and just a few regrown hair at this location. In contrast, group 1 and 3 showed a high density of regrown fur, which made it painful and hard to remove the tape and the patch in all these animals.

Skin condition after patch removal: Group 2 showed no signs of skin reaction after patch removal, whereas in group 1 and 3, a diffuse erythema underlying the drug-delivery portion of the patches was recognized in all the animals (Figure 6).

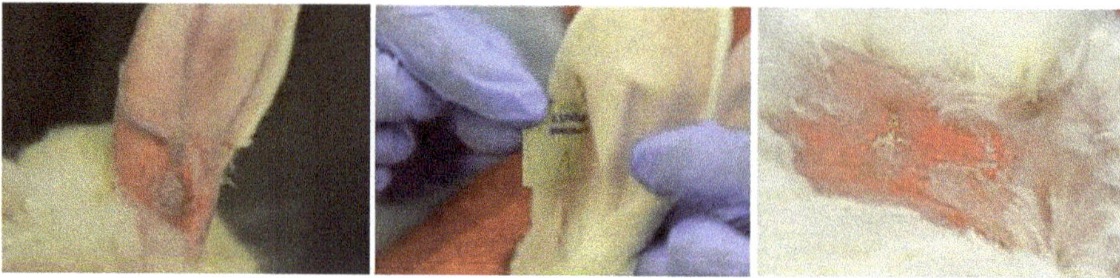

Figure 6. Representative images of skin condition after patch removal (**left**: group 1, **middle**: group 2 and **right**: group 3).

No differences in general condition, food or water intake, body weight, defecation, coprophagy or temperature were noticed among the three groups.

4. Discussion

The aim of the study was to determine fentanyl plasma concentrations in rabbits from patches applied at three different locations with the final aim to refine (postoperative) analgesia. A standard skin preparation and patch application protocol was applied to avoid inconsistent results as previously reported [16]. For the different locations, fentanyl plasma concentrations were measured, and practicability of patch application, maintenance and removal were assessed. To the authors' knowledge, the current study is the first to report fentanyl plasma concentrations in rabbits following patch application at different locations.

The results of this study indicate that the patch placed on the neck provided fentanyl plasma concentrations higher than 0.5 ng/mL from 6 to 72 h after application. This plasma concentration was reached earlier and peak concentrations were higher when the patch was applied on the neck compared to both ear sites. However, preparation of this location was the most complex due to the high density of fur. Additionally, animal handling was needed to check the patch adhesion on the skin. Furthermore, patch removal was accompanied by aversive reactions and lead to mild skin erythema. The cause of the erythema associated with patch application seemed to be due to the occlusive nature of the patch systems [24]. Cutaneous irritation was often encountered and limits the duration of time, a patch can be worn at a single site. However, the adhesive agent on the tape may be causally related, too.

For patches applied on the outer ear surface analgesic plasma fentanyl concentrations were measured from 9 to 48 h, whereas skin preparation was easier than in group 3. We hypothesize, that a reason for the difference in fentanyl absorption might be owed to the influence of a difference in body temperature, as it is described in several former studies. In fact, a temperature of 40 °C can raise drug delivery up to one third [8,25]. Rabbits can regulate their ear blood flow, which might lead to a lower drug delivery due to a lower body temperature in contrast to the neck. Depending on how long potent analgesia is requested, the application on the outer surface of the ear pinna might still be relevant for the attachment of fentanyl patches. There is a study published by Christou et al. using sheep, in which the change of fentanyl patches after 24 and 72 h is performed to investigate the fentanyl plasma concentrations thereafter [26]. It was shown that a patch change lead to an extended period of analgesia without a lack of insufficient pain management in between and a higher peak concentration due to pre-loading. Additionally, the peak concentration increased in that study, although we assume that this does not represent an advantage, as long as a certain threshold plasma concentration is exceeded permanently. Most likely the incidence of side effects will increase with rising values.

An obvious finding of our study is that it is inappropriate to apply the fentanyl patch on the inner surface of the ear. While practicability was good and the upcoming burden for the rabbit low, fentanyl plasma concentrations were below the predefined threshold, which implies the unsuitability of this location. Low plasma concentrations in this group might be attributable to poor drug penetration due to endogenous and exogenous influences such as differing skin conditions as described in former studies [24,27]. Additionally, rate of manipulation was higher in this group, which probably favored detaching of the patches in 5 of 6 animals. Furthermore, fixation of the patch with tape was more difficult due to the concave anatomy of the ear pinna, which probably also facilitated detachment. However, frequent checks to ensure that the patch is still in place were necessary in all groups.

Individual peak values of animals of group 1 and 3 showed a high variability within the study period. Probably several individual factors, like different pH, variable depot formation in the stratum corneum and dermis, different body temperatures and differences in cutaneous blood flow, influenced this outcome as previously reported [24].

The neck has been already described for patch application in rabbits, but a detailed description of skin preparation and patch application process have not been reported [15]. In the present study, the method of application was accurately planned and described to avoid variations in drug absorption due to differences in site preparation. Riviere and Papich showed that abrasion removes the stratum corneum barrier and can alter the transdermal flux [24]. Foley and coworkers investigated plasma fentanyl concentrations in rabbits using different methods of hair removal. For rabbits with clipped hair, comparable to the study described here, they reported a peak of 1.11 ± 0.32 ng/mL at 24 h and a decrease to 0.77 ± 0.21 ng/mL 72 h after patch application [15]. This is lower than the peak values in this study, even though fentanyl patches with a half dose rate were used: 3.60 µg/kg/h in this study versus 6.67 µg/kg/h reported by Foley and coworkers. One possible explanation for the better uptake is the use of alcohol to degrease the skin and ameliorate the drug uptake in this study. Two other studies reported the use of alcohol to degrease the skin prior to patch application in sheep [8,26]. In other studies involving dogs and cats, no degreasing with alcohol was performed [18,28,29].

The main limitation of this study is that rabbit-specific fentanyl plasma concentration able to provide analgesia is unknown. In humans, the minimum effective plasma concentration for analgesia is assumed to range between 0.5 and 2.0 ng/mL [11]. Previous veterinary studies extrapolated this range to animals such as sheep, rabbits and minipigs [15,16,23]. In former human studies, wide ranges of 0.23 to 3 ng/mL were described [13,30]. In humans, contrasting evidence has been reported about the relationship between respiratory adverse effects and fentanyl plasma concentration. While in one study concentrations above 2.0 ng/mL were suggested to cause respiratory depression [31]. In another study no correlation between fentanyl plasma concentrations and side effects was found [32]. In our study, no obvious adverse effects were observed, although concentrations above 2.0 ng/mL were measured. However, there was no evaluation of heart rate and respiratory frequency, which are known among others to be influenced at an early stage [33]. In cats, a plasma concentration higher than 2.2 ng/mL was observed to cause dysphoric behavior [17]. No dysphoria was observed in the rabbits of the present study.

Further limitations of this study are the use of only female rabbits. In humans, no difference in fentanyl uptake between male and female patients has been shown. To our knowledge, this has not been investigated yet in animals [34]. Another limitation is the use of a commercially available enzyme-linked immunosorbent assay (ELISA) developed to analyze human samples and no second test was performed to verify the results. In the literature the use of several different tests for analysis of fentanyl plasma concentration has been reported. A human ELISA was used in sheep, while no details were provided about the ELISA method applied in the study by Gilbert et al. [16,26]. In contrast to this, tests as the liquid chromatography–MS and gas chromatography–MS methods served for analysis in humans as well as in sheep [16,30,35]. Besides these tests, radioimmunoassays (RIA) are commonly used for analysis of plasma or serum fentanyl concentrations, too [29,36].

Concerning the timing of fentanyl patch application in experimental settings, some additional aspects need to be considered. Based on the findings of the present study, at least 6 h are needed for the fentanyl patch to provide a potentially analgesic plasma concentration. This implies that additional analgesics are needed to cover this period if the patch is applied during preparation for surgery. Some authors recommend patch application 12–24 h prior to surgery in dogs and sheep to achieve sufficient analgesia for surgery [7,12,37]. However, earlier application also means more stress for the rabbit, as an additional handling session accompanied by sedation would be needed. Another important aspect is the desired duration of pain treatment with opioids, which will depend on the invasiveness of surgery as well as administration of other pain medication (e.g., NSAIDs). This has to be defined for each study or patient and adapted as needed. Patches can be removed earlier or a second patch could be applied to extend the analgesic effect. As already mentioned, the effects of a patch change were investigated in sheep [26], while no comparable studies have been performed in rabbits so far.

5. Conclusions

The present findings support the use of transdermal fentanyl patches as a method to provide long-lasting analgesia in NZW rabbits. Depending on how long potent analgesia is required, either the neck or the outer surface of the ear are suitable for patch application. Summarizing, the administration of the patch can simplify postoperative pain management in laboratory rabbits and improve animal welfare in sense of the 3R.

Author Contributions: Conceptualization, V.M., C.S., H.R. and S.Z.; methodology, V.M., C.v.D., A.H. and S.Z.; validation, V.M., C.v.D. and A.H.; formal analysis, V.M., C.v.D. and A.H.; investigation, V.M., C.v.D., A.H. and S.Z.; resources, S.Z.; data curation, V.M. and S.Z.; writing—original draft preparation, V.M.; writing—review and editing, V.M., C.v.D., A.H., C.S., H.R. and S.Z.; visualization, V.M. and A.H., supervision, C.S., H.R. and S.Z.; project administration, V.M. and S.Z.; funding acquisition, S.Z. All authors have read and agreed to the published version of the manuscript.

Funding: Funding for this study was provided by the AO Foundation (grant number FS2017_02). No external funds were received in support of this work.

Acknowledgments: The authors would like to thank the whole team of the Preclinical Facility at the AO Research Institute Davos for their support with the implementation of the project.

Conflicts of Interest: The authors declare no conflict of interest.

References

1. Bottagisio, M.; Coman, C. Animal models of orthopaedic infections. A review of rabbit models used to induce long bone bacterial infections. *J. Med. Microbiol.* **2019**, *68*, 506–537. [CrossRef] [PubMed]

2. Mapara, M.; Thomas, B.S. Rabbit as an animal model for experimental research. *Dent. Res. J. (Isfahan)* **2012**, *9*, 111–118.

3. Permuy, M.; López-Peña, M. Rabbit as model for osteoporosis research. *J. Bone Miner. Metab.* **2019**, *37*, 573–583. [CrossRef] [PubMed]

4. Coulter, C.A.; Flecknell, P.A. Reported analgesic administration to rabbits undergoing experimental surgical procedures. *BMC Vet. Res.* **2011**, *7*, 12. [CrossRef] [PubMed]

5. Hofmeister, E.H.; Egger, C.M. Transdermal fentanyl patches in small animals. *J. Am. Anim. Hosp. Assoc.* **2004**, *40*, 468–478. [CrossRef]

6. Benato, L.; Rooney, N.J.; Murrell, J.C. Pain and analgesia in pet rabbits within the veterinary environment: A review. *Vet. Anaesth. Analg.* **2019**, *46*, 151–162. [CrossRef]

7. Weaver, L.A.; Blaze, C.A. A model for clinical evaluation of perioperative analgesia in rabbits (Oryctolagus cuniculus). *J. Am. Assoc. Lab. Anim. Sci.* **2010**, *49*, 845–851.

8. Ahern, B.J.; Soma, L.R. Comparison of the analgesic properties of transdermally administered fentanyl and intramuscularly administered buprenorphine during and following experimental orthopedic surgery in sheep. *Am. J. Vet. Res.* **2009**, *70*, 418–422. [CrossRef]

9. Mathews, K.A. Pain assessment and general approach to management. *Vet. Clin. N. Am. Small Anim. Pract.* **2000**, *30*, 729–755. [CrossRef]

10. Mayer, J. Use of behavior analysis to recognize pain in small mammals. *Lab. Anim. (N. Y.)* **2007**, *36*, 43–48. [CrossRef]

11. Flecknell, P.A.; Roughan, J.V. Assessing pain in animals- putting research into practice. *J. Anim. Welf.* **2004**, *13*, S71–S76.

12. Keating, S.C.; Thomas, A.A.; Flecknell, P.A.; Leach, M.C. Evaluation of EMLA cream for preventing pain during tattooing of rabbits: Changes in physiological, behavioural and facial expression responses. *PLoS ONE* **2012**, *7*, e44437. [CrossRef] [PubMed]

13. Calis, K.A.; Kohler, D.R. Transdermally administered fentanyl for pain management. *Clin. Pharm.* **1992**, *11*, 22–36. [PubMed]

14. Schultheiss, P.J.; Morse, B.C. Evaluation of a transdermal fentanyl system in the dog. *Contemp. Top. Lab. Anim. Sci.* **1995**, *34*, 75–81.

15. Foley, P.L.; Henderson, A.L. Evaluation of fentanyl transdermal patches in rabbits: Blood concentrations and physiologic response. *Comp. Med.* **2001**, *51*, 239–244.

16. Gilbert, D.B.; Motzel, S.L. Postoperative pain management using fentanyl patches in dogs. *Contemp. Top. Lab. Anim. Sci.* **2003**, *42*, 21–26.

17. Freise, K.J.; Newbound, G.C. Pharmacokinetics and the effect of application site on a novel, long-acting transdermal fentanyl solution in healthy laboratory Beagles. *J. Vet. Pharmacol. Ther.* **2012**, *35* (Suppl. 2), 27–33. [CrossRef]

18. Egger, C.M.; Duke, T. Comparison of plasma fentanyl concentrations by using three transdermal fentanyl patch sizes in dogs. *Vet. Surg.* **1998**, *27*, 159–166. [CrossRef]

19. Scherk-Nixon, M. A study of the use of a transdermal fentanyl patch in cats. *J. Am. Anim. Hosp. Assoc.* **1996**, *32*, 19–24. [CrossRef]

20. Davidson, C.D.; Pettifer, G.R. Plasma fentanyl concentrations and analgesic effects during full or partial exposure to transdermal fentanyl patches in cats. *J. Am. Vet. Med. Assoc.* **2004**, *224*, 700–705.

21. Ahern, B.J.; Soma, L.R. Pharmacokinetics of fentanyl administered transdermally and intravenously in sheep. *Am. J. Vet. Res.* **2010**, *71*, 1127–1132. [CrossRef] [PubMed]

22. Mills, P.C.; Cross, S.E. Regional differences in transdermal penetration of fentanyl through equine skin. *Res. Vet. Sci.* **2007**, *82*, 252–256. [CrossRef] [PubMed]

23. Wilkinson, A.C.; Thomas, M.L. Evaluation of a transdermal fentanyl system in yucatan miniature pigs. *Contemp. Top. Lab. Anim. Sci.* **2001**, *40*, 12–16. [PubMed]

24. Riviere, J.E.; Papich, M.G. Potential and problems of developing transdermal patches for veterinary applications. *Adv. Drug Deliv. Rev.* **2001**, *50*, 175–203. [CrossRef]

25. Pettifer, G.R.; Hosgood, G. The effect of rectal temperature on perianesthetic serum concentrations of transdermally administered fentanyl in cats anesthetized with isoflurane. *Am. J. Vet. Res.* **2003**, *64*, 1557–1561. [CrossRef]

26. Christou, C.; Oliver, R.A. Transdermal fentanyl and its use in ovine surgery. *Res. Vet. Sci.* **2015**, *100*, 252–256. [CrossRef]

27. Varvel, J.R.; Shafer, S.L. Absorption characteristics of transdermally administered fentanyl. *Anesthesiology* **1989**, *70*, 928–934. [CrossRef]

28. Lee, D.D.; Papich, M.G. Comparison of pharmacokinetics of fentanyl after intravenous and transdermal administration in cats. *Am. J. Vet. Res.* **2000**, *61*, 672–677. [CrossRef]

29. Reed, F.; Burrow, R. Evaluation of transdermal fentanyl patch attachment in dogs and analysis of residual fentanyl content following removal. *Vet. Anaesth. Analg.* **2011**, *38*, 407–412. [CrossRef]

30. Gourlay, G.K.; Kowalski, S.R. Fentanyl blood concentration-analgesic response relationship in the treatment of postoperative pain. *Anesth. Analg.* **1988**, *67*, 329–337. [CrossRef]

31. Peng, P.W.; Sandler, A.N. A review of the use of fentanyl analgesia in the management of acute pain in adults. *Anesthesiology* **1999**, *90*, 576–599. [CrossRef] [PubMed]

32. Grond, S.; Radbruch, L. Clinical pharmacokinetics of transdermal opioids: Focus on transdermal fentanyl. *Clin. Pharm.* **2000**, *38*, 59–89. [CrossRef] [PubMed]

33. Streisand, J.B.; Varvel, J.R. Absorption and bioavailability of oral transmucosal fentanyl citrate. *Anesthesiology* **1991**, *75*, 223–229. [CrossRef] [PubMed]

34. Solassol, I.; Caumette, L.; Bressolle, F.; Garcia, F.; Thézenas, S.; Astre, C.; Culine, S.; Coulouma, R.; Pinguet, F. Inter-and intra-individual variability in transdermal fentanyl absorption in cancer pain patients. *Oncol. Rep.* **2005**, *14*, 1029–1036. [PubMed]

35. Nomura, M.; Inoue, K. Serum concentration of fentanyl during conversion from intravenous to transdermal administration to patients with chronic cancer pain. *Clin. J. Pain* **2013**, *29*, 487–491. [CrossRef] [PubMed]

36. Thomasy, S.M.; Slovis, N. Transdermal fentanyl combined with nonsteroidal anti-inflammatory drugs for analgesia in horses. *J. Vet. Intern. Med.* **2004**, *18*, 550–554. [CrossRef]

37. Robinson, T.M.; Kruse-Elliott, K.T. A comparison of transdermal fentanyl versus epidural morphine for analgesia in dogs undergoing major orthopedic surgery. *J. Am. Anim. Hosp. Assoc.* **1999**, *35*, 95–100. [CrossRef]

Article

Transdermal Fentanyl Uptake at Two Different Patch Locations in Swiss White Alpine Sheep

Tim Buchholz [1], Maria Hildebrand [1], Anja Heider [2], Valentina Stenger [1], Daniel Arens [1], Claudia Spadavecchia [3] and Stephan Zeiter [1,*]

[1] AO Research Institute Davos, 7270 Davos Platz, Switzerland; tim.buchholz@aofoundation.org (T.B.); maria.hildebrand@aofoundation.org (M.H.); valentina.stenger@aofoundation.org (V.S.); daniel.arens@aofoundation.org (D.A.)

[2] Swiss Institute of Allergy and Asthma Research (SIAF), University Zurich, 7265 Davos Wolfgang, Switzerland; anja.heider@siaf.uzh.ch

[3] Vetsuisse Faculty, Department for Veterinary Medicine, University of Bern, 3012 Bern, Switzerland; claudia.spadavecchia@vetsuisse.unibe.ch

* Correspondence: stephan.zeiter@aofoundation.org; Tel.: +41-814-142-311

Received: 20 August 2020; Accepted: 11 September 2020; Published: 17 September 2020

Simple Summary: Providing adequate and prolonged pain relief to sheep undergoing invasive orthopedic surgery while keeping side effects and stress for the animals at a minimum is challenging. Transdermal patches continuously releasing the synthetic opioid fentanyl through the skin, are a frequently used method in veterinary and human medicine. To refine the current analgesia protocol, we compared the uptake of fentanyl from a transdermal fentanyl patch applied at two different skin locations in sheep. The fentanyl plasma levels were measured at different time points over five days. The patch applied on the foreleg resulted in a faster fentanyl uptake with higher peaks and a longer time within or above the target fentanyl plasma concentration when compared to the one on the thorax. Additionally, it was easier to apply the patch at the foreleg than at the thorax. Our findings suggest that the fentanyl patch should be applied to the foreleg 3–6 h before the painful insult and that its effect should last at least 48 h.

Abstract: When using animals in biomedical research, investigators have the responsibility to ensure adequate analgesia. Currently, transdermal fentanyl patches (TFP) are often used to provide postoperative analgesia in large laboratory animals. The aim of this study was to compare the fentanyl uptake resulting from TFP applied at two different locations, namely the foreleg and the thorax, in healthy adult sheep. Twelve sheep received a TFP with an intended dosage of 2 ug/kg/h. Blood samples were taken at different time points over a period of five days and the fentanyl plasma levels were measured. The TFP applied on the foreleg allowed a faster fentanyl uptake with higher peaks and a longer time within or above the target concentration of 0.6–1.5 ng/mL, shown to be analgesic in humans, when compared to the one on the thorax. Assuming that the effective plasma concentration described for humans is providing analgesia in sheep as well, the present findings suggest that it should be sufficient to apply the TFP 3–6 h before the painful insult and that its effect should last at least 48 h. Furthermore, when TFP are used to provide postoperative analgesia in sheep, they should be placed on the foreleg rather than on the thorax.

Keywords: 3R principles; refinement; analgesia; transdermal fentanyl patch; sheep

---

1. Introduction

Animals are frequently used in research, in particular in biomedical studies [1]. The selection of the appropriate animal species and a model for the specific scientific question under investigation are the first steps towards a successful study [2,3]. If the species and model utilized have not been tailored to the study goal, it might prove impossible to answer the research question. Furthermore, when planning an animal study, it is necessary to consider the 3R principles. An animal study should only be considered if no alternative method exists to answer the research question (replace). Further, by careful planning, it should also be ensured that a maximum of relevant information is obtained from as few animals as possible (reduce), and last but not least, the burden of the animals is alleviated as much as possible by adapting standard protocols and procedures to the chosen species and model (refine) [2].

Sheep are commonly used as a preclinical model in orthopedic research [1,4–9]. They are chosen due to their similarities to humans in terms of body weight, rate of bone healing [7], and characteristics of mineralization of long bones [1]. Additionally, the size of several bones can be compared with the one of humans [10]. For example, sheep were employed by Yamamuro et al. to test a prosthesis for the replacement of the lumbar vertebrae [6], and by den Boer et al. to develop a new segmental long bone defect model [7]. In an implant-associated infection study, Boot et al. used a sheep model to investigate different treatment strategies after implantation of a intramedullary implant in the tibia [8]. In all these models and the majority of orthopedic research, surgical intervention is essential to mimic the clinical problem. As in clinical cases, surgical interventions require the implementation of an appropriate analgesic protocol. In research, the analgesic protocol is not only a tool but can also be an important determinant for the study outcome since pain can majorly impact the reproducibility of results [11]. Therefore, an analgesic protocol tailored to the species and model is a necessity not only for animal welfare reasons, but also to ensure the acquisition of high-quality data.

In order to control pain produced by invasive surgeries, analgesic drugs are typically administered pre-, intra- and a postoperatively. Before and during surgery, the sheep is under general anesthesia with a venous access allowing easy administration of analgesic medication intravenously. On the contrary, after surgery, the sheep is awake, and administration of pain medication can be stressful. In addition, pain assessment during the postoperative phase is very challenging and insufficient analgesia could easily be unrecognized [12,13]. An adequate postoperative analgesic protocol should therefore guarantee constant analgesia adapted to the level of pain inflicted with minimal side effects and stress.

Opioids are analgesics for moderate to severe pain, often used to obtain sufficient pain relief due to invasive surgical procedures [14]. Fentanyl is a strong synthetic opioid commonly used in analgesia protocols for humans and animals [15–19]. Due to the short half-life of this drug, systems allowing continuous administration are required [16]. During surgery, fentanyl can be given by continuous rate infusion (CRI) via infusion pumps. Such tools can be heavy and can disturb the freedom of movement of sheep when awake. For this reason, fentanyl, as an injectable medication, is not the method of choice during the postoperative phase. However, fentanyl can also be applied via a transdermal fentanyl patch (TFP), which consists of a polyacrylate adhesive layer allowing absorption of fentanyl through the skin (according to the information provided by Mepha® Pharma AG). With such a TFP providing a continuous and prolonged fentanyl administration, the sheep can be left undisturbed, neither stressful injections nor infusion pumps are needed [20,21]. Therefore, the transdermal application of fentanyl is a suitable method to provide analgesia for a time period of several days [16].

In sheep, the foreleg is a commonly used location for TFP since the application is very simple [20,22]. However, not only the ease of application but also the onset of action as well as the overall achieved fentanyl plasma concentration should be considered. In a study by Ahern et al., the TFP was applied 12 h prior to surgical intervention to compensate the slow uptake and achieve analgesic levels intraoperatively [22]. In dogs, TFP application has been recommended even 24 h before surgery [23]. In horses however, it has been shown that the application of a TFP at the foreleg results in a lower absorption rate and lower fentanyl plasma levels when compared to the application at the groin region or thorax [24]. In another study in horses by Orsini et al., the TFP applied at the thorax showed a fast fentanyl uptake [25]. Considering these findings in horses, it is logical to hypothesize that the thorax might be preferable in sheep as well, because of better skin perfusion, less affected by temperature variation and lower mobility when compared to the limb. Additionally, the onset of action might be faster, making it unnecessary to apply the TFP many hours before its desired effect.

The comparison of different locations for TFP application in sheep has not been described yet.

The aim of this study was to compare the fentanyl uptake resulting from a TFP applied at two different locations, namely the foreleg and the thorax, in healthy adult sheep.

Following TFP application, fentanyl plasma levels were measured at different time points. It was hypothesized that with the TFP applied at the thorax the fentanyl uptake would be faster and a higher peak would be reached than with the TFP applied at the foreleg.

2. Materials and Methods

2.1. Study Design

This experimental, prospective, randomized study was approved by the Cantonal Committee for Animal Experiments of Graubünden (GR-TVB No. 35_2018) and conducted in an AAALAC International approved facility. The sheep were randomly assigned to one of two equal groups ($n = 6$ per group) by drawing lots. An ear tag with an internal consecutive number was used as identification. In one group, the TFP (fentanyl Mepha® Matrix patches/Mepha Pharma AG/57362) was applied to the foreleg and in the other group to the thorax (Figure 1)

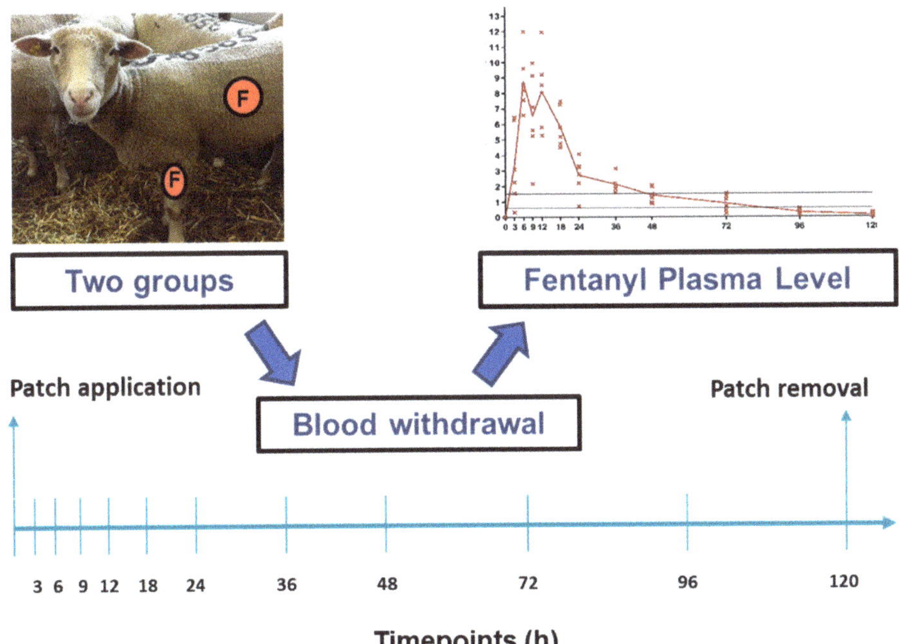

Figure 1. Study Design Overview—The letter "F" in the first picture shows the locations of the TFP on the sheep. However, only one location was used per sheep. After application of a transdermal fentanyl patch (TFP) to one of two possible locations depending on group allocation, fentanyl plasma levels were measured at different time points.

The foreleg was defined as the middle of the lateral surface starting from the elbow joint proximally and ending at carpal joint distally. The region of the thorax was defined as approx. 5 cm caudal from the scapula, at the thorax (Figure 2). The primary outcome, the fentanyl plasma level, was measured at 12 different time points.

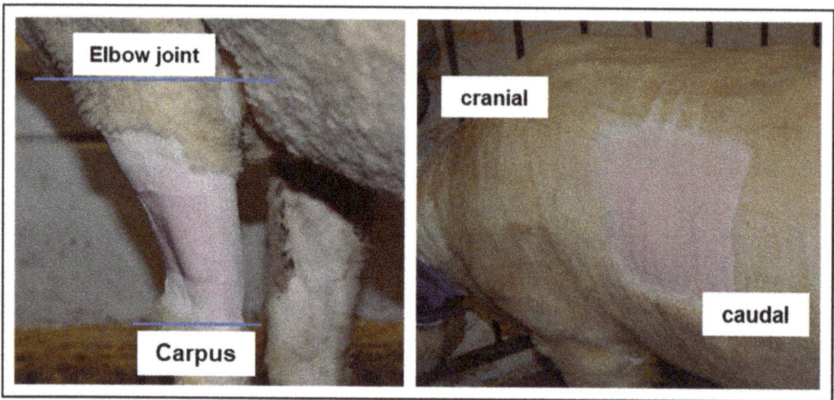

Figure 2. Transdermal fentanyl patch location—left picture: foreleg, right picture: thorax.

2.2. Animals

Sheep included in this study were all females and more than 2 years old and determined to be healthy based on clinical examination, hemogram (hematocrit and white blood cell count determined by a "Vet ABC" (ABX Diagnostics, France) and biochemistry analysis (total proteins measured with a refractometer). None of the sheep had undergone anesthesia or surgery before to exclude any prior effects. In total, 13 female White Swiss Alpine sheep were part of this study (mean age 3.2, range 2 to 6.5 years; mean weight 76.2 kg, range 59.5 to 99.5 kg). The sheep were acclimatized to the research facility and the daily routine for 2 weeks in shared pens prior to TFP application. The stable had a 12 h day/night cycle and all sheep had access to daylight in form of a window. Temperature was maintained between 15 and 20 °C, with relative humidity above 30%, and 10–15 air changes per hour were performed. Maintenance diet consisting of a mixture of straw, hay, silage, maize, and salt was fed twice a day. Water was available ad libitum in an automatic water drinker.

2.3. Transdermal Fentanyl Patch Application and Long-Term Catheter

A fentanyl dosage of 2 µg/kg/h was targeted based on previous literature [22,26]. For each sheep, the required dose was approximated using the available TFP sizes of 12 µg/h, 25 µg/h, 50 µg/h and 100 µg/h.

The sheep were sedated with an intramuscular injection of detomidine (0.04 mg kg^{-1}; Equisedan; Dr. E. Graeub AG, Bern, Switzerland). After 15 min, the left jugular vein was prepared aseptically and a permanent catheter (Certofix® Mono, B.Braun, Sempach, Switzerland) was placed using a Seldinger method [27]. The catheter was sutured with a 2-0 Ethilon II (Ethicon Inc., Somerville, NJ, USA) on three points and covered with adhesive retention dressing. Additionally, a cotton bandage and cohesive bandage was applied to prevent the catheter from loosening. The TFP location was then prepared according to the group. The area was clipped (Figure 2) (foreleg group: 10 cm wide band around the foreleg; thorax group: 15 × 15 cm square on left thorax) and carefully shaved with warm water and soap (Softaskin®, B.Braun). Then, the area was degreased by scrubbing three times with Ethanol 80% and allowed to dry for 5 min. The TFP were applied 75 min after the sedation. This time was chosen so all sheep had the same sedation effect when the TFP was applied. The TFP was covered with a self-adhesive "vet polster bandage" by HENRY SCHEIN® (product code: 900-9911). For the foreleg, this bandage was cut to fit once around the foreleg while for the thorax a square was cut out about 2–3 cm larger than the area covered by the TFP. Above the polster bandage, a cohesive bandage ("cohesive elastic crepe bandage", 20 m × 10 cm, HENRY SCHEIN®, product code: 900-8588) was applied. Depending on the TFP location, this bandage was either wrapped around the foreleg or around the entire thorax (Figure 3).

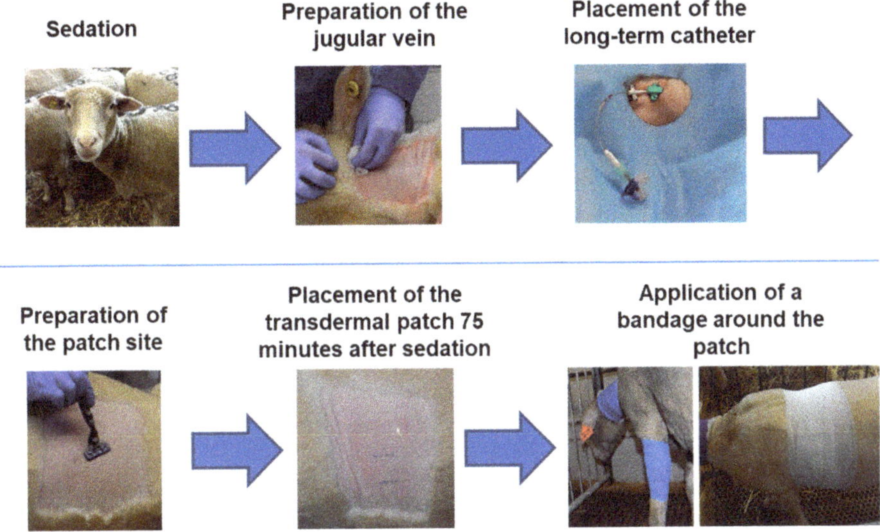

Figure 3. Work flow of the long-term catheter placement and transdermal fentanyl patch application.

2.4. Blood Withdrawal

Blood samples were taken at 12 different time points: 0 (before TFP application), and then 3, 6, 9, 12, 18, 24, 36, 48, 72, 96, and 120 h after the application of the TFP (Figure 1). All blood samples were taken using the long-term catheter. The first 5–6 mL were discarded and afterwards an EDTA blood tube (9 mL S-Monovette®, Sarstedt AG&Co.KG) was filled. Then, the catheter was flushed with 10 mL Saline with 500 IE Heparin/mL (Heparinum natricum, FRESENIUS Medical Care (Schweiz) AG). Within two hours after withdrawal, the blood samples were centrifuged for 10 min with 2500 rpm. The plasma was aliquoted in Eppendorf tubes (Vaudaux-Eppendorf AG, Schönenbuch, Switzerland) and frozen at −80 °C until analysis.

2.5. Blood Analysis

Fentanyl plasma levels were analyzed by a human enzyme-linked immunoabsorbent assay (Forensic ELISA-Kit, Neogen Toxicology). Standard values were created by adding known fentanyl concentrations (8 in total, fentanyl 0.5 mg/10 mL solution for injection, Sintetica S.A., Switzerland) to plasma of fentanyl-naive sheep. Fentanyl-naïve plasma was used as the blank. The absorbance was read at 450 nm using a Mithras microplate reader (Berthold Technologies). Both standards and samples were measured in duplicates and the assay was performed according to the protocol.

2.6. Study Outcomes

The primary outcome of this study was the fentanyl plasma level. As an analgesic fentanyl plasma level for sheep is not described in the literature, the minimum analgesic fentanyl plasma level described for opioid-naive human patients and corresponding to the concentration range of 0.6–1.5 ng/mL was used as a reference in the present study [28]. Time to reach the concentration threshold of 0.6 ng/mL and duration of effect, defined as fentanyl concentration > 0.6 ng/mL, were evaluated.

Secondary outcomes were recorded twice a day from TFP application until the end of the study (5 days) to detect potential fentanyl related side effects. Physiological parameters, including rectal body temperature and respiratory rate, water and food uptake, as well as amount and texture of feces were evaluated at regular intervals. Furthermore, the condition of the bandage and the patch application site were evaluated. The TFP was removed after the last blood sample was taken. Pictures of the skin from every sheep were taken to assess the skin reaction to the TFP or to the bandage. General behavior, and in particular the occurrence of sedation or excitation, was observed and described. Additionally, the body weight was taken before TFP application and after TFP removal, always before sheep were fed. These data were collected by the same observer for all time points. This person was also responsible for handling of the sheep before the study commenced, clinical examination, and the blood work.

2.7. Statistics

A descriptive statistical analysis was performed for the data of this study as a statistically significant difference is not necessarily connected to an appropriate fentanyl level. The number of sheep that achieved the predefined fentanyl plasma level of 0.6–1.5 ng/mL was determined. The goal was to define the time period during which the expected minimum fentanyl plasma level was achieved in every individual sheep.

3. Results

Thirteen sheep were part of this study. One sheep of the thorax group was excluded as the TFP detached 18 h after placement. Data from this sheep were excluded from the study. All sheep recovered well after sedation.

3.1. Primary Outcome: Fentanyl Plasma Level

Based on the selected TFP size, the overall average expected fentanyl administration rate was 2.08 µg/h/kg (range 2.01 to 2.16 µg/h/kg). The average expected fentanyl administration rate was 2.01 µg/h/kg (range 2.01 to 2.16 µg/h/kg) in the foreleg group and 2.07 µg/h/kg (range 2.01 to 2.1 µg/h/kg) in the thorax group.

In the foreleg group, five out of six sheep achieved a fentanyl plasma level threshold of >0.6 ng/mL after 3 h, and all sheep within 6 h. At least this level was maintained in all sheep for a minimum of 48 h and in four out of six sheep for a maximum of 72 h (Table 1, Figure 4). In the thorax group, only two sheep had fentanyl levels above 0.6 ng/mL after 3 h, four sheep after 6 h, five sheep after 9 h, and all six sheep after 12 h (Table 1, Figure 5). At 24 h, all six sheep still had a concentration above 0.6 ng/mL. However, thereafter it started to decline so that, by 36 h, the level of 0.6 ng/mL was achieved in only five sheep, at 72 h in three sheep, and by 96 h two sheep had level above 0.6 ng/mL. When comparing both groups, the fentanyl uptake was more consistent in terms of the onset of the desired plasma concentration as well as duration of effect (Figure 6), when applied to the foreleg than to the thorax. At 120 h, no sheep from either group were above the predefined threshold levels.

All sheep in the foreleg group were above 1.5 ng/mL at least once during the study, with 5/6 sheep reaching this level already after 3 h. In contrast, two out of six sheep of the thorax group never attained a fentanyl plasma level of 1.5 ng/mL and only one sheep reached this value after 3 h. A fentanyl plasma level > 1.5 ng/mL was achieved in all sheep of the foreleg group between 6 h and 18 h. In the thorax group, only four out of six sheep reached a plasma level of 1.5 ng/mL. These sheep achieved the predefined fentanyl plasma level between 9 h and 36 h. At 96 h, no sheep from either group was in the target plasma level range.

For the individual fentanyl plasma concentration peaks see Table 1.

In the foreleg group, one sheep displayed a striking drop in plasma levels after 24 h with values well above the 1.5 ng/mL threshold before (18 h) and after (36 h).

Table 1. Fentanyl plasma level of both groups (values marked in yellow are above 0.6 ng/mL; values marked with green are above 1.5 ng/mL; every column represents one sheep).

	Fentanyl Concentration ng/ml											
	Foreleg							Thorax				
Timepoint	1	2	3	4	5	6	7	8	9	10	11	12
0 h	0.01	0.01	0.01	0.01	0.01	0.01	0.02	0.01	0.01	0.01	0.01	0.01
3 h	3.11	6.46	1.56	0.32	2.27	6.28	5.67	0.01	1.43	0.01	0.48	0.23
6 h	8.56	7.57	6.58	8.20	11.99	9.60	5.80	0.37	3.78	0.00	3.10	1.21
9 h	9.12	9.96	5.63	7.13	5.26	2.17	7.87	0.48	3.85	0.87	3.15	2.21
12 h	9.20	5.29	8.54	8.01	11.95	5.82	7.33	0.65	3.94	1.24	3.83	2.38
18 h	7.29	4.51	4.75	5.20	7.48	5.82	5.00	0.90	3.91	1.33	3.92	4.69
24 h	2.21	3.30	4.08	0.72	2.77	3.23	4.05	0.71	3.52	1.30	3.43	4.78
36 h	2.03	1.84	3.13	1.60	2.19	2.04	2.09	0.38	3.38	0.85	2.02	4.55
48 h	1.26	0.88	1.97	1.48	0.96	2.08	1.20	0.21	2.36	0.83	1.36	3.31
72 h	1.37	0.71	1.14	0.43	0.21	1.54	1.17	0.01	1.53	0.40	0.28	2.11
96 h	0.51	0.08	0.32	0.56	0.11	0.26	0.52	0.01	0.77	0.11	0.01	0.94
120 h	0.19	0.01	0.20	0.07	0.01	0.28	0.30	0.01	0.18	0.01	0.01	0.11

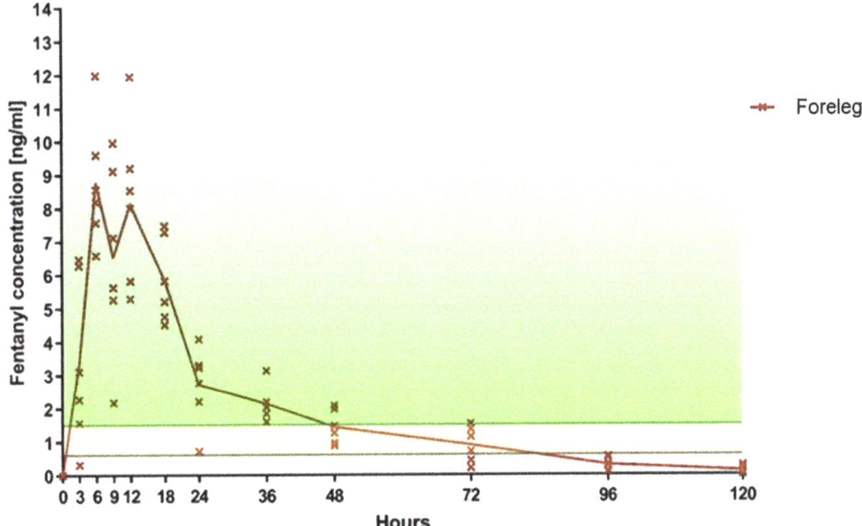

Figure 4. Foreleg—Fentanyl plasma level over time (the fentanyl level within the area marked in yellow, correspond to the minimal analgesic plasma level of 0.6–1.5 ng/mL described for humans; the area above marked in green represents values higher than 1.5 ng/mL).

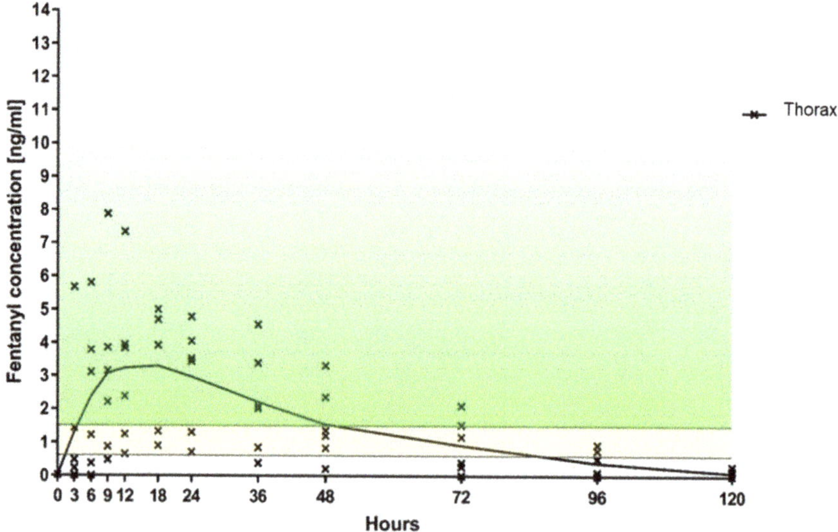

Figure 5. Thorax—Fentanyl plasma level over time (the fentanyl level within the area marked in yellow, correspond to the minimal analgesic plasma level of 0.6–1.5 ng/mL described for humans; the area above marked in green represents values higher than 1.5 ng/mL).

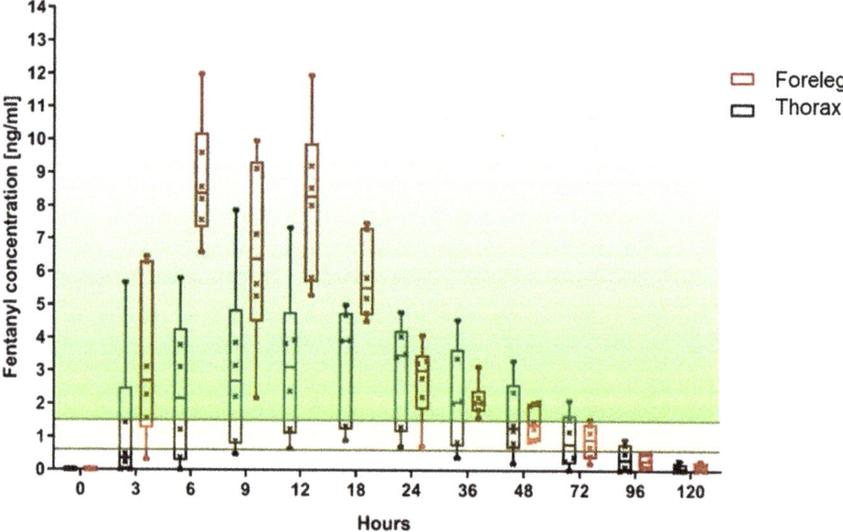

Figure 6. Fentanyl plasma levels measured over time for TFP applied at the foreleg (**red**) and thorax (**black**) displayed in a box plot graph (the fentanyl level within the area marked in yellow, corresponds to the minimal analgesic plasma level of 0.6–1.5 ng/mL described for humans; the area above marked in green represents values higher than 1.5 ng/mL).

3.2. Secondary Outcomes

In all sheep, the general behavior and bodyweight were affected by the TFP. All sheep lost weight during the study. The average weight loss was 3.7 kg (range from 1–9 kg) at 120 h. The average weight loss was 4 kg for the foreleg group and 3.3 kg for the thorax group. The parameters that remained unchanged were respiratory rate (12–40/min), rectal body temperature (38.0–39.5 °C), water uptake and amount and texture of feces, which remained within normal limits for all sheep at all time points. Food uptake was reduced for the first three days for all sheep.

For all sheep, the bandage stayed in place regardless of the location. After TFP removal, the skin was slightly reddened and slightly moist in half of the sheep. These observations were made in both groups. In one sheep of the foreleg group, a part of the bandage rubbed the skin and created a 2 × 0.3 cm superficial wound.

During the first 36 h after TFP application, all sheep showed similar alternating phases, of being apathetic or excited. Apathy was observed as a reduced response to environmental influences (observer, food etc.). Excitement was shown as nervous behavior or pressing their head against the door of the fence when the observer entered the pen. Three sheep were mainly extremely nervous, 3 others head-pressed, and during this time period two sheep were observed frequently walking back and forth along the fence.

Eight out of 12 sheep displayed hypersalivation. These behavioral abnormalities started in six sheep after 3 h, in four sheep after 6 h, and in two after 12 h. In seven sheep, the abnormal behavior disappeared 72 h after TFP application. For the rest after 36 h. These behavioral changes occurred in both groups equally and were not related to a specific fentanyl plasma level.

4. Discussions

Aim of this study was to quantify and compare the characteristics of fentanyl uptake following the application of a TFP at two different locations, namely the foreleg and the thorax, in healthy adult sheep. It was hypothesized that the absorption of fentanyl at the thorax is better, leading to a faster onset of action and a higher peak of fentanyl plasma levels than at the foreleg. Contrary to our hypothesis, the faster absorption and highest plasma concentration values were observed for TFP applied on the foreleg. For this patch location, the lowest peak measured was 8.01 ng/mL and five out of six sheep reached fentanyl plasma levels > 1.5 ng/mL after 3 h, whereas for the patch applied to the thorax, the highest concentration reached was 7.87 ng/mL, only one sheep achieving levels above 1.5 ng/mL after 3 h.

It has been shown that the transdermal uptake rate of fentanyl is influenced by the characteristics of the skin. Especially the subcutaneous fat content [29], but also other characteristics of the skin such as skin thickness, corneal layer, temperature, blood vessel architecture, as well as perfusion [30,31] have an impact on absorption in transdermal drug delivery. The specific skin characteristics of the foreleg seemed to favor the fentanyl uptake at this location compared to the thorax. Lyne et al. described significant differences of the skin composition and thickness between sheep breeds and differences depending on the skin location. Skin areas with wool were thinner compared to skin areas with no or less hair [32].

In line with this, a fast and efficient uptake of fentanyl from the foreleg has been previously reported by Ahern et al. in sheep [22]. However, in horses it has been shown that the application of a TFP to the foreleg results in a lower absorption rate and lower fentanyl plasma level when compared to the application to the groin region or thorax [21]. This was confirmed in another study in horses where the TFP applied at the thorax resulted in a fast uptake of fentanyl [22]. The discrepancy between the reported finding in horses and the present study might be due to differences in skin composition between species. According to Riviere et al. [31], differences in the pharmacokinetics of fentanyl uptake between species is linked to differences in the thickness of the stratum corneum. Bouclier et al. described morphological differences between species (rabbit, rat, mice, minipig) and humans [33]. In addition, Bartek et al. showed differences in terms of skin permeability in rat, rabbit, pig, and human [34].

In sheep, other locations for TFP application have also been described in the literature. Jen et al. [35] applied a TFP at the intrascapular region resulting in lower values and a slower uptake than the one achieved in the foreleg group of the present study. In another study, Musk et al. reported a sufficient analgesia applying the TFP at the groin region of pregnant sheep undergoing a hysterotomy and laparotomy. The efficacy of analgesia was assessed via a pain score postoperatively [36]. These two examples as well as the present study highlight differences in fentanyl uptake when patches are applied at different locations in sheep and thereby enabling scientists to use these differences to refine their analgesic protocols. Other possibilities to refine postoperative pain management in sheep are discussed in an article by Lane et al. [37] mentioning a transdermal patch releasing buprenorphine for seven days. A buprenorphine patch was applied in minipigs in a study by Thiede et al., in which a therapeutic plasma levels were reached at least for 72 h [38]. Considering the behavioral changes observed in our study after fentanyl patch application, it could be of interest to compare clinical efficacy and side effects of fentanyl and buprenorphine administered via transdermal route, buprenorphine being potentially associated with lower incidence of side effects.

Behavioral changes (inter alia dysphoria, hypersalivation) were observed after patch application in both groups. These behavioral changes were also described by Marchionatti et al. in a Holstein calve [39]. The TFP dosage of 2 µg/kg/h used in this study is based on the literature [22,26]. In sheep undergoing surgery and treated with the dosage of 2 µg/kg/h, this dysphoria is normally not seen at our institution. In dogs and cats it is described that pain from a surgery can hide side effects like dysphoria due to opioids [40]. Most likely, the "missing" pain in this study conducted in healthy, pain free animals could be a reason for the observed behavior. These side effects were very likely caused by fentanyl since all sheep underwent an uneventful clinical examination by a veterinarian at the beginning of the study. The sheep were normal again after the first 36 h. However, a reliable correlation of the side effects to fentanyl would only be possible with a control group undergoing the same procedure (sedation, long-term catheter, bandage, blood taking) without a TFP application.

Hypersalivation caused by opioids is to our knowledge not described in sheep. However nausea and vomiting are described in small animals [14] and might have caused hypersalivation in the sheep of the present study. In addition, hypersalivation is described in dogs after application of a TFP [41]. The sheep in this study lost weight most likely due to the behavioral changes in the first days leading to reduced food uptake. This weight loss as an additional side effect indicates again the need of an adequate dosage of fentanyl adapted to the surgical procedure. No other side effects of fentanyl described in literature (i.e., respiratory depression, reduced food uptake, reduced gut motility, and hyperkinesia [14,42]) were observed in this study

From a practical point of view, the application of the TFP to the foreleg was slightly easier. However, the risk of tightening the bandage too much and creating a wound or edema, is higher at the foreleg. Regular controls of the bandage after TFP application are highly recommended.

The major limitation of this study is assuming the same fentanyl plasma level being analgesic in sheep as in humans. To our knowledge, no study identifying the minimal analgesic level in sheep has been published and even for humans a wide range is reported in the literature. Frequently referenced by other veterinary studies [38,39,43] is the range between 0.5 and 2 ng/mL as the minimal analgesic level published in a review in the early nineties [44]. In an older study, 0.6 ng/mL has been reported [45]. In a more recent article by Grape et al., the analgesic plasma level for humans of 0.6–1.5 ng/mL was described and used as a reference in this study [28]. Even though it is common practice to apply the human minimal analgesic level for different species in veterinary medicine, it would clearly be beneficial to establish these levels for frequently treated species. However, as it is already challenging to define the minimal analgesic level in humans, it is even more so in animals. Therefore, the present study was designed without surgical intervention or other painful stimuli to avoid interaction with the additionally necessary medication and surgery as well as individual differences in pain sensitivity between animals. Thus, evaluation of the analgesic effectiveness using a pain score was not feasible. Further studies are needed to describe the minimum analgesic level for fentanyl in sheep, using quantitative assessment methods like for example the nociceptive withdrawal reflex [46].

Fentanyl plasma levels can be measured by different methods, which may result in some discrepancies regarding the reported levels between studies. In this study a commercial ELISA-Kit was used. Ahern et al. [22] analyzed samples using liquid and gas chromatography mass spectrometry. Even though the same dose was applied to the animals, the measured fentanyl plasma levels were substantially lower compared to the present study. Using an ELISA-Kit from a different company, Christou et al. [20] measured lower values as well. However, Jen et al. [35] showed that both methods mentioned are positively correlated. Therefore, the difference between the two locations described here is valid even though the peak values may differ if a different detection method would have been used. In the foreleg group, one sheep displayed a striking drop in plasma levels after 24 h. This sample might be diluted due to unknown reasons. No other explanation can be given.

Only female sheep were used in this study, as it was shown in a recent European survey that female sheep are used significantly more often in research [1,4–9,47] most likely due to the fact that males are usually castrated shortly after birth. However, Solassol et al. showed that sex does not influence fentanyl uptake in humans [48] and to our knowledge, information about the effect of sex on transdermal fentanyl uptake in sheep is not recorded. Suárez-Morales et al. compared fentanyl doses needed by male and female human patients undergoing general anesthesia and showed a higher dose level was required in females [49]. Many other studies investigating the influence of sex on the response to opioids in rodents [50] showed the same tendency with males having a higher sensitivity to opioids [51–53]. However, this result could not be shown in a study using fentanyl in Sprague-Dawley rats [54].

In an experimental animal study, the analgesia protocol should provide reliable and appropriate pain relief in all sheep, and therefore decreasing the need for additional, individual pain medication which could impact the outcome. In this study, the TFP applied to the foreleg demonstrated a reliable onset of fentanyl uptake in all animals. After 6 h, all sheep of the foreleg group reached the level of 1.5 ng/mL fentanyl plasma level and five of these after 3 h. Therefore, an application of the TFP 12 h before the desired onset of effect as done by Ahern [22] may not be needed.

5. Conclusions

A TFP, applied either on the foreleg or on the thorax, induced an effective fentanyl uptake in healthy sheep. However, a faster uptake, longer duration of action and easier application were observed for the foreleg compared to the thorax. Assuming that the effective plasma concentration described for humans is providing analgesia in sheep as well, the present findings suggest that it should be sufficient to apply the TFP 3–6 h before the painful insult and that its effect should last at least 48 h.

Author Contributions: Conceptualization, S.Z., T.B., and C.S.; data curation, T.B., S.Z., D.A., V.S., and A.H.; methodology, T.B., S.Z., M.H., V.S., D.A., and A.H.; formal analysis, T.B., M.H., and A.H.; writing—original draft preparation, T.B., M.H., D.A., and V.S.; writing—review and editing, S.Z., C.S., and M.H.; supervision, S.Z. and C.S.; project administration, T.B.; funding acquisition, S.Z. All authors have read and agreed to the published version of the manuscript.

Funding: Funding for this study was provided by the AO Foundation (grant number FS2018_08). No external funds were received in support of this work.

Acknowledgments: The authors want to thank the whole team of the Preclinical Facility of the AO Research Institute for their support. Another thanks to the Swiss Institute of Allergy and Asthma Research (SIAF) in Davos for letting us analyze the blood samples in their laboratory.

Conflicts of Interest: The authors declare no conflict of interest.

References

1. Martini, L.; Fini, M.; Giavaresi, G.; Giardino, R. Sheep model in orthopedic research: A literature review. *Comp. Med.* **2001**, *51*, 292–299. [PubMed]
2. Russell, W.M.S.; Burch, R.L. *The Principles of Humane Experimental Technique*; Methuen: London, UK, 1959.

3. Pearce, A.; Richards, R.; Milz, S.; Schneider, E.; Pearce, S. Animal models for implant biomaterial research in bone: A review. *Eur. Cell Mater.* **2007**, *13*, 1–10. [CrossRef] [PubMed]

4. Zeiter, S.; Montavon, P.; Schneider, E.; Ito, K. Plate Stabilization With Bone Rivets: An Alternative Method for Internal Fixation of Fractures. *J. Orthop. Trauma* **2004**, *18*, 279–285. [CrossRef] [PubMed]

5. Tate, M.L.K.; Ritzman, T.F.; Schneider, E.; Knothe, U.R. Testing of a new one-stage bone-transport surgical procedure exploiting the periosteum for the repair of long-bone defects. *JBJS* **2007**, *89*, 307–316. [CrossRef]

6. Yamamuro, T.; Shikata, J.; Okumura, H.; Kitsugi, T.; Kakutani, Y.; Matsui, T.; Kokubo, T. Replacement of the lumbar vertebrae of sheep with ceramic prostheses. *J. Bone Jt. Surg. Br.* **1990**, *72*, 889–893. [CrossRef]

7. den Boer, F.C.; Patka, P.; Bakker, F.C.; Wippermann, B.W.; van Lingen, A.; Vink, G.Q.; Boshuizen, K.; Haarman, H.J.T.M. New segmental long bone defect model in sheep: Quantitative analysis of healing with dual energy X-ray absorptiometry. *J. Orthop. Res.* **1999**, *17*, 654–660. [CrossRef]

8. Boot, W.; D'Este, M.; Schmid, T.; Zeiter, S.; Richards, R.G.; Eglin Moriarty, T.F.D. Local antibiotic delivery from a thermoresponsive hyaluronan hydrogel successfully treats a chronic implant-related infection in a single stage revision in sheep. In *Orthopaedic Proceedings Volume 99-B, Issue Supp 22, Proceedings of the European Bone and Joint Infection Society (EBJIS), Nantes, France, 7–9 September 2017*; The Bone & Joint Journal: Nantes, France, 2017; p. 99.

9. Pobloth, A.M.; Checa, S.; Razi, H.; Petersen, A.; Weaver, J.C.; Schmidt-Bleek, K.; Windolf, M.; Tatai, A.; Roth, C.P.; Schaser, K.D.; et al. Mechanobiologically optimized 3D titanium-mesh scaffolds enhance bone regeneration in critical segmental defects in sheep. *Sci. Transl. Med.* **2018**, *10*, 423. [CrossRef]

10. Newman, E.; Turner, A.; Wark, J. The potential of sheep for the study of osteopenia: Current status and comparison with other animal models. *Bone* **1995**, *16*, S277–S284. [CrossRef]

11. Carbone, L.; Austin, J. Pain and laboratory animals: Publication practices for better data reproducibility and better animal welfare. *PLoS ONE* **2016**, *11*, e0155001. [CrossRef]

12. Flecknell, P. Analgesia from a veterinary perspective. *Br. J. Anaesth.* **2008**, *101*, 121–124. [CrossRef] [PubMed]

13. Huxley, J.; Whay, H. Current attitudes of cattle practitioners to pain and the use of analgesics in cattle. *Vet. Rec.* **2006**, *159*, 662–668. [CrossRef] [PubMed]

14. Hammond, R.; Christie, M.; Nicholson, A. *14—Opioid Analgesics, in Small Animal Clinical Pharmacology*, 2nd ed.; Maddison, J.E., Page, S.W., Church, D.B., Eds.; W.B. Saunders: Edinburgh, UK, 2008; pp. 309–329.

15. Janssen, P.A. A review of the chemical features associated with strong morphine-like activity. *BJA Br. J. Anaesth.* **1962**, *34*, 260–268. [CrossRef] [PubMed]

16. Dyson, D.H. Perioperative pain management in veterinary patients. *Vet. Clin. N. Am. Small Anim. Pract.* **2008**, *38*, 1309–1327. [CrossRef] [PubMed]

17. Liu, S.S.; Allen, H.W.; Olsson, G.L. Patient-controlled Epidural Analgesia with Bupivacaine and Fentanyl on Hospital Wards Prospective Experience with 1030 Surgical Patients. *Anesthesiol. J. Am. Soc. Anesthesiol.* **1998**, *88*, 688–695.

18. Twycross, R.G. *Palliative Care Formulary*; Radcliffe Publishing: Oxford, UK, 2002.

19. Vadivelu, N.; Mitra, S.; Narayan, D. Recent advances in postoperative pain management. *Yale J. Biol. Med.* **2010**, *83*, 11–25.

20. Christou, C.; Oliver, R.A.; Rawlinson, J.; Walsh, W.R. Transdermal fentanyl and its use in ovine surgery. *Res. Vet. Sci.* **2015**, *100*, 252–256. [CrossRef]

21. Holley, F.; Van Steennis, C. Postoperative analgesia with fentanyl: Pharmacokinetics and pharmacodynamics of constant-rate iv and transdermal delivery. *BJA Br. J. Anaesth.* **1988**, *60*, 608–613. [CrossRef]

22. Ahern, B.J.; Soma, L.R.; Boston, R.C.; Schaer, T.P. Comparison of the analgesic properties of transdermally administered fentanyl and intramuscularly administered buprenorphine during and following experimental orthopedic surgery in sheep. *Am. J. Vet. Res.* **2009**, *70*, 418–422. [CrossRef]

23. Robinson, T.M.; Kruse-Elliott, K.T.; Markel, M.D.; Pluhar, G.E.; Massa, K.; Bjorling, D.E. A comparison of transdermal fentanyl versus epidural morphine for analgesia in dogs undergoing major orthopedic surgery. *J. Am. Anim. Hosp. Assoc.* **1999**, *35*, 95–100. [CrossRef]

24. Mills, P.C.; Cross, S.E. Regional differences in transdermal penetration of fentanyl through equine skin. *Res. Vet. Sci.* **2007**, *82*, 252–256. [CrossRef]

25. Orsini, J.A.; Moate, P.J.; Kuersten, K.; Soma, L.R.; Boston, R.C. Pharmacokinetics of fentanyl delivered transdermally in healthy adult horses—Variability among horses and its clinical implications. *J. Vet. Pharm.* **2006**, *29*, 539–546. [CrossRef] [PubMed]

26. Pugh, D.G.; Baird, N. *Sheep & Goat Medicine*; Elsevier Health Sciences: Philadelphia, PA, USA, 2012.

27. Ivar Seldinger, S. Catheter replacement of the needle in percutaneous arteriography: A new technique. *Acta Radiol.* **2008**, *49* (Suppl. 434), 47–52. [CrossRef] [PubMed]

28. Grape, S.; Schug, S.A.; Lauer, S.; Schug, B.S. Formulations of fentanyl for the management of pain. *Drugs* **2010**, *70*, 57–72. [CrossRef] [PubMed]

29. Gilbert, D.B.; Motzel, S.L.; Das, S.R. Postoperative pain management using fentanyl patches in dogs. *Contemp. Top. Lab. Anim. Sci.* **2003**, *42*, 21–26.

30. Pettifer, G.R.; Hosgood, G. The effect of rectal temperature on perianesthetic serum concentrations of transdermally administered fentanyl in cats anesthetized with isoflurane. *Am. J. Vet. Res.* **2003**, *64*, 1557–1561. [CrossRef]

31. Riviere, J.E.; Papich, M.G. Potential and problems of developing transdermal patches for veterinary applications. *Adv. Drug Deliv. Rev.* **2001**, *50*, 175–203. [CrossRef]

32. Lyne, A.G.; Hollis, D.E. The skin of the sheep: A comparison of body regions. *Aust. J. Biol. Sci.* **1968**, *21*, 499–527. [CrossRef]

33. Bouclier, M.; Cavey, D.; Kail, N.; Hensby, C. Experimental models in skin pharmacology. *Pharmacol. Rev.* **1990**, *42*, 127–154.

34. Bartek, M.J.; LaBudde, J.A.; Maibach, H.I. Skin permeability in vivo: Comparison in rat, rabbit, pig and man. *J. Investig. Dermatol.* **1972**, *58*, 114–123. [CrossRef]

35. Jen, K.Y.; Dyson, M.C.; Lester, P.A.; Nemzek, J.A. Pharmacokinetics of a Transdermal Fentanyl Solution in Suffolk Sheep (*Ovis aries*). *J. Am. Assoc. Lab. Anim. Sci.* **2017**, *56*, 550–557.

36. Musk, G.C.; Catanchin, C.S.M.; Usuda, H.; Woodward, E.; Kemp, M.W. The uptake of transdermal fentanyl in a pregnant sheep model. *Vet. Anaesth. Analg.* **2017**, *44*, 1382–1390. [CrossRef] [PubMed]

37. Lane, M.E. The transdermal delivery of fentanyl. *Eur. J. Pharm. Biopharm.* **2013**, *84*, 449–455. [CrossRef] [PubMed]

38. Thiede, A.J.; Garcia, K.D.; Stolarik, D.F.; Ma, J.; Jenkins, G.J.; Nunamaker, E.A. Pharmacokinetics of sustained-release and transdermal buprenorphine in Göttingen minipigs (*Sus scrofa domestica*). *J. Am. Assoc. Lab. Anim. Sci.* **2014**, *53*, 692–699. [PubMed]

39. Marchionatti, E.; Lardé, H.; Steagall, P.V. Opioid-induced adverse effects in a Holstein calf. *Vet. Anaesth. Analg.* **2015**, *42*, 229–230. [CrossRef]

40. Grimm, K.A.; Lamont, L.A.; Tranquilli, W.J.; Greene, S.A.; Robertson, S.A. *Veterinary Anesthesia and Analgesia: The Fifth Edition of Lumb and Jones*; John Wiley & Sons: Hoboken, NJ, USA, 2015.

41. Linton, D.; Wilson, M.; Newbound, G.; Freise, K.; Clark, T. The effectiveness of a long-acting transdermal fentanyl solution compared to buprenorphine for the control of postoperative pain in dogs in a randomized, multicentered clinical study. *J. Vet. Pharmacol. Ther.* **2012**, *35*, 53–64. [CrossRef]

42. Otto, K.A.; Short, C.E. Pharmaceutical control of pain in large animals. *Appl. Anim. Behav. Sci.* **1998**, *59*, 157–169. [CrossRef]

43. Varvel, J.R.; Shafer, S.L.; Hwang, S.S.; Coen, P.A.; Stanski, D.R. Absorption characteristics of transdermally administered fentanyl. *Anesthesiol. J. Am. Soc. Anesthesiol.* **1989**, *70*, 928–934. [CrossRef]

44. Calis, K.A.; Kohler, D.R.; Corso, D.M. Transdermally administered fentanyl for pain management. *Clin. Pharm.* **1992**, *11*, 22–36.

45. Gourlay, G.K.; Kowalski, S.R.; Plummer, J.L.; Cousins, M.J.; Armstrong, P.J. Fentanyl blood concentration-analgesic response relationship in the treatment of postoperative pain. *Anesth. Analg.* **1988**, *67*, 329–337. [CrossRef]

46. Rohrbach, H.; Andersen, O.K.; Zeiter, S.; Wieling, R.; Spadavecchia, C. Repeated electrical stimulations as a tool to evoke temporal summation of nociceptive inputs in healthy, non-medicated experimental sheep. *Physiol. Behav.* **2015**, *142*, 85–89. [CrossRef]

47. Berset, C.M.; Lanker, U.; Zeiter, S. Survey on Sheep Usage in Biomedical Research. *Animals* **2020**, *10*, 1528. [CrossRef] [PubMed]

48. Solassol, I.; Caumette, L.; Bressolle, F.; Garcia, F.; Thézenas, S.; Astre, C.; Culine, S.; Coulouma, R.; Pinguet, F. Inter- and intra-individual variability in transdermal fentanyl absorption in cancer pain patients. *Oncol. Rep.* **2005**, *14*, 1029–1036. [CrossRef]

49. Suárez-Morales, M.; Mendoza-Popoca, C.Ú. Intravenous total anesthesia in neuroanesthesia: Influence of gender and gender/age on fentanyl consumption. *Rev. Mex. Anestesiol.* **2008**, *31*, 160–165.

50. Kest, B.; Sarton, E.; Dahan, A. Gender Differences in Opioid-mediated Analgesia Animal and Human Studies. *Anesthesiol. J. Am. Soc. Anesthesiol.* **2000**, *93*, 539–547.
51. Badillo-Martinez, D.; Kirchgessner, A.L.; Butler, P.D.; Bodnar, R. Monosodium glutamate and analgesia induced by morphine: Test-specific effects. *Neuropharmacology* **1984**, *23*, 1141–1149. [CrossRef]
52. Craft, R.; Stratmann, J.; Bartok, R.; Walpole, T.; King, S. Sex differences in development of morphine tolerance and dependence in the rat. *Psychopharmacology* **1999**, *143*, 1–7. [CrossRef]
53. Kasson, B.G.; George, R. Endocrine influences on the actions of morphine: IV. Effects of sex and strain. *Life Sci.* **1984**, *34*, 1627–1634. [CrossRef]
54. Bartok, R.E.; Craft, R.M. Sex differences in opioid antinociception. *J. Pharmacol. Exp. Ther.* **1997**, *282*, 769–778.

Article

Differential Transcription of Selected Cytokine and Neuroactive Ligand-receptor Genes in Peripheral Leukocytes from Calves in Response to Cautery Disbudding

Kavitha Kongara [1,*], Venkata Sayoji Rao Dukkipati [1,2], Hui Min Tai [3], Axel Heiser [1,4], Alan Murray [1], James Webster [5] and Craig Brian Johnson [1]

[1] School of Veterinary Science, Massey University, Palmerston North 4410, New Zealand; R.Dukkipati@massey.ac.nz (V.S.R.D.); axel.heiser@agresearch.co.nz (A.H.); a.murray@massey.ac.nz (A.M.); C.B.Johnson@massey.ac.nz (C.B.J.)
[2] School of Agriculture and Environment, Massey University, Palmerston North 4410, New Zealand
[3] Town and Country Vets, 257 Great South Road, Drury, Auckland 2113, New Zealand; hmin_93@hotmail.com
[4] AgResearch, Hopkirk Research Institute, Palmerston North 4410, New Zealand
[5] AgResearch, Ruakura, Hamilton 3214, New Zealand; jim.webster@agresearch.co.nz
* Correspondence: K.Kongara@massey.ac.nz; Tel.: +64-6-951-8235

Received: 13 May 2020; Accepted: 6 July 2020; Published: 14 July 2020

Simple Summary: Calf disbudding is a painful husbandry practice on dairy and beef cattle farms. Continuing efforts to enhance the accuracy of pain assessment can aid in the application of effective anti-nociceptive (analgesic) agents in non-verbal animals. The aim of this study was to evaluate the changes in the expression of genes involved in inflammation and pain sensitisation in response to removal of horn buds in calves, using hot-iron cauterization. The efficacy of an analgesic, meloxicam, was also tested in attenuating the changes in expression of the studied genes post-disbudding. It was revealed that cautery disbudding induces significant changes in the expression of genes involved in inflammation. Meloxicam was able to blunt the increased expression of some of the genes at 4 h and 24 h after disbudding, while it could not attenuate the increased expression of a few other genes associated with inflammation.

Abstract: Calf disbudding is a painful husbandry practice on dairy and beef cattle farms. An objective measurement of pain is useful to reliably evaluate the pain intensity and anti-nociceptive (analgesic) efficacy of therapeutic agents. The aim of this study was to investigate the changes in peripheral leucocyte inflammatory cytokine gene expression in calves after disbudding, and to assess whether the changes in cytokine gene expression could be an indicator of the efficacy of analgesic drugs. In a randomised controlled study, 16 calves (aged 31 to 41 days and weighing 58 to 73 kg), undergoing routine disbudding, were randomly allocated into two groups ($n = 8$ in each group). Calves in the control group received no analgesic, while those in the treatment group received 0.5 mg kg^{-1} meloxicam subcutaneously prior to disbudding. Disbudding was performed using an electric debudder. Blood (10 mL) was sampled from the jugular vein just before and 4 and 24 h post-disbudding, RNA was extracted from leukocytes, and the transcription of 12 genes of interest was assessed using nCounter gene expression assay. The results showed significantly higher transcription (compared to baseline values) of the studied genes (except *CRH*, *IFNγ*, and *IL10*) in the control group calves at either 4 or 24 h post-disbudding. The administration of meloxicam one hour before disbudding significantly attenuated the upregulation of *IL6*, *PGHS2*, *TAC1*, *NOS1*, and *CRH* gene transcription post-disbudding, while it did not suppress the elevated transcription of acute and pro-inflammatory cytokines such as *IL1β*, *IFNγ*, *IL8*, and *TNFα* genes. In conclusion, nCounter gene expression assay seems to be a promising tool to study the expression of cytokine genes and thus could be used for the pre-clinical evaluation of novel analgesics.

Keywords: calves; disbudding; gene transcription; cytokines; pain; meloxicam; nCounter

---

1. Introduction

Calf disbudding is a routine management practice on dairy and beef cattle farms. The veterinary medical associations in several countries recommend this procedure to be performed at an early age [1]. Since late 2019, the disbudding of calves in New Zealand requires the administration of a local anaesthetic [2]. However, it is still performed without the provision of analgesia in most countries. Pain is a subjective state, and it can only be measured indirectly in non-verbal animals [3]. Previous studies have used a variety of pain assessment methods following calf disbudding. These include changes in behaviour [4] and/or physiological variables such as plasma cortisol [4–7], substance P levels [8,9], heart rate, eye temperature [7,8,10], nociceptive thresholds [7,8], and performance measures such as activity and weight gain [11].

During the recent decade, several studies have focussed on evaluating the effect of local anaesthetics and nonsteroidal anti-inflammatory drugs (NSAID) in mitigating the acute pain and distress following the disbudding/dehorning of calves [3,9,11]. The findings of those studies varied regarding the effects of local anaesthetics and/or analgesics on measures of pain and welfare following disbudding in calves. While some studies demonstrated that pre-procedural administration of local anaesthetics and/or analgesics resulted in benefits such as acute pain relief, increased weight gains, and growth rates [3,11], a few other studies did not find any significant reduction of physiological and behavioural changes [8,9,12]. None of the pain assessment methods have been described as the gold standard for measuring the nociceptive blocking ability of analgesics.

The molecular mechanisms behind nociception and the resultant peripheral and central sensitisation are gradually being unravelled. Cytokines are small proteins secreted by different cell types at the site of injury and by systemic immune cells in response to injury [13]. They appear rapidly following injury due to active gene transcription and translation by the injured cells. They initiate the acute phase response and induce the production of other pro-inflammatory cytokines resulting in amplified response to injury. Increased expression of inflammatory cytokine genes has been demonstrated post-surgery in humans and rodents [14], while perioperative analgesia attenuated the rise in inflammatory cytokine levels and postoperative pain scores in humans [14]. In addition, a significant increase in the expression of inflammatory cytokine mRNA was evident in tissue samples from the testes, epididymis, and scrotum after the castration of calves [15]. Important pro-inflammatory cytokines include interleukin 1 beta (*IL1B*), interleukin 6 (*IL6*), interleukin 8 (*IL8*), tissue necrosis factor α (*TNFα*) and interferon gamma (*IFNγ*) [13], while interleukin 10 (*IL10*) is an anti-inflammatory cytokine, and its induction has been shown to attenuate systemic inflammatory response [16]. Apart from pro-inflammatory cytokines, several molecules involved neuroactive ligand–receptor interaction, such as the nitric oxide synthase 1 (*NOS1*), prostaglandin-endoperoxide synthase 2 (*PGHS2*), angiotensin II receptor type 2 (*AGTR2*), corticotropin-releasing hormone (*CRH*), nerve growth factor (*NGF*), and tachykinin precursor 1 (*TAC1*), have an important role in nociception pathways [17]. Very few studies looked at the expression of inflammatory cytokine genes in response to calf disbudding [18,19]. A precise understanding of the cytokine response to injury can be exploited therapeutically to improve animal welfare.

Quantitative reverse transcriptase polymerase chain reaction (RT-qPCR) is routinely used to investigate gene expression. However, the enzymatic reactions (reverse transcription and polymerization) used in this method could contribute to variability. nCounter gene expression assay enables the amplification-free multiplex detection of nucleic acids by the molecular barcoding of target molecules using a colour-coded probe pair [20].

The aim of this study was to investigate the changes in peripheral leucocyte inflammatory cytokine gene expression, using nCounter assay, in calves after disbudding, and to assess whether the changes in cytokine gene expression could be an indicator of the efficacy of analgesic drugs. It was hypothesised that the cautery disbudding of calves would induce significant changes in systemic inflammatory cytokine gene transcription, and that meloxicam (an NSAID analgesic) administered before disbudding would significantly attenuate the changes in cytokine gene expression.

2. Materials and Methods

2.1. Animals, Groups and Ethics Approval

This trial was undertaken on Holstein–Friesian Jersey crossbred calves born at the AgResearch farm, Tokanui, Hamilton, New Zealand, during the July to September calving season. The study protocol (#15/61) was approved by the Massey University Animal Ethics Committee, Palmerston North, New Zealand. Sixteen calves (aged 31 to 41 days and weighing 58 to 73 kg), undergoing routine disbudding, were randomly allocated into two groups (n = 8 in each group). Randomisation was performed by an online software QuickCalcs (GraphPad Software, San Diego, CA, USA). Calves in the control group received no analgesic before disbudding, and those in the treatment group received meloxicam prior to disbudding. During the study period, the calves were housed together indoors in a calf-rearing unit floored with wood shavings and maintained under normal farm practice.

2.2. Disbudding and Blood Sampling

Calves in the treatment group received 0.5 mg kg^{-1} meloxicam (Metacam 20 mg/mL for injection, Boehringer Ingelheim, NZ Ltd., Manukau, New Zealand) subcutaneously (SC) 1 h prior to disbudding. No analgesic was administered to the control group calves. Calves in both groups were sedated with 0.1 mg kg^{-1} Xylazine (Xylazine 2% injection, Phoenix Pharm Distributors, Auckland, New Zealand) intravenously 10 min before disbudding. Disbudding was performed by a single veterinarian in all calves by cautery, using an electric debudder (Shoof International Ltd., Cambridge, New Zealand). The contact time between the cautery iron and each horn bud was maintained for 12–15 s [11]. Blood (10 mL) was sampled from the jugular vein into heparinised vacutainers just before and 4 and 24 h post-disbudding and transported to the laboratory at ambient temperature for processing. The 4 and 24 h sampling time points post-disbudding were chosen with a view to study the kinetics of the early and late transcription of cytokine and neuroactive ligand-receptor genes. Since calves in the control group received no analgesic either prior to disbudding, meloxicam (0.5 mg kg^{-1} SC) was administered soon after the final sampling at 24 h post-disbudding to alleviate pain.

2.3. RNA Extraction and Purification

Two ml of heparinised blood contained in a 50 mL conical tube was mixed with 20 mL pre-warmed tris-buffered ammonium chloride (TAC) buffer (46 mM Tris-Cl, pH 8.1, and 1 mM CaCl$_2$), incubated at 37 °C for 10 min, and centrifuged at 350× g for 7 min at room temperature. The supernatant was discarded, and the pellet was re-suspended in one mL of RNeasy Lysis Buffer (Qiagen GmbH, Hilden, Germany). RNA in the lysates was purified using a QIAamp RNA Blood Mini Kit (Qiagen GmbH, Hilden, Germany), and its quantity and quality were assessed using a Nanodrop spectrophotometer (Thermo Fisher Scientific Inc., Waltham, MA, USA). RNA concentrations in the samples were adjusted to ≥25 ng/µL and stored at −80 °C.

2.4. Enumeration of Gene-Specific RNA

The detection and enumeration of RNA specific to 15 genes (12 of interest and 3 reference) was performed using nCounter gene expression assay [20]. Forty-eight RNA samples pertaining to the 16 animals, over three time points, were analysed using a custom-designed probe panel for 15 genes. Details of the genes, their Genebank accession numbers, and the position of target sequences are shown in Table 1. The simultaneous detection of mRNA of the 15 genes (Table 1) was undertaken using this assay as per the protocol outlined in the nCounter™ Gene Expression Assay Manual, v.20090807

(NanoString, Seattle, WA, USA). In brief, the gene-specific capture and reporter probes were hybridised to complementary target mRNA (around 100 ng) in solution by overnight incubation at 65 °C in a thermocycler, followed by washing off the excess probes and non-target transcripts in the solution by a two-step magnetic bead-based purification on the robotic nCounter™ Prep Station (NanoString, Seattle, WA, USA). Subsequently, the tripartite molecules were eluted and immobilised on cartridge for the enumeration of different colour-coded probe-mRNA hybrids, using a nCounter™ Digital Analyzer (NanoString, Seattle, WA, USA) with maximal sensitivity (555 fields of view, FOV).

Table 1. Details of genes analysed for transcription in peripheral leukocytes.

Gene	Gene Category	Genebank Accession	Target Sequence Position
IFNγ	Pro-inflammatory cytokine	NM_174086.1	503–602
IL1B	Pro-inflammatory cytokine	NM_174093.1	331–430
IL6	Pro-inflammatory cytokine	NM_173923.2	293–392
IL8 (CXCL8)	Pro-inflammatory cytokine	NM_173925.2	278–377
TNFα	Pro-inflammatory cytokine	NM_173966.2	950–1049
IL10	Anti-inflammatory cytokine	NM_174088.1	145–244
AGTR2	Neuroactive ligand-receptor	XM_001249373.2	1206–1305
CRH	Neuroactive ligand-receptor	NM_001013400.1	443–542
NGF	Neuroactive ligand-receptor	NM_001099362.1	558–657
NOS1	Neuroactive ligand-receptor	XM_867630.5	2657–2756
PGHS2	Neuroactive ligand-receptor	NM_174445.2	881–980
TAC1	Neuroactive ligand-receptor	NM_174193.1	317–416
GAPDH	Reference	NM_001034034.1	213–312
GUSB	Reference	NM_001083436.1	1815–1914
YWHAZ	Reference	NM_174814.2	147–246

2.5. Processing mRNA Expression Data

Tabulated data in comma-separated value (CSV) format, obtained from the nCounter™ Digital Analyzer as reporter code count (RCC) file, was input into nSolver™ Analysis Software, version 4.0 (https://www.nanostring.com/products/analysis-software/nsolver) and analysed as per the "nCounter Gene Expression Data Analysis Guide", MAN-C0011-04 (NanoString, Seattle, USA). The reporter library file containing the CodeSet information specific for the genes in this study was used to undertake the quality control routine. Default quality control (QC) settings were used: an Imaging QC (a measure of the percentage of requested fields of view successfully scanned in each cartridge lane) of <75; a Binding Density QC (a measure of reporter probe density in each cartridge lane) range of 0.1 to 2.25; a Positive Control Linearity QC (a measure of correlation between the counts observed for the positive control probes and the concentrations of the spike-in synthetic target nucleic acids) of <0.95; and a Positive Control Limit of Detection QC (indicates whether the counts for the positive control E probe and target sequence spiked in at 0.5 fM, assumed to be the system's limit of detection, are significantly above the counts of the negative control probes) of ≤2 SD above the mean of the negative controls. All samples (except a 24 h sample for an animal in the meloxicam group) passed the quality control. A background minimisation of counts was undertaken by subtracting the number of counts for the highest negative control +2 SD from all the mRNA counts. Subsequently, a positive control normalisation of RNA counts was then performed using the geometric mean of the 6 positive controls included in the nCounter assay. Finally, a biological normalisation of gene-specific RNA counts was undertaken based on the RNA counts of the chosen three mRNA reference genes (glyceraldehyde 3-phosphate dehydrogenase (*GAPDH*), glucuronidase beta (*GUSB*), and 14-3-3 protein zeta/delta (*YWHAZ*), and the final counts were exported into an Excel workbook.

2.6. Statistical Analyses

Differences in normalised mRNA counts with respect to the 12 genes of interest were tested using a mixed model analysis in SAS® 9.4 (SAS Institute Inc. Cary, NC, USA). The employed model included the fixed effects of group (control versus meloxicam), time (0, 4, and 24 h), and their interaction, and the random effect of animals. An autoregression 1 (AR1) model [21] was used to account for the covariance between the repeated measures within the individuals. In addition, to account for minor differences in the baseline (0 h) mRNA counts within each group, baseline values were considered as a covariate in the mixed model. The normality of residuals of data was checked by Shapiro–Wilk and Anderson–Darling tests using the CAPABILITY Procedure in SAS® 9.4 and the residuals were found to be normally distributed. Probability (*p*) values of ≤0.05 were considered statistically significant.

A priori power analysis could not be undertaken since there were no prior studies in cattle that used nCounter gene expression assay for pain-related cytokines. Hence, an indicative post-hoc power analysis was performed to estimate the power of detecting a significant difference between the two groups at a given time point as well as between two time points within a group. Power was estimated for two genes (IL8 and PGHS2) that exhibited significant differential expression between groups at 4 and 24 h post-disbudding, respectively. Power was calculated using G*Power software [22], assuming a simple t-test for differences between two independent means. The mean and standard deviation of mRNA counts, as well as sample size of the groups, were used as input values to estimate the realised power. The level of significance was set at 5%. Based on the observed effect size, the number of individuals required to achieve power values up to 1 were extrapolated for each of those two genes. Similarly, using the same software, the power of detecting a significant difference in mRNA counts (compared to baseline values) of *IL8* and *PGHS2* genes at 4 and 24 h, respectively, was also determined. A paired t-test for differences between two dependent means (matched pairs) was assumed. The observed mean and standard deviation of mRNA counts at the two time points as well as sample size were inputted, and a correlation of 0.5 between the readings at the two time points was assumed.

3. Results

Least square means ± standard errors (LSM ± SE) for normalised mRNA counts in peripheral leukocytes with respect to the 12 genes of interest in control and meloxicam-administered calves, prior to as well as at 4 and 24 h post-disbudding, are presented in Figure 1a,b. Significant differences in mRNA counts, between groups as well as between time points, were observed with regard to *IL8*, *IFNγ*, *IL1B*, *NOS1*, and *PGHS2* genes. The *IL8* gene exhibited significantly higher transcription (compared to baseline mRNA counts) at 4 and 24 h post-disbudding in both groups (Figure 1a). The *IL8* mRNA counts were particularly high in the control group calves at 4 h post-disbudding, which significantly differed from those in the meloxicam-administered calves. *IL1B* transcription was also significantly higher post-disbudding in both groups, with the mRNA counts at 4 and 24 h in the control group and at 4 h in the meloxicam group being significantly higher compared to the respective baseline values (Figure 1a). Furthermore, the *IL1B* mRNA counts at 4 h post-disbudding in the meloxicam group were significantly higher compared to those in the control group.

A slight but significant increase in *IFNγ* gene transcription, compared to time point 0, was evident in the meloxicam group at 4 and 24 h post-disbudding (Figure 1a). In addition, the mRNA counts for this gene at 24 h post-disbudding in the meloxicam group were significantly higher than those in the control group. The mRNA counts for *NOS1* and *PGHS2* genes were significantly higher in the control group at 24 h post-disbudding compared to their respective baseline values as well as the meloxicam-administered group (Figure 1b).

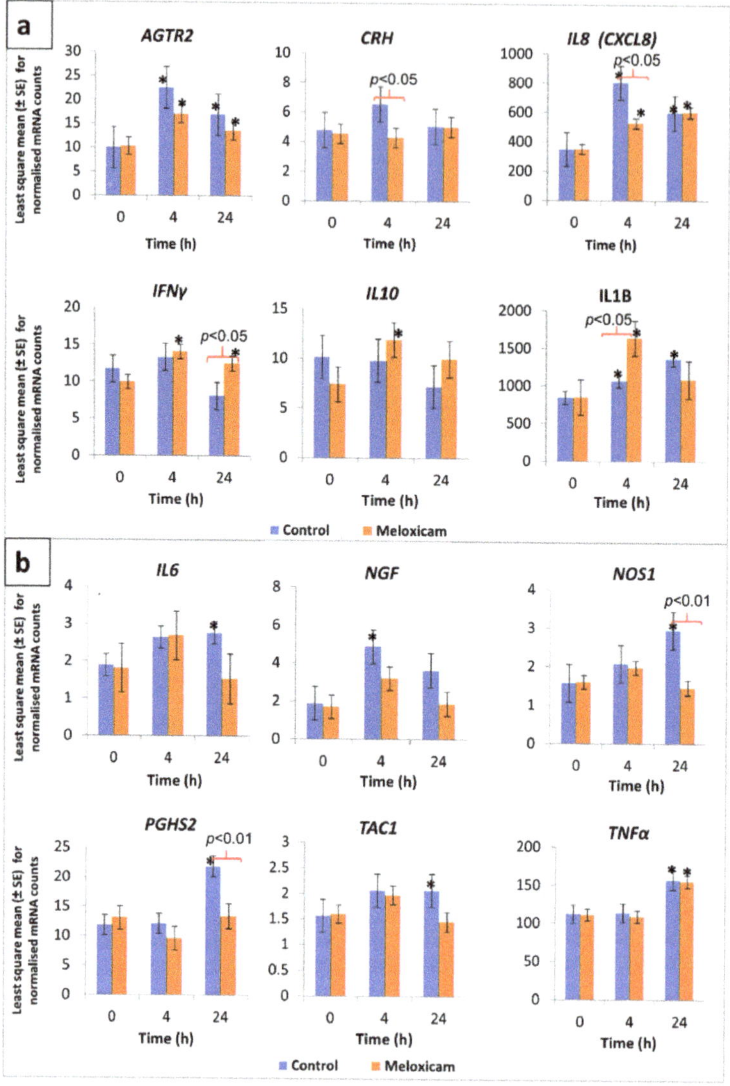

Figure 1. Transcription profile of selected cytokine and neuroactive ligand-receptor genes in peripheral leukocytes from calves, in response to cautery disbudding. *Two groups of calves (aged 31 to 41 days, n = 8 per group), that were either administered meloxicam or no analgesic (control), were disbudded using an electric debudder. mRNA counts in peripheral leukocytes, as determined by amplification-free nCounter gene expression assay (NanoString, Seattle, WA, USA), were normalised based on the RNA counts of three reference genes included in the assay. An asterisk (*) over the mean bars indicates a significant (p < 0.05) difference of transcription compared with respective pre-disbudding mean, within treatment, while significant differences between the treatment and control group means, within each time point, were denoted with braces and respective p-values on top of the bars. (a) Data pertaining to angiotensin II receptor type 2 (ATGR2), corticotropin-releasing hormone (CRH), interleukin 8 (IL8), interferon gamma (IFNγ), interleukin 10 (IL10), and interleukin 1 beta (IL1B) are shown in this graph. (b) Data pertaining to interleukin 6 (IL6), nerve growth factor (NGF), nitric oxide synthase 1 (NOS1), prostaglandin-endoperoxide synthase 2 (PGHS2), tachykinin precursor 1 (TAC1), and tissue necrosis factor α (TNFα) are shown in this graph.*

A significantly higher transcription, compared to respective group baseline mRNA counts, was observed in the case of *AGTR2* (in meloxicam as well as control groups at both 4 and 24 h post-disbudding), *IL10* (at 4 h in meloxicam group), *IL6* (at 24 h in control group), *NGF* (at 4 h in control group), *TAC1* (at 24 h in control group), and *TNFα* (at 24 h in both control and meloxicam groups) genes (Figure 1a,b). In the case of the *CRH* gene, although the mRNA counts were relatively low, a between-group difference was observed at 4 h post-disbudding (Figure 1a). The mRNA counts for this gene in the control group calves were significantly higher compared to the meloxicam-administered calves.

Post-hoc power analysis results indicated a power of 0.706 for detecting a significant difference in the *IL8* mRNA counts between the meloxicam and control groups at 4 h post-disbudding, and 0.922 in the case of *PGHS2* mRNA counts between the two groups at 24 h post-disbudding. A sample size of 10 and 16 animals per group would increase the power of detection to 0.8 and 0.9, respectively, in case of the *IL8* gene. Similarly, powers of 0.966 and 0.999 were realised in the case of detecting a within-group (control) difference (compared to baseline values) in *IL8* mRNA at 4 h and *PGHS2* mRNA at 24 h post-disbudding, respectively.

4. Discussion

Peripheral leukocyte inflammatory cytokine gene expression profiles were investigated in calves in response to cautery disbudding in this study. Both control and meloxicam groups showed significant changes in a variety of pro- and anti-inflammatory cytokine mRNA after disbudding. There was a significantly higher transcription of *TNFα* gene at 24 h post-disbudding in both groups, but no significant difference was detected at 4 h after disbudding, compared to respective baseline values. *TNFα* is mainly secreted by activated macrophages after tissue injury and plays a pivotal role in the initiation of acute phase response along with other early pro-inflammatory cytokines such as *IL1B*, *IL6*, and *IFNγ* [23]. Disbudding with hot iron involves the burning of a ring of tissue containing horn bud cells. The systemic inflammatory response following a burn injury in mice has been demonstrated to induce the release of a myriad of pro- and anti-inflammatory cytokines, with a peak of serum *TNFα* detected at 24 h and 48 h post-burn [24]. Only a few studies [18,19] have assessed the expression of cytokines and other inflammatory mediators in response to calf disbudding. In the present study, it is likely that the mRNA levels of *TNFα* have started rising in the window between 4 and 24 h and by 24 h after disbudding, a significant increase from the basal level was found. Nonsteroidal anti-inflammatory drugs have been shown to upregulate the production of *TNFα* from human peripheral blood leucocytes [25,26]. Results from the current study appear to support this as the *TNFα* mRNA of meloxicam group, similar to the control group, significantly increased from baseline by 24 h post injury.

IL1B, *TNFα*, and *IFNγ* have been postulated to be specific and early markers of inflammation and nociception. No significant differences in *IL1B*, *TNFα*, and *IL6* mRNA abundance between sham-handled and disbudded calves treated with a lidocaine block and IV meloxicam have been reported in one study [18]. In the present study, the mRNA transcripts of both *IL1B* and *INFγ* were found to be significantly elevated in the meloxicam group post-disbudding, compared to the respective baseline values and values of the control group at 4 h and 24 h. A previous study in calves reported a similar elevation of these early pro-inflammatory cytokine levels as early as 15 min after disbudding [19]. In the current study, it is likely that the mRNA of these pro-inflammatory cytokines had begun to rise at an earlier time point than 4 h, which was the first sampling time point after disbudding in our study. The finding that meloxicam did not attenuate the rise in early pro-inflammatory cytokine mRNA in this study appears to support the previous literature that the administration of only a systemic NSAID without local anaesthetic prior to disbudding/dehorning did not completely mitigate the acute phase response [9,19,27].

Calves in both groups showed a significantly elevated transcription of the *IL8* gene at 4 and 24 h post-disbudding (Figure 1a). *IL8* is a pro-inflammatory cytokine produced by a variety of tissues and blood cells in response to inflammation, and it exhibits chemotaxic activities against neutrophils and T lymphocytes, drawing these cells to the site of inflammation [28]. *IL8* has been shown to evoke hyperalgesia in rats by a prostaglandin-independent mechanism [29]. More recently, plasma *IL8* levels were found to be positively correlated with the intensity of burning mouth syndrome pain in humans [30]. In the current study, the significantly lowered transcription levels of this gene in the meloxicam group compared to those in the control group at 4 h post-disbudding potentially indicate the ability of meloxicam to reduce the pain caused by noxious stimuli via its anti-inflammatory effect. This is supported by the observed significantly higher transcription of the *IL10* gene in the meloxicam group at this time point (Figure 1a). In the current study, the significantly lowered transcription levels of this gene in the meloxicam group compared to those in the control group at 4 h post-disbudding potentially indicate the ability of meloxicam to reduce the pain caused by noxious stimuli via its anti-inflammatory effect. This is supported by the observed significantly higher transcription of *IL10* gene in the meloxicam group at this time point (Figure 1a). A similar negative regulatory effect of *IL10* on *IL8* expression in human monocytes has been documented [31]. It is interesting to see markedly elevated transcription levels of *IL8* in the leukocytes of calves in both groups prior to disbudding. This could be due to the age of the calves, as it has been found that *IL8* levels in healthy infant humans are significantly higher than those in adults [32]. The high *IL8* levels in infants have been attributed to be a major T cell effector function that has the potential to activate antimicrobial neutrophils and γδ T cells.

IL6 is a Janus-faced complex cytokine. It is one of the pro-inflammatory cytokines released early in the cascade [33] and induces the production of acute phase proteins. It also acts as an anti-inflammatory cytokine. *IL6* production has been shown to be significantly upregulated in skin cells close to a heat-induced injury site in rats at 24 h post-injury [34]. The anti-inflammatory effect of IL6 is thought to be mediated, in part, through the induction of prostaglandin E2 (PGE$_2$) synthesis, which in turn leads to the inhibition of *TNFα* and *IL1* receptors and subsequently their production [35,36]. In the present study, *IL6* mRNA counts did not significantly change in the meloxicam group, but the control group exhibited a significant increase at 24 h after disbudding. The precise time point at which the anti-inflammatory effect of *IL6* initiates in the course of inflammation relative to tissue injury is unknown. Treatment with NSAIDs for 3–6 days after injury in humans has been shown to result in lesser concentrations of serum *IL6* compared to a control group [37]. Thus, it appears that meloxicam had stabilized the *IL6* mRNA levels compared to the control group post-disbudding in the current study.

In the present study, the mRNA counts of *PGHS2, NOS1,* substance P *(TAC1), CRH,* and *NGF* showed no significant differences in transcription between pre-and post-disbudding time points in the meloxicam group. The control group showed a significant increase in the mRNA counts of these neuroactive ligand–receptor interaction molecules at 24 h post-disbudding. The *PGHS2* (aka cyclooxygenase 2, cox-2) gene encodes the inducible isozyme cox-2, which is predominantly involved in inflammatory prostanoid biosynthesis in response to injury [38]. Meloxicam is a preferential cox-2 inhibitor [39] and thus inhibits inflammatory prostanoids to produce its therapeutic effects in injured animals. Our finding of lower *PGHS2* mRNA counts in the meloxicam group could possibly reflect decreased ex vivo prostaglandin E2 synthesis in the plasma after cautery dehorning [40].

TAC1 (a precursor of Substance P, SP) is a neuropeptide member of the tachykinin family and is widely distributed in the central, peripheral, and enteric nervous systems. Neuronal SP is released from sensory neurons on noxious stimulation and activates several immune cells such as macrophages, mast cells, and T lymphocytes [41]. Sensory neuropeptide activated immune cells release inflammatory mediators such as histamine, arachidonic acid derivatives, and cytokines/chemokines. The synergistic interplay between SP and prostaglandins has been demonstrated in various inflammation and pain models [42]. Lower plasma SP concentrations compared with a control group have been reported in meloxicam-treated calves following scoop dehorning without local anaesthesia [9], which is in line with the lower mRNA counts of *TAC1* in the meloxicam group at 24 h after disbudding in our study.

Nitric oxide synthases (*NOS1*) catalyse the oxidation of the amino acid L-arginine to produce the free radical, nitric oxide (NO). Nitric oxide has been shown to activate the cox enzymes for prostanoid biosynthesis [43] and facilitate prostaglandin-induced hyperalgesia in rats [44]. The inducible isoform of NOS (iNOS) is released from macrophages in response to inflammatory stimuli [45]. Nonsteroidal anti-inflammatory drugs have been shown to inhibit the expression of iNOS mRNA in rat in vitro studies [45]. Likewise, the co-administration of meloxicam and iNOS inhibitor produced a synergistic anti-inflammatory effect in carrageenan-induced acute inflammation in rats [46]. The findings from the current study support those of previous studies demonstrating the efficacy of NSAIDs in inhibiting the NO synthesis, with significantly lower mRNA of *NOS1* in the meloxicam-treated calves compared to the control calves.

The *CRH* gene encodes a member of the corticotropin-releasing hormone family and is mainly expressed by the hypothalamic paraventricular nucleus and secreted into the hypophyseal portal system [47]. It plays a crucial role in eliciting the stress response through stimulation of the hypothalamic, pituitary, adrenal axis and the secretion of cortisol in response to noxious stimulation [48]. In the present study, post-disbudding *CRH* mRNA were significantly higher in the control group than the meloxicam group at 4 h, and no difference was found in both groups at 24 h after disbudding. Plasma cortisol concentration peaks within 30 min and returns to pre-treatment levels 6–8 h after cautery disbudding [6]. It is likely that more obvious changes in *CRH* mRNA have occurred earlier than 4 h post-disbudding, and only a slight increase could be detected in the control group at 4 h in the current study.

NGF, a member of the neurotrophin family, is essential for the development and maintenance of both central and peripheral nervous systems [49]. Rats injected with *NGF* into paws showed rapid and prolonged hypersensitivity to noxious thermal stimulation, confirming its role in inflammatory pain [50]. This inflammatory hyperalgesic effect of *NGF* is thought to be primarily mediated via tropomyosin receptor kinase A (TrkA) receptors [51]. In the current study, it is interesting to note that while there was a significantly higher transcription of NGF in the control group calves at 4 h post-disbudding, its transcription in meloxicam-administered calves at 4 and 24 h after disbudding remained relatively unchanged compared to baseline values.

Two recent studies [52,53] revealed the *AGTR2* to be a promising target for therapeutics aimed at treating neuropathic pain. The studies showed that damage to the peripheral nerve might lead to pain hypersensitivity as a result of signalling through *AGTR2* found on peripheral macrophages infiltrating the site of injury, rather than those on sensory neurons. The activation of *AGTR2* on peripheral macrophages triggers the release of reactive oxygen/nitrogen, which in turn activates the transient receptor potential ankyrin 1 ion channel, thus leading to nociceptive signaling in sensory neurons [52]. The current study also revealed the significantly higher transcription of this gene at 4 and 24 h post-disbudding in the calves of both groups (Figure 1a), indicating that meloxicam administration could not attenuate the transcription of this gene in the peripheral macrophages.

IL10 is a potent anti-inflammatory cytokine, reducing the expression of pro-inflammatory cytokines to balance the inflammatory response to injury [54]. A significant increase in *IL10* gene expression has been reported 12 h after the burdizzo castration of cattle [15]. The pre-emptive administration of an NSAID, carprofen, has been shown to have no effect on *IL10* mRNA expression [55]. Meloxicam has been shown to have no effect on the production of *IL10* from bovine lymphocytes [56] despite the involvement of cox-2 (target of meloxicam) in the production of *IL10* [57]. Similarly, in the current study, meloxicam did not affect the upregulation of IL10 mRNA. Taken together, it appears that NSAIDs such as meloxicam and carprofen do not inhibit cox-2 activity to the extent of suppression of different populations of immune cells that produce *IL10* [56].

In the current study, a representative *post-hoc* power analysis pertaining to the mRNA data of two genes, *IL8* (at 4 h) and *PGHS2* (at 24 h), revealed that there was adequate power (>0.9, except for between-group differences in the case of *IL8* at 4 h, which was 0.71) of detecting between-group and between time-point differences. In the case of *IL8*, a sample size of 10 per group would provide

0.8 power for between-group comparison. It is to be noted that these power analyses were based on assuming a simple t-test, but for the actual analysis of the study data, a more power linear mixed model, which would better account for correlation between repeated measures, was employed.

It has been shown in humans [58] that cytokine increases in cerebrospinal fluid (CSF) during peripheral surgery are more marked compared to those in blood, indicating the role of pro-inflammatory cytokines in increased central nervous system sensitivity to surgical pain. Hence, to check if a similar trend was evident in the current study, a 1–2 mL CSF was obtained from the atlanto-occipital joint of all calves prior to and 4 and 24 h post-disbudding. However, attempts to quantify the transcription of cytokine and neuroactive ligand-receptor genes (as well as housekeeping genes) in the RNA purified from those samples were unsuccessful due to the very low yield of RNA (<20 ng/sample) being inadequate for the employed nCounter gene expression assay. A volume of 8–10 mL of CSF might contain an adequate number of cells, yielding the required mRNA quantity for the nCounter gene expression assay.

nCounter gene expression assay (NanoString, Seattle, WA, USA) has been employed in this study to explore the amplification-free expression of peripheral leukocyte inflammatory cytokine genes. The transcription of genes, as actual mRNA counts in relation to the inbuilt positive controls as well as the three selected reference genes (*GAPDH, GUSB, and YWHAZ*) was successfully quantified. Although the expression of only 12 genes (plus 3 reference genes) was investigated in the current study, the expression of as many as 96 genes (of choice) in total could be simultaneously investigated using this assay. Using this technique, differential transcription of a few cytokine and neuroactive ligand-receptor genes was detected in the peripheral leucocytes of claves post-disbudding. Disbudding resulted in an increased transcription of pro-inflammatory cytokine genes (such as *IL1β, IFNγ, IL8,* and *TNFα*) in all calves (control as well as meloxicam groups), while meloxicam administration attenuated the upregulation of a few other genes (*IL6, PGHS2, TAC1, NOS1,* and *CRH*) involved in pain sensitisation pathways. These findings indicate that meloxicam alone would not be able to completely reduce the pro-inflammatory response following cautery disbudding. However, the nCounter assay seems to be an efficient tool to screen combinations of different analgesics for their ability to attenuate the pro-inflammatory response of a variety of genes. However, further studies need to be undertaken to validate the findings of this study. It would be interesting to see if this differential transcription of genes reflects in terms of actual protein levels, using MILLIPLEX® Cytokine/Chemokine panels (EMD Millipore Corporation, Billerica, MA, USA) that can simultaneously quantify up to 15 different cytokines. Similarly, corroboration of the differential expression of cytokines with either behaviour-based pain scores or other objective methods such as electroencephalogram variables would be useful.

5. Conclusions

Hot iron disbudding induced significant changes in the expression of a complex network of inflammatory cytokine mRNA in the peripheral blood leukocytes of calves. The subcutaneous administration of meloxicam one hour before disbudding significantly attenuated the upregulation of IL6, PGHS2, TAC1, NOS1, and CRH gene transcription post-disbudding. The mRNA expression levels of acute and specific pro-inflammatory cytokines such as IL1β, IFNγ, IL8, and TNFα were significantly increased after disbudding, and meloxicam did not suppress the elevation of these cytokine mRNA compared to the control group. The current study results indicate that the administration of only a systemic NSAID could not completely reduce the acute inflammatory response following disbudding. nCounter gene expression assay (NanoString, Seattle, WA, USA) was used in the current study, which is a promising tool to study the expression of cytokine genes and thus could be used for the pre-clinical evaluation of novel analgesics. However, it would be useful to corroborate these gene transcription results at protein levels in further studies.

Author Contributions: K.K., V.S.R.D. and C.B.J. conceived and supervised the study, K.K., V.S.R.D., C.B.J. and J.W. planned and carried out disbudding of calves and sample collection. V.S.R.D., H.M.T., A.H. and A.M. were involved in planning and execution of gene expression analysis. K.K. and V.S.R.D. prepared the paper drafts. All authors contributed to editing the article and approved the final manuscript.

Funding: This research received no external funding and was funded by the McGeorge research fund, Massey University, New Zealand.

Acknowledgments: We acknowledge the help of Alison Cullum, Veterinarian and Animal Welfare Officer, AgResearch, Ruakura, New Zealand, with disbudding the calves.

Conflicts of Interest: The authors declare no conflict of interest. The funders had no role in the design of the study; in the collection, analyses, or interpretation of data; in the writing of the manuscript, or in the decision to publish the results.

References

1. Welfare Implications of Dehorning and Disbudding Cattle. Available online: https://www.avma.org/resources-tools/literature-reviews/welfare-implications-dehorning-and-disbudding-cattle (accessed on 25 April 2020).
2. Animal Welfare (Care and Procedures) Regulations 2018 (LI 2018/50). Available online: http://www.legislation.govt.nz/regulation/public/2018/0050/latest/whole.html (accessed on 29 June 2020).
3. Heinrich, A.; Duffield, T.F.; Lissemore, K.D.; Millman, S.T. The effect of meloxicam on behavior and pain sensitivity of dairy calves following cautery dehorning with a local anesthetic. *J. Dairy Sci.* **2010**, *93*, 2450–2457. [CrossRef]
4. Faulkner, P.M.; Weary, D.M. Reducing pain after dehorning in dairy calves. *J. Dairy Sci.* **2000**, *83*, 2037–2041. [CrossRef]
5. McMeekan, C.M.; Stafford, K.J.; Mellor, D.J.; Bruce, R.A.; Ward, R.N.; Gregory, N.G. Effects of regional analgesia and/or a non-steroidal anti-inflammatory analgesic on the acute cortisol response to dehorning in calves. *Res. Vet. Sci.* **1998**, *64*, 147–150. [CrossRef]
6. Stafford, K.J.; Mellor, D.J. Dehorning and disbudding distress and its alleviation in calves. *Vet. J.* **2005**, *169*, 337–349. [CrossRef]
7. Stock, M.L.; Millman, S.T.; Barth, L.A.; Van Engen, N.K.; Hsu, W.H.; Wang, C.; Gehring, R.; Parsons, R.L.; Coetzee, J.F. The effects of firocoxib on cautery disbudding pain and stress responses in preweaned dairy calves. *J. Dairy Sci.* **2015**, *98*, 6058–6069. [CrossRef]
8. Stock, M.L.; Barth, L.A.; Van Engen, N.K.; Millman, S.T.; Gehring, R.; Wang, C.; Voris, E.A.; Wulf, L.W.; Labeur, L.; Hsu, W.H.; et al. Impact of carprofen administration on stress and nociception responses of calves to cautery dehorning. *J. Anim. Sci.* **2016**, *94*, 542–555. [CrossRef] [PubMed]
9. Coetzee, J.F.; Mosher, R.A.; KuKanich, B.; Gehring, R.; Robert, B.; Reinbold, J.B.; White, B.J. Pharmacokinetics and effect of intravenous meloxicam in weaned Holstein calves following scoop dehorning without local anesthesia. *BMC Vet. Res.* **2012**, *8*, 153. [CrossRef]
10. Stewart, M.; Stafford, K.J.; Dowling, S.K.; Schaefer, A.L.; Webster, J.R. Eye temperature and heart rate variability of calves disbudded with or without local anaesthetic. *Physiol. Behav.* **2008**, *93*, 789–797. [CrossRef]
11. Bates, A.J.; Eder, P.; Laven, R.A. Effect of analgesia and anti-inflammatory treatment on weight gain and milk intake of dairy calves after disbudding. *N. Z. Vet. J.* **2015**, *63*, 153–157. [CrossRef]
12. Van der Saag, D.; Lomax, S.; Windsor, P.A.; Taylo, C.; Whit, P.J. Evaluating treatments with topical anaesthetic and buccal meloxicam for pain and inflammation caused by amputation dehorning of calves. *PLoS ONE* **2018**, *13*, e0198808. [CrossRef]
13. Lin, E.; Calvano, S.E.; Lowry, S.F. Inflammatory cytokines and cell response in surgery. *Surgery* **2000**, *127*, 117–126. [CrossRef]
14. Beilin, B.; Bessler, H.; Mayburd, E.; Smirnov, G.; Dekel, A.; Yardeni, I.; Shavit, Y. Effects of preemptive analgesia on pain and cytokine production in the postoperative period. *Anesthesiology* **2003**, *98*, 151–155. [CrossRef]
15. Pang, W.; Earley, B.; Sweeney, T.; Gath, V.; Crowe, M.A. Temporal patterns of inflammatory gene expression in local tissues after banding or burdizzo castration in cattle. *BMC Vet. Res.* **2009**, *5*, 36. [CrossRef]

16. Van der Poll, T.; Marchant, A.; Buurman, W.A.; Berman, L.; Keogh, C.V.; Lazarus, D.D.; Nguyen, L.; Goldman, M.; Moldawer, L.L.; Lowry, S.F. Endogenous IL-10 protects mice from death during septic peritonitis. *J. Immunol.* **1995**, *155*, 397–401.

17. Julius, D.; Basbaum, A.I. Molecular mechanisms of nociception. *Nature* **2001**, *413*, 203–210. [CrossRef]

18. Mirra, A.; Spadavecchia, C.; Bruckmaier, R.; Gutzwiller, A.; Casoni, D. Acute pain and peripheral sensitization following cautery disbudding in 1- and 4-week-old calves. *Physiol. Behav.* **2018**, *184*, 248–260. [CrossRef]

19. Korkmaz, M.; Saritaş, Z.K.; Bülbül, A.; Demirkan, I. Effect of pre-emptive dexketoprofen trometamol on acute cortisol, inflammatory response and oxidative stress to hot-iron disbudding in calves. *Kafkas Univ. Vet. Fak. Derg.* **2015**, *21*, 563–568.

20. Geiss, G.K.; Bumgarner, R.E.; Birditt, B.; Dahl, T.; Dowidar, N.; Dunaway, D.L.; Fell, H.P.; Ferree, S.; George, R.D.; Grogan, T.; et al. Direct multiplexed measurement of gene expression with color-coded probe pairs. *Nat. Biotechnol.* **2008**, *26*, 317–325. [CrossRef]

21. Littell, R.C.; Henry, P.R.; Ammerman, C.B. Statistical analysis of repeated measures data using SAS procedures. *J. Anim. Sci.* **1998**, *76*, 1216–1231. [CrossRef]

22. Faul, F.; Erdfelder, E.; Lang, A.; Buchner, A. G*Power 3: A flexible statistical power analysis program for the social, behavioral, and biomedical sciences. *Behav. Res. Methods* **2007**, *39*, 175–191. [CrossRef] [PubMed]

23. Van Miert, A.S.J.P.A.M. Pro-inflammatory cytokines in a ruminant model: Pathophysiological, pharmacological, and therapeutic aspects. *Vet. Q.* **1995**, *17*, 41–50. [CrossRef] [PubMed]

24. Finnerty, C.C.; Przkora, R.; Herndon, D.N.; Jeschke, M.G. Cytokine expression profile over time in burned mice. *Cytokine* **2009**, *45*, 20–25. [CrossRef] [PubMed]

25. Tsuboi, I.; Tanaka, H.; Nakao, M.; Shichijo, S.; Itoh, K. Nonsteroidal anti-inflammatory drugs differentially regulate cytokine production in human lymphocytes: Up-regulation of TNF, IFN-y and IL-2, in contrast to down-regulation of IL-6 production. *Cytokine* **1995**, *7*, 372–379. [CrossRef] [PubMed]

26. Page, T.H.; Turner, J.J.; Brown, A.C.; Timms, E.M.; Inglis, J.J.; Brennan, F.M.; Foxwell, B.M.; Ray, K.P.; Feldmann, M. Nonsteroidal Anti-Inflammatory Drugs Increase TNF Production in Rheumatoid Synovial Membrane Cultures and Whole Blood. *J. Immunol.* **2010**, *185*, 3694–3701. [CrossRef]

27. Ballou, M.A.; Sutherland, M.A.; Brooks, T.A.; Hulbert, L.E.; Davis, B.L.; Cobb, C.J. Administration of anesthetic and analgesic prevent the suppression of many leukocyte responses following surgical castration and physical dehorning. *Vet. Immunol. Immunopathol.* **2013**, *151*, 285–293. [CrossRef]

28. Poznanski, S.M.; Lee, A.J.; Nham, T.; Lusty, E.; Larché, M.J.; Lee, D.A.; Ashkar, A.A. Combined stimulation with interleukin-18 and interleukin-12 potently induces interleukin-8 production by natural killer cells. *J. Innate Immun.* **2017**, *9*, 211–525. [CrossRef]

29. Cunha, F.Q.; Lorenzetti, B.B.; Poole, S.; Ferreira, S.H. Interleukin-8 as a mediator of sympathetic pain. *Br. J. Pharmacol.* **1991**, *104*, 765–767. [CrossRef]

30. Barry, A.; O'Halloran, K.D.; McKenna, J.P.; McCreary, C.; Downer, E.J. Plasma IL-8 signature correlates with pain and depressive symptomatology in patients with burning mouth syndrome: Results from a pilot study. *J. Oral Pathol. Med.* **2018**, *47*, 158–165. [CrossRef] [PubMed]

31. Méndez-Samperio, P.; García, E.; Vázquez, A.; Palma, J. Regulation of interleukin-8 by interleukin-10 and transforming growth factor beta in human monocytes infected with *Mycobacterium bovis*. *Clin. Diagn. Lab. Immunol.* **2002**, *9*, 802–807.

32. Gibbons, D.; Fleming, P.; Virasami, A.; Michel, M.L.; Sebire, N.J.; Costeloe, K.; Carr, R.; Klein, N.; Hayday, A. Interleukin-8 (CXCL8) production is a signatory T cell effector function of human newborn infants. *Nat. Med.* **2014**, *20*, 1206–1210. [CrossRef]

33. Hirano, T. Interleukin 6 and its Receptor: Ten Years Later. *Int. Rev. Immunol.* **1998**, *16*, 249–284. [CrossRef] [PubMed]

34. Summer, G.J.; Romero-Sandoval, E.A.; Bogen, O.; Dina, O.A.; Khasar, S.G.; Levine, J.D. Proinflammatory cytokines mediating burn-injury pain. *Pain* **2008**, *135*, 98–107. [CrossRef] [PubMed]

35. Kunkel, S.L.; Chensue, S.W.; Phan, S.H. Prostaglandins as endogenous mediators of interleukin 1 production. *J. Immunol.* **1986**, *136*, 186–192. [PubMed]

36. Kunkel, S.L.; Spengler, M.; May, M.A.; Spengler, R.D.; Larrick, J.; Remick, D.G. Prostaglandin E2 regulates macrophage-derived tumor necrosis factor gene expression. *J. Biol. Chem.* **1988**, *263*, 5380–5384.

37. Lisboa, F.A.; Bradley, M.J.; Hueman, M.T.; Schobel, S.A.; Gaucher, B.J.; Styrmisdottir, E.L.; Potter, B.K.; Forsberg, J.A.; Elster, E.A. Nonsteroidal anti-inflammatory drugs may affect cytokine response and benefit healing of combat-related extremity wounds. *Surgery* **2016**, *161*, 1164–1173. [CrossRef]

38. Furst, D.E. Meloxicam: Selective COX-2 inhibition in clinical practice. *Semin. Arthritis Rheum.* **1997**, *26*, 21–27. [CrossRef]

39. Engelhardt, G. Pharmacology of meloxicam, a new non-steroidal anti-inflammatory drug with an improved safety profile through preferential inhibition of COX-2. *Br. J. Rheumatol.* **1996**, *35* (Suppl. 1), 4–12. [CrossRef]

40. Allen, K.A.; Coetzee, J.F.; Edwards-Callaway, L.N.; Glynn, H.; Dockweiler, J.; KuKanich, B.; Lin, H.; Wang, C.; Fraccaro, E.; Jones, M.; et al. The effect of timing of oral meloxicam administration on physiological responses in calves after cautery dehorning with local anesthesia. *J. Dairy Sci.* **2013**, *96*, 5194–5205. [CrossRef]

41. Nicolette, M.; Teri, G.; Maccauro, G.; Tripodi, D.; Varvara, G.; Saggini, A.; Potalivo, G.; Castellani, M.L.; Fulcheri, M.; Rosati, M.; et al. Impact of neuropeptide substance P an inflammatory compound on arachidonic acid compound generation. *Int. J. Immunopathol. Pharmacol.* **2012**, *25*, 849–857. [CrossRef]

42. Zhang, J.M.; An, J. Cytokines, inflammation, and pain. *Int. Anesthesiol. Clin.* **2007**, *45*, 27–37. [CrossRef]

43. Salvemini, D.; Misko, T.P.; Masferrer, J.L.; Seibert, K.; Currie, M.G.; Needle-man, P. Nitric oxide activates cyclooxygenase enzymes. *Proc. Natl. Acad. Sci. USA* **1993**, *90*, 7240–7244. [CrossRef] [PubMed]

44. Aley, K.O.; McCarter, G.; Levine, J.D. Nitric Oxide Signaling in Pain and Nociceptor Sensitization in the Rat. *J. Neurosci.* **1998**, *18*, 7008–7014. [CrossRef] [PubMed]

45. Aeberhard, E.E.; Henderson, S.A.; Arabolos, N.S.; Griscavage, J.M.; Castro, F.E.; Barrett, C.T.; Ignarro, L.J. Non-steroidal anti-inflammatory drugs inhibit expression of inducible nitric oxide synthase gene. *Biochem. Biophys. Res. Commun.* **1995**, *208*, 1053–1059. [CrossRef] [PubMed]

46. Dudhgaonkar, S.P.; Tandon, S.K.; Bhat, A.S.; Jadhav, S.H.; Kumar, D. Synergistic anti-inflammatory interaction between meloxicam andaminoguanidine hydrochloride in carrageenan-inducedacute inflammation in rats. *Life Sci.* **2006**, *78*, 1044–1048. [CrossRef]

47. Zhou, J.N.; Fang, H. Transcriptional regulation of corticotropin-releasing hormone gene in stress response. *IBRO Rep.* **2018**, *5*, 137–146. [CrossRef] [PubMed]

48. Pacak, K.; Palkovit, M. Stressor Specificity of Central Neuroendocrine Responses: Implications for Stress-Related Disorders. *Endocr. Rev.* **2001**, *22*, 502–548. [CrossRef]

49. Farinas, I. Neurotrophin actions during the development of the peripheral nervous system. *Microsc. Res. Tech.* **1999**, *45*, 233–242. [CrossRef]

50. Lewin, G.R.; Ritter, A.M.; Mendell, L.M. Nerve growth factor-induced hyperalgesia in the neonatal and adult rat. *J. Neurosci.* **1993**, *13*, 2136–2148. [CrossRef]

51. McMahon, S.B.; Armanini, M.P.; Ling, L.H.; Phillips, H.S. Expression and coexpression of Trk receptors in subpopulations of adult primary sensory neurons projecting to identified peripheral targets. *Neuron* **1994**, *12*, 1161–1171. [CrossRef]

52. Shepherd, A.J.; Copits, B.A.; Mickle, A.D.; Karlsson, P.; Kadunganattil, S.; Haroutounian, S.; Tadinada, S.M.; de Kloet, A.D.; Valtcheva, M.V.; McIlvried, L.A.; et al. Angiotensin II Triggers Peripheral Macrophage-to-Sensory Neuron Redox Crosstalk to Elicit Pain. *J. Neurosci.* **2018**, *38*, 7032–7057. [CrossRef]

53. Shepherd, A.J.; Mickle, A.D.; Golden, J.P.; Mack, M.R.; Halabi, C.M.; de Kloet, A.D.; Samineni, V.K.; Kim, B.S.; Krause, E.G.; Gereau, R.W., IV; et al. Macrophage angiotensin II type 2 receptor triggers neuropathic pain. *Proc. Natl. Acad. Sci. USA* **2018**, *115*, E8057–E8066. [CrossRef] [PubMed]

54. Saraiva, M.; O'Garra, A. The regulation of IL-10 production by immune cells. *Nat. Rev. Immunol.* **2010**, *10*, 170–181. [CrossRef]

55. Pang, W.Y.; Earley, B.; Murray, M.; Sweeney, T.; Gath, V.; Crowe, M.A. Banding or Burdizzo castration and carprofen administration on peripheral leukocyte inflammatory cytokine transcripts. *Res. Vet. Sci.* **2011**, *90*, 127–132. [CrossRef]

56. Maślanka, T.; Jaroszewski, J.J. In vitro effects of meloxicam on the number, Foxp3 expression, production of selected cytokines, and apoptosis of bovine CD25+CD4+ and CD25-CD4+ cells. *J. Vet. Sci.* **2013**, *14*, 125–134. [CrossRef] [PubMed]

57. Harizi, H.; Juzan, M.; Pitard, V.; Moreau, J.F.; Gualde, N. Cyclooxygenase-2-issued prostaglandin E2enhances the production of endogenous IL-10, which down-regulates dendritic cell functions. *J. Immunol.* **2002**, *168*, 2255–2263. [CrossRef] [PubMed]

58. Bromander, S.; Anckarsäter, R.; Kristiansson, M.; Blennow, K.; Zetterberg, H.; Anckarsäter, H.; Wass, C.E. Changes in serum and cerebrospinal fluid cytokines in response to non-neurological surgery: An observational study. *J. Neuroinflammation* **2012**, *9*, 242. [CrossRef]

Review

Optimal Methods of Documenting Analgesic Efficacy in Neonatal Piglets Undergoing Castration

Meredith Sheil [1,*] **and Adam Polkinghorne** [2,3]

1   Animal Ethics Pty. Ltd., Yarra Glen, VIC 3775, Australia
2   Department of Microbiology and Infectious Diseases, NSW Health Pathology, Nepean Hospital, Penrith, NSW 2750, Australia; adam.polkinghorne@health.nsw.gov.au
3   Faculty of Medicine and Health, Nepean Clinical School, The University of Sydney Medical School, University of Sydney, Penrith, NSW 2750, Australia
*   Correspondence: mlksheil@me.com.au

Received: 27 July 2020; Accepted: 13 August 2020; Published: 19 August 2020

Simple Summary: Surgical castration in piglets is widely used in commercial pig production systems; however, it may cause pain and stress to the animal. There is an urgent need to develop effective pain-relieving medications to use for this procedure. Such products must meet high standards of proof confirming that they are effective. This requires undertaking trials to determine the duration and severity of pain that piglets experience during and after castration, and the extent of pain reduction in anaesthetic/analgesic treated piglets. Unfortunately, responses to pain may be transient, subtle, or variably expressed. Furthermore, there is no simple "gold standard" method to measure pain in neonatal piglets. Instead, researchers must rely on using a range of indirect measures of pain of varying reliability. Without understanding the nature of expression of piglet pain, and the reliability of test measures to detect it, there is the potential of misinterpreting trial outcomes. Although there is a high degree of variability in the literature of test methods employed and outcomes obtained, there is nevertheless a growing body of evidence to suggest that some piglet responses to pain induced by castration, are more consistently reproduced and specific to the pain experienced during castration than others. In this narrative review, we examine the potential indicators of pain in neonatal piglets undergoing castration to determine the optimal methods currently available to most accurately detect pain and assess pain mitigation.

Abstract: Analgesic products for piglet castration are critically needed. This requires extensive animal experimentation such as to meet regulatory-required proof of efficacy. At present, there are no validated methods of assessing pain in neonatal piglets. This poses challenges for investigators to optimize trial design and to meet ethical obligations to minimize the number of animals needed. Pain in neonatal piglets may be subtle, transient, and/or variably expressed and, in the absence of validated methods, investigators must rely on using a range of biochemical, physiological and behavioural variables, many of which appear to have very low (or unknown) sensitivity or specificity for documenting pain, or pain-relieving effects. A previous systematic review of this subject was hampered by the high degree of variability in the literature base both in terms of methods used to assess pain and pain mitigation, as well as in outcomes reported. In this setting we provide a narrative review to assist in determining the optimal methods currently available to detect piglet pain during castration and methods to mitigate castration-induced pain. In overview, the optimal outcome variables identified are nociceptive motor and vocal response scores during castration and quantitative sensory-threshold response testing and pain-associated behaviour scores following castration.

Keywords: piglet; castration; pain; behaviour; peri-operative; vocalisation; nociception; neonate; anaesthesia; analgesia

---

1. Introduction

A variety of animal husbandry procedures that cause pain to the animal are routinely employed in livestock species as a part of effective animal management systems. A primary example of such a procedure is castration, a technique that involves the removal of the testicles or the removal of testicular function [1]. In pigs, castration is employed in commercial swine facilities for several purposes, including improving meat flavour, preventing unwanted breeding and modifying animal behaviour. It is generally performed in the first week of life in male piglets intended to be kept past sexual maturity. Meat quality is improved by reducing the potential for 'boar taint', an unpleasant odour and flavour associated with the presence of androstenone (5α-androst-16-ene-3-one), produced in the testes of intact male pigs following sexual maturity [2]. Castration also reduces the risk of unwanted breeding that can interfere with the maintenance of genetic lines, and assists with management of boars by reducing the presence of aggressive behaviours that pose a welfare risk to other animals and also to the safety of humans interacting with them [3].

Traditional methods involve the use of surgical castration, a rapid (<1 min) method commonly performed by farmers in piglets between 2–7 days of age. The procedure involves restraining the piglet, incising the skin of the scrotum, extracting the testes, and severing the spermatic cords. Antiseptic is commonly sprayed onto the wound, and, less commonly, antibiotics are administered with the piglet finally returned to its sow. The wound is left to heal by secondary intention [1,4–6]. As an alternative to this procedure, there is growing interest in raising entire males, and/or the use of immunocastration by an anti-GnRH vaccine, which has shown to be effective in reducing boar taint and increasing growth performance in male pigs [7,8]. Whilst this review focusses on methods of assessing pain mitigation for surgical castration, the reader is referred to comprehensive review articles regarding surgical and non-surgical options and pig welfare [3,6,9].

Surgical procedures induce pain via a number of mechanisms [10]. The acute phase is primarily neurally mediated. Tissue incision causes trauma to keratinocytes and nerve fibres at the incision site, resulting in a barrage of nociceptive neural transmission from the damaged tissue to the central nervous system (nociception) inducing spinal reflexes such as the nociceptive withdrawal reflex, and, on reaching the cerebrum, the perception of acute pain and induction of the neuroendocrine response [11]. A second, sub-acute or prolonged inflammatory phase arises, primarily due to local release of various mediators in response to tissue damage, that promote ongoing pain or pain hypersensitivity against thermal, mechanical, and chemical stimuli [12,13]. Pro-nociceptive mediators such as ATP, glutamate, kinins, cytokines, tropic factors, and prostaglandins activate primary afferent neurons directly or indirectly to enhance nociceptive signal transmission to the central nervous system [14–17]. Prostaglandins derived from the arachidonic acid cascade are implicated in the production of inflammatory pain, and in sensitising nociceptors to the actions of other mediators. Bleeding and coagulation due to tissue injury are closely associated with the initiation of inflammation resulting in reflex erythema and acute pain responses. Kallikrein released during coagulation produces bradykinin, a strong allogenic factor [18]. Degranulation of activated mast cells results in the release of proteases, cytokines, serotonin, and histamine into the extracellular space. These substances sensitize primary afferent neurons to produce hyperalgesia [19]. Sensitization of peripheral and central neuronal structures amplifies and sustains postoperative pain [10,15,19].

Consistent with this, piglet castration is reported to cause pain and stress to the animal involving (i) discomfort and stress prior to the procedure due to handling and restraint; (ii) acute pain and stress during the procedure itself associated with incision of the scrotum, separation of the tissue to release each testicle, followed by severing of the spermatic cord; and (iii) post-operative pain and/or discomfort in the hours and days following the procedure [1,6]. Despite this, historically, castration has been typically performed without any pain relief, including in North America [20] and the EU [5]. In a detailed survey of 26 European countries, undertaken as part of the PIGCAS project (Attitudes, Practices and State-of-the-Art regarding Piglet Castration in Europe) in 2009, in the European Union [5] it was estimated that 79.3% of the about 98 million male pigs were castrated and analgesic use was reported as "very rare" or "never" in most EU member countries surveyed. Over the past decade, however, welfare concerns and ethical objectives have led to a drive to develop effective pain relief strategies for piglet castration, along with strategies to support the phasing out of the procedure where possible. In 2010, for example, the 'European Declaration on alternatives to surgical castration of pigs' was agreed, stipulating the intention that from 1 January 2012, surgical castration of pigs should only be performed with prolonged analgesia and/or anaesthesia. From 2018, the declaration stipulates that the surgical castration of pigs should be phased out altogether. This has seen progress with non-surgical alternatives along with exploration of a range of different anesthetic/analgesic options for piglet castration. These include the use of general anaesthesia (with CO_2, isoflurane, or injectable agents); the use of injectable local anaesthesia (such as lignocaine or procaine) administered by a combination of subcutaneous scrotal and intra-testicular (i.t.), or infundibular injection 5–15 min prior to the procedure; and/or the use of non-steroidal anti-inflammatory (NSAID) medications generally also administered 20 min prior to castration, via intra-muscular (i.m.) injection or oral administration [3,6]. An updated survey of 24 different European countries in 2016 [21] identified significant progress; however, concluded that the deadlines were far from being met. Whilst 6 of the countries had the practice of raising entire males, an average 80% of pigs continued to be surgically castrated in the remainder. The average percentage of piglets receiving immunocastration was 2.7%, 5% of the male pigs surgically castrated received anaesthesia and analgesia while 41% received analgesia alone [21] and 54% received no anaesthesia or analgesia. As analgesia alone ameliorates post-operative but not acute procedural pain, the development of practical and effective anaesthesia for the procedure was identified as an urgent priority.

The challenge faced by stakeholders in this field is to identify options that are effective in mitigating pain but are also safe, practical and economically sustainable for use in commercial swine facilities. Few medications are specifically approved for this use in piglets, and many must be used off-label under veterinary prescription [3]. General and local anaesthesia may be effective to provide pain relief during the procedure, but not after [22–25], and may require specialized equipment or veterinary administration, precluding practicality or commercial viability in many situations. Although some countries allow farmers to administer injectable local anaesthesia, this is not widespread [21]. Furthermore, injected sedatives or anaesthetics often require time to take effect resulting in negative welfare impacts, such as due to the pain of injection and/or the need for double handling. There may also be negative consequences if agents induce post-operative sedation due to interference with feeding and increased risk of crushing [3,6]. Although data on this is conflicting, NSAIDs may assist to mitigate post-operative inflammatory pain [26], however, they appear to offer little effective alleviation of pain during the procedure or in the early minutes and hours following the procedure [24–29] when pain is most acute. As NSAIDs take time to reach therapeutic effect, they commonly require administration well before castration, thus also resulting in negative welfare impacts due to pain of injection and the need for double handling of piglets. Hence, currently there is a critical therapeutic gap in availability of practical farmer-applied methods of delivering safe and effective peri-operative anaesthesia. Our group is investigating the use of a combination topical anaesthetic and antiseptic formulation, which may be farmer-applied during the procedure, (administered via intra-operative wound instillation), as a method to mitigate acute peri-operative pain in piglets. This has proven effective to alleviate castration pain in lambs and calves and is now widely used on farms in Australia [30,31]. Administered immediately following skin incision, the topical anesthetic formulation (containing 5% lidocaine, 0.5% bupivacaine, cetrimide, and 1:2000 adrenalin), can act rapidly, within 30 s, to anaesthetise the wound and the exposed cordal tissues prior to severing the spermatic cords [32], which is considered the most painful part of the procedure. The longer acting local anaesthetic bupivacaine is included in the formulation to assist in providing extended post-operative sensory analgesia [33]. Extensive animal experimentation, such as to confirm safety and efficacy, is required for regulatory approval and authorization for use in piglets. Prior to commencing such studies, we performed a review of methods of assessing analgesic efficacy in neonatal piglets to identify those most valid, sensitive, and specific for the assessment of pain and the efficacy of analgesic medications.

Proof of anaesthetic/analgesic efficacy is challenging in neonatal piglets. There is no one gold standard or validated measure of pain in piglets. Signs of pain in neonatal piglets can be subtle and variably expressed, and readily confounded by extraneous variables, particularly when required to be examined in the field setting (as opposed to in a laboratory) as is a standard requirement for regulatory approvals. Nevertheless, it is generally accepted that piglets react to stimuli in a number of ways including: physiologically, behaviorally, and through resistance movements and vocalization [1,6]. On this basis, a range of outcome variables have been used to assess piglet pain during and following castration, and to assess amelioration of pain due to use of local anaesthetics or analgesia. These include; (a) physiological responses during the procedure [22,25,28,29,34–58], (b) nociceptive motor responses during the procedure [28,29,32,48,50,55,59,60]; (c) vocal responses during the procedure [22,25,28,29,32,47,50,60–68]; (d) mechanical sensory testing in the minutes and hours following the procedure [32,33,45] and; (e) post-operative pain-related behaviours in the minutes and hours following the procedure [22–24,27–29,35,45,47,51,54,58,63,65–72]. More recently, newer technologies have been explored including (e) facial expression [45,65,71], and (f) infra-red thermography (IRT) [29,40,46,52]. Unfortunately, the methods used to examine analgesic efficacy in the reported literature have varied considerably between investigators, and the detail and quality of reporting has been highly variable, precluding the ability to make standardized assessments of the validity of each measure. As highlighted in previous reviews on this topic [26,73,74], this variation in the methods has impeded efforts to develop science-based guidelines for pain management protocols for castration.

To be valuable as indicators of pain mitigation, measures must be capable of consistently detecting a significant difference in pain-associated responses during and/or following castration as compared with pre-operative values, and/or as compared between castrated and non-castrated piglets. Secondly, variables must optimally be physiologically and/or clinically relevant to the evaluation of the type of pain being measured e.g., intraoperative pain or post-operative pain. Ideally, these measures (i) must be practically measured within the study without being confounded by the assessment of other variables; and; (ii) have the ability to be measured using an analytical method or measurement device/subjective assessment tool that has sufficient validation.

In the current review, we summarise literature on the currently available methods for assessing peri-operative pain in surgically castrated neonatal piglets and provide a critical analysis of the outcome variables identified to ascertain those that most closely meet these criteria. It is anticipated that this critical analysis may assist the future development of more standardized methods and optimise (reduce and refine) future analgesic efficacy trials in this field.

2. Physiological Measurements of Pain in Piglets

Physiological responses occur in response to pain and stress, including activation of the hypothalamus-pituitary-adrenal axis (HPA-axis) and sympathetic nervous system (SNS), and release of opiate neuropeptides. This acts to increase the metabolic rate in preparation for flight or fight as well as mediate the inflammatory response and mitigate pain. Adrenalcorticotrophic hormone (ACTH) is released by the pituitary and acts on the adrenal gland. Cortisol and adrenalin are released and, in turn, result in an increase in the level of glucose and lactate in the blood. Activation of the SNS may result in an increased heart rate and blood pressure and reduced skin temperature as blood is diverted to muscles and vital organs. β-endorphins (endogenous opioid-neuropeptides) are released from the anterior pituitary and act on opiate receptors in the peripheral and central nervous system to induce analgesia principally through effects on mu-opioid receptors. Indicators of the HPA axis and SNS activation, or β-endorphin release are thus often used as indirect measures of pain.

These physiological responses; however, are not specific to pain. They may be triggered by stress alone, and/or by tissue trauma (such as induced by surgical incision), even in the absence of pain. Surgical studies reveal that animals under a general anaesthetic increased cortisol and ACTH production, irrespective of the animal's sensation of pain [75,76]. Haemorrhage alone is known to result in an increase in ACTH, cortisol, β-endorphin concentration, as well as tissue content of pro-inflammatory cytokines; (including tumour necrosis factor-alpha (TNF-a) and interleukin-1alpha (IL-1a), IL-6 and IL10), and opiates have a proposed role in regulating the hemodynamic response to blood loss [77]. In a porcine model of abdominal surgery, for example, a standardized laparotomy without visceral involvement was performed on 24 anaesthetized pigs. Surgery gave rise to dramatic increases in plasma ACTH and cortisol ($p < 0.01$ and $p < 0.001$, respectively) within 15 min of incision, while animals were still under full general anaesthesia [75]. The activation of the HPA axis, and inflammatory cascade in response to surgical tissue trauma is generally termed the surgical stress response [78], and plays an important role in haemostasis and fluid homeostasis, immune defence, endogenous pain mitigation, and wound healing [76].

Similar to other surgical procedures, piglet castration results in an acute physiological response with activation of the HPA-axis and SNS, and opiate neuropeptide release. Prunier et al. [4] reported that castration of piglets induced significant ($p < 0.05$) increases in ACTH from 5 to 60 min, cortisol (from 15 to 90 min), and lactate (from 5 to 30 min) following the procedure, although no significant changes in blood glucose were observed. These authors hypothesised that glucose may not increase in neonatal piglets due to lack of glycogen stores. There is also a very rapid and transient increase in plasma adrenaline, followed by a longer lasting increase in plasma noradrenaline [4] as well as an increase in heart rate, blood pressure, and other signs of activation of the SNS such as reduced skin temperature that have also been reported [4,53,66]. Elevated β-endorphin levels have been reported in piglets castrated via cutting, but not via tearing the spermatic cord, despite equivalent rises in cortisol, as well as motor and vocal responses during the procedure [50]. This was hypothesised to be due to the increased risk of blood loss when cutting as opposed to tearing the cordal tissues.

Highlighting concerns over interpreting such physiological markers as being indicative of pain rather than in response to surgical tissue trauma, comparisons of anaesthetised and non-anaesthetised castrated piglets have found no significant difference in stress hormone responses [48,49]. Plasma cortisol, ACTH, and β-endorphins did not differ significantly between the anaesthetised and non-anesthetised castration groups indicating that tissue trauma (with inflammatory mediator release) and/or blood loss, rather than pain, is primarily responsible for the physiological HPA-activation and opiate neuropeptide response. Cortisol was reported as "not a sensitive tool to judge castration stress" in piglets castrated under general anaesthesia [49]. This indicates that variability in wound size, blood loss, and a piglet's neuroendocrine and immune response to wounding may all have a greater impact on cortisol levels than pain in piglets undergoing castration.

Furthermore, activation of the HPA axis and SNS may occur simply through handling and restraining piglets. Marchant-Forde et al. [50] reported that cortisol and β-endorphin levels were increased 45 min following the procedure in castrated piglets versus sham-handled controls ($p < 0.1$), however this was associated with a significant difference in the duration of handling and restraint, and was no longer evident when these factors were taken into account. Hay et al. [51] did not find differences in urinary levels of corticosteroids and catecholamines over the 4 days following surgical castration of piglets, as compared with sham-handled controls. This was considered most likely due to the short-lived activity of the adrenal and sympathetic axes [4]. Lonardi et al. [52] reported a short-lived increase in cortisol levels in castrated versus sham-handled animals at 20 min but not at 3–24 h following the procedure. Lactate and glucose levels were not significantly different between the two groups. Sutherland et al. [22] reported increased cortisol levels in castrated versus sham-handled piglets 30–120 min, but not 180 min or 24 h following procedure, however the study involved prolonged handling of piglets for blood collection and/or administration of anaesthetic treatments prior to castration, and the actual duration of restraint and handling was not documented for each piglet to allow group comparisons. Substance P (SP), however, was not significantly different between groups. SP is a neurotransmitter released directly from damaged nerve fibres at the site of tissue damage and is associated with increased pain perception and, hence, used as a biomarker of pain [79]. Other studies have reported that castrated piglets tended to have higher cortisol levels than sham-handled pigs, however this did not reach statistical significance at the $p < 0.05$ level [47,50]. Interestingly, where duration of restraint was controlled to be equivalent between groups, there were also no significant differences between castrated and sham-handled piglets in plasma levels of pro-inflammatory cytokines; TNF-α and interleukin-1beta (IL-1β), or on acute phase proteins C-reactive protein (CRP), serum amyloid A (SAA) and haptoglobin (Hp) and Moya et al. [54] concluded that pro-inflammatory cytokines and acute phase proteins did not provide relevant information on the physiological consequences of castration in neonatal piglets. Together, these data suggest that handling alone may induce a physiological response similar to that of castration in neonatal piglets. Despite the significant impact that the duration of restraint and handling may have on results, this variable is not always detailed in study reports or included as a variable in analyses.

Local anaesthetics and NSAIDS act to block pain via different mechanisms. This has important implications regarding interpreting the validity of biomarkers of HPA axis, neuroendocrine and/or inflammatory cascade activation as indicators of pain in this setting. NSAIDs mitigate pain via blockade of the conversion of arachidonic acid to prostaglandins by cyclooxygenase enzymes (COX), preventing activation of the inflammatory cascade and release of pain-inducing inflammatory mediators. Prostaglandins also directly stimulate ACTH and cortisol release. Separate from mitigating pain, NSAIDs thus also may directly mitigate the humoral aspect of the surgical stress response to tissue trauma [80,81]. A reduction in cortisol following NSAID administration, may be anticipated to indicate a collateral reduction in production of prostaglandins and other associated pain-inducing inflammatory mediators in piglets post castration, and hence also an associated decrease in pain. Hence, cortisol or ACTH levels may provide an indirect biomarker of pain in piglets following NSAID administration. This is not the case for local or general anaesthetics, however.

Local anaesthetics act by blocking nerve fibre conduction of pain signals. These prevent pain sensation via local or central nervous system effects, without primary effect on the humoral/inflammatory response to tissue trauma or associated HPA-axis activation. Biomarkers associated with the surgical stress response may thus be elevated, even although pain induced by them is blocked. Such variables are thus unlikely to be reliable indicators of pain in animals administered local or general anaesthesia. An additional confounding factor in the case of local anaesthetics is that, in many cases, these are administered in combination with adrenalin. This is to enhance local anaesthetic effects and minimize risks of systemic absorption. Adrenalin and nor-adrenalin, may have centrally and/or peripheral effects to stimulate corticotrophin releasing hormone and increase the breakdown of proopiomelanocortins into ACTH and β-endorphins [82–84]. Exogenously administered adrenalin may thus confound markers of endogenous HPA-axis and SNS activation and opiate-peptide production in castrated piglets.

In view of these factors, it is not surprising that studies investigating the impact of local anaesthesia or analgesia on physiological parameters in piglet castration have shown highly variable and, at times, apparently conflicting results (Table 1). The more consistent results are seen with the use of NSAIDs. Compared with piglets castrated without analgesic treatment, significantly reduced plasma cortisol and/or ACTH levels have been documented in NSAID-treated piglets at 30 min [35,38,41,45], 60 min [35,37,45,46], or up to 4 h post-procedure [35,37]. Others however, have reported no significant ($p < 0.05$) effect of NSAIDs administered prior to [25,36,39] or at the time of the procedure [22,42], on cortisol and/or ACTH, nor acute phase reactants, Hp, SAA, and/or CRP. Bates et al. [40] reported significantly greater amount of prostaglandin E_2 (PGE$_2$) inhibition at 10h, and from 30–100 h post castration in piglets which had nursed from meloxicam- as opposed to placebo treated sows prior to procedures. Cortisol and SP concentrations, however, were not significantly different ($p < 0.05$) between the two groups. O'Connor [74] and associates concluded a weak recommendation for use of NSAIDS for pain alleviation in piglets 1–24 h post-castration following a systematic review of available trial data, based principally on impact on cortisol. In the same review, NSAIDs were not found to have any impact on vocalisation to suggest an effect to mitigate procedural pain, which is discussed further below. Together, these data support the conclusion that some NSAIDs may have activity to reduce the inflammatory response and HPA-axis activation resulting from tissue trauma in piglets in the hours following castration, consistent with their known mechanism of action. Where cortisol and ACTH levels are reduced post castration (despite equivalent handling duration between treatment and control groups), this may be indicative of the efficacy of NSAIDs to mitigate post-operative inflammatory pain.

Table 1. Summary of studies investigating physiological (neuro-endocrine and inflammatory) responses during piglet castration.

Authors	Piglets N, Age	Castration Experimental Groups	Significant Findings
Prunier et al. [57]	18, 7–9 days	Castrated without analgesia/anesthesia (CAST), sham-handled (SHAM) or no handling	↑ ACTH; (5 to 60 min), cortisol (15 to 90 min), and lactate (5 to 30 min) in CAST animals. No effect on glucose.
Marchant-Forde et al. [50]	328, 2–3 days	CAST (cut or tear), SHAM	Blood sampling immediately before and at 45 min, 4 h, 48 h, 1- and 2-weeks post procedure. 45 min post-castration—↑ cortisol (trend) in CAST vs. SHAM piglets. And ↑β-endorphin (trend) in cut vs. tear and SHAM piglets. Significantly longer duration of procedure noted in CAST piglets vs. SHAM piglets, however.
Moya et al. [54]	40, 5 days	CAST, SHAM (controlled for time of restraint)	Blood sampling before (0 h) and 1, 2, 3, and 4 h after procedures (cortisol, TNF-a, and IL-1b) and before (0 h) and 12, 24, 48, and 72 h after procedures (CRP, SAA, and Hp). ↑cortisol trend only ($p < 0.1$) in CAST vs. SHAM and no statistically significant difference between groups (NSD) for TNF-a, IL-1b, CRP, SAA, or Hp.
Lonardi et al. [52]	32, 4 days	CAST, SHAM	Blood sampling 1 h before and at 20 min, 3, 5, and 24 h after procedures. ↑ cortisol in CAST vs. SHAM animals 20 min but not 3–24 h post-castration; ↓ lactate and glucose (SHAM and CAST) 3–24 h post-castration.
Carroll et al. [58]	90, 3, 6, 9, and 12 days	CAST, SHAM	Blood sampling before and at 0.5, 1, 1.5, 2, 24, and 48 h after castration. ↑cortisol for 0.5–2 h after procedure CAST > SHAM and ↑cortisol in older versus 3-day-old piglets.
Hay et al. [51]	84; 5 days	CAST; SHAM; (Animals previously tail-docked)	NSD between CAST vs. SHAM animals during 4 days of urinary measurements
Keita et al. [28]	90; mean 5 days	CAST; SHAM; NSAID (NSAID = Meloxicam (M) i.m. 10–30 min prior to castration).	30 min post castration—↑ cortisol in CAST and M versus SHAM. ↓ cortisol and ACTH in M vs. CAST group, (ACTH in M group similar to SHAM). NSD for Hp at 24 h.
Langhoff et al. [35]	245; 4–6 days	CAST, SHAM, NSAID (NSAID = M, flunixin (F), metamizole (MET), carprofen (C)), or saline i.m. 15–30 min prior)	Blood sampling before and at 30 min, 1, 4 and 24 h following procedures. ↑ cortisol in CAST piglets 1 and 4 h post castration; ↓ cortisol in all NSAID vs. CAST piglets; (↓ cortisol in M and F vs. CAST at 30 min, 1 h and 4 h; NSD vs. SHAM treated animals at 1 h).
Reiner et al. [36]	N/A	SHAM, NSAID (M or F)	↑ cortisol in NSAID vs. SHAM piglets 30 min post-castration

Table 1. *Cont.*

Authors	Piglets N, Age	Castration Experimental Groups	Significant Findings
Zöls et al. [37]	78; 4–6 days	CAST, SHAM, NSAID (M) i.m. prior	↑ cortisol in CAST vs. NSAID and SHAM piglets 1, 4 (but not 28) h post castration.
Schwab et al. [38]	130; <7 days	CAST, SHAM, NSAID (Ketoprofen, (K) i.m. 10–30 min prior)	30 min post-castration—↑cortisol and ACTH CAST > NSAID > SHAM piglets.
Wavreille et al. [39]	66; 5–6 days	CAST, SHAM, NSAID (Tolfenamic acid (T) or M)	NSD CAST vs. SHAM or M; ↑cortisol 30 min post-castration in T-pigs.
Bates et al. [40]	10 sows; 60 piglets; 5 days	CAST(M)-(piglets from M-treated sows), CAST(p)-(piglets from placebo-treated sows)	↑ PGE_2 inhibition, 10 h and 30–100 h post (castration and tail docking and iron injection) in CAST-M vs. CAST-p piglets. NSD between groups for plasma cortisol and SP. (Peak cortisol occurred 1 h post procedures).
Marsálek et al. [34]	36, 4 days	CAST, SHAM, local anaesthesia (LA) (LA = Lignocaine(L) with Noradrenalin (N-adr), administered i.t. 3 min prior)	↑ cortisol CAST and LA vs. SHAM at 1 h after castration. (L with N-adr did not modify cortisol concentrations).
Saller et al. [55]	54; 3–7 days	CAST, ±NaCl, L2%, Procaine (P) 4%, Bupivacaine (B) 0.5%, Mepivacaine 2% 20 min prior; SHAM (all with low flow isoflurane)	CAST: ↑ Heart rate + blood pressure. NSE on cortisol, Adr, Nor-Adr or chromogranin A. (LA did not modify cortisol or catecholamine concentrations, despite significant reduction in heart rate, blood pressure and nociceptive motor responses)
Zöls et al. [59]	124; 4–6 days	CAST, SHAM, LA (LA = p i.t. 15 min prior)	↑ cortisol in CAST and LA vs. SHAM piglets 1, 4 (but not 28) h post-castration. (P did not modify cortisol concentrations)
Courboulay et al. [41]	96	CAST, SHAM, NSAID (K), LA (L).	↑ cortisol at 30 min in L and CAST vs. K and SHAM.
Kluivers-Poodt et al. [25]	160, 3–5 days	CAST, SHAM, NSAID (M), LA(L), L and M (L-i.t.and s.c. M-i.m. administered 15mins prior)	Cortisol, lactate glucose and creatinine kinase (CK) measured before and 20 min following procedures. ↑ cortisol all grp vs. SHAM. ↓ cortisol L vs. CAST and M. NSD any treatment groups, for lactate, glucose or CK.
Hansson et al. [29]	564; 1–7 days	CAST; NSAID (M); LA (L with adr), LA and NSAID (Administration L with adr -i.t. 3–30 min prior, M-i.m. post castration).	Trend to reduced SAA in NSAID-treated piglets.

Table 1. *Cont.*

Authors	Piglets N, Age	Castration Experimental Groups	Significant Findings
Bonastre et al. [42]	120; 4–7 days	CAST, SHAM, SHAM and NSAID (M), CAST with M, CAST with LA(L), CAST with L and M, CAST with L and B, CAST with L, B and M (Administration; L and B i.t. 20 min prior; M i.m. immediately post-castration).	↑ cortisol (20 min) in all groups except SHAM and CAST with L+M; ↑ glucose (20 min) in all groups except SHAM and CAST and L.
Nyborg et al. [43]	NA	CAST, LA. (LA= L and B administered intrafunicularly (bilateral) and subcutaneously prior to castration)	↑ cFos protein (spinal cord) in CAST vs. LA piglets
Svendsen [44]	20	CAST, CAST and LA, CAST and CO_2/O_2 general anaesthesia (GA)	↑ cFos protein (spinal cord) in CAST vs. LA and GA piglets
Gottardo et al. [45]	196; 4 days	CAST; SHAM; NSAID (M, K or T); CAST with topical anaesthesia (TA) (TA = 2% or 6% topical tetracaine hydrochloride prior and applied to wound immediately post-procedure);	↑ cortisol and ACTH at 30 and 60 min in CAST vs. NSAID, TA and SHAM groups.
Sutherland et al. [47]	36; 3 days	CAST; SHAM; TA (tetracaine); TA (L, B and adr). (TA administered post-incision, to spermatic cords and skin edge immediately prior to castration).	Trend ($p = 0.06$) ↑cortisol in CAST and TA piglets 0.5–1 h post castration but not at 90–180 min; ↑cortisol ($p < 0.05$) in TA (L,B and adr) piglets between 30–180 min post-castration.
Sutherland et al. [22]	70; 3 days	CAST; SHAM; SHAM with NSAID, SHAM with GA(CO_2), CAST with NSAID, CAST with GA(CO_2), CAST with both (NSAID = F, i.m. immediately prior to procedure)	Blood sampled before, and 30, 60, 120, and 180 min, 24 h, and 3 d after castration for cortisol, Substance P (0–180 min) and CRP (24 h–3 days). ↑ cortisol (30 min) in all CAST grps vs. SHAM grps. ↑cortisol (60–120 min) in CAST and CAST with NSAID vs SHAM grp. ↑CRP in CAST(trend) and CAST with GA (CO_2) piglets. (↓CRP CAST with GA (CO2) vs. CAST piglets). ↑ SP in all piglet groups receiving GA (CO_2).
Walker et al. [48]	85; 2–12	CAST; CAST with GA (Isofluorane)	↑ cortisol, ACTH and β-endorphins in CAST animals; NSD between anaesthetised and non-anesthetised groups despite obvious behavioural differences.
Kohler et al. [49]	21–28 days	CAST, CAST + GA (CO_2/O_2), CAST + GA(Halothane)	↑ cortisol, ACTH, β-endorphin; NSD between groups despite obvious behavioural differences.

By contrast, as expected, the majority of studies have found little or no impact of either local or general anaesthesia on markers of the tissue trauma/inflammatory response to piglet castration and resulting activation of the HPA axis. Pre-emptive use of local anaesthesia via intra-testicular (i.t.) or infundibular injection, or via topical wound instillation, has been associated with reduced cortisol levels as compared with untreated animals in some trials [25,40,45], while not in others [25,47,55,56], or only where local anaesthetics and NSAIDs have been used in combination [20]. As detailed above, the lack of efficacy of local or general anaesthesia to reduce cortisol or ACTH does not, however, represent lack of efficacy to mitigate pain. These agents act via a different mechanism and mitigate pain via blockade of neural transmission. Neural markers of pain mitigation, such as the expression of the c-fos gene and its protein product, Fos, in neurons of the spinal cord [85], are significantly reduced when piglets are castrated under effective local or general anaesthesia, as compared with piglets castrated without anaesthesia [43,44]. Furthermore, this is associated with a dramatic reduction in the nociceptive physiological [4,53], motor and vocal response to castration [22,25,29,32,48,55,59,60,66,67]. Additionally, reduced post-operative hyperalgesia has been documented in local anaesthetic-treated piglets [32,33]. Together, these factors are considered to indicate that biomarkers of activation of the HPA axis, and inflammatory response lack specificity for pain mitigating effects of local and general anaesthetics, and are poor indicators of pain in piglets castrated under general or local anaesthesia [1]. They are similarly not suited to comparative efficacy trials with NSAIDs.

Based on this review, it is concluded that biomarkers of activation of the HPA axis, SNS, opiate neuropeptides and immune response, lack specificity as indicators of pain associated with neonatal piglet castration, and are confounded by the physiological response to restraint and to tissue trauma. They may provide some indication regarding the efficacy of NSAIDs to reduce post-operative inflammatory pain; however, are very poor markers of potential pain mitigating effects of local or general anaesthetics.

3. Nociceptive Motor Responses during Piglet Castration

Piglet castration without anaesthesia induces protracted violent struggling and escape behaviour in piglets during the procedure [48]. This piglet motor response is usually accompanied by a loud vocal response and is attributable to the nociceptive withdrawal response to acute pain induced during the procedure. It is referred to in the literature by a variety of terms including 'escape attempts' [50]; 'defense behaviour' [55,60] or; 'resistance movements' [29,59]. Measurement of the nociceptive motor response is typically conducted by use of a variety of methodologies [86] including (i) ordinal scales [60] (ii) focal assessments [28,55], (iii) a visual analogue scale (VAS) [29], or; (iv) the use of a numerical rating scale (NRS) [32,48]. Regardless of the methods used, analysis of the nociceptive motor responses of piglets consistently detects a marked and significant increase in castrated versus sham-handled animals, and successful mitigation of this response through use of general or local anaesthesia, indicative of sensitivity to detect pain mitigating effects (Table 2).

Table 2. Summary of studies measuring motor response movements during castration.

Authors	Piglets N, Age	Castration Experimental Groups	Method	Significant Findings ($p < 0.05$)
Marchant-Forde et al. [50]	32; 2–8 days	Castration without anaesthesia (CAST); Cutting or tearing spermatic cord; sham-handled animals (SHAM)	Number of escape attempts (sequential kicks) during procedure	↑ escape attempts CAST vs. sham groups; no significant difference (NSD) in response between castration method (cut versus tear)
Horn et al. [59]	36; 10–14 days	CAST, local anaesthesia (LA) (LA = Lignocaine (L) administered i.t. +/− intrafunicularly prior to castration)	Relative proportion of resistance movements	↑ resistance movements in CAST, particularly prominent during spermatic cord cutting. ↓ in L-treated group
Leidig et al. [60]	61; 3–4 days	CAST; SHAM; LA; (LA = L or Procaine (P) i.t. prior to castration)	Ordinal scale measuring duration and intensity	↑ scores in CAST animals; ↓ scores in SHAM, L and P-treated animals
Saller et al. [55]	54; 3–7 days	(All - GA with minimum alveolar concentration isoflurane): CAST +/- saline or LA (L, P, bupivacaine or mepivacaine); SHAM	Ordinal scale measuring defensive limb movements	↑ scores in CAST+saline animals; ↓ scores in SHAM and all LA treated animals
Sheil et al. [32]	40; 3–7 days	CAST; topical wound anesthetic (TA), applied by wound instillation 30 s prior to excising testes.	Numerical rating scale	↑ scores in CAST piglets with traction on each teste and cutting of each spermatic cord; significantly reduced in TA-treated group
Walker et al. [48]	85; 2–12	CAST; CAST under general anaesthesia (GA) (Isofluorane)	Numerical rating scale	↑ scores in CAST piglets with skin incision and testis excision; significantly reduced in GA group
Keita et al. [28]	90; mean 5 days	CAST; SHAM; NSAID (NSAID = Meloxicam (M) i.m. 10–30 min prior to castration).	"Global" behaviour score (GBS) calculated from presence or absence of: foreleg; or hind leg; or other body movements; urine or faeces emission; tremors.	GBS was similar in the meloxicam and placebo groups. There was a behavioural response (i.e., global score of 1 or more) in more than 95% of all piglets in the study during castration
Hansson et al. [29]	564; 1–7 days	CAST; LA (L + adrenalin); NSAID(M); LA + M (Administration L and adr -i.t. 3–30 min prior, M-i.m. post castration)	Visual analogue scale	↑ scores in CAST animals; ↓ scores in L- and LM-treated animals

Numerous studies have demonstrated that the piglet nociceptive motor response to castration is significantly increased in piglets undergoing castration as compared with sham-handled controls and/or following the application of effective local or general anaesthesia (Table 2). Marchant–Forde et al. [50] reported that castration triggered significant escape attempts in piglets undergoing castration compared to sham-handled controls. Focal sampling observations revealed that the piglet's nociceptive motor response often involved a sequence of sequential leg kicks in an attempt to escape, followed by a pause. Injectable anaesthesia (i.e., 2% Lignocaine) applied via intra-testicular or infundibular injection with an effective wait time has been shown to reduce the relative proportion of resistance movements from the entire period of fixation, including during the cutting of the spermatic cords, which elicits the greatest response and is considered to be the most painful step of the procedure [59]. A subsequent study investigating lignocaine effectiveness also confirmed less resistance movements during castration in piglets pre-injected with 10 mg/mL lignocaine into each testicle as compared to untreated animals. By contrast, pre-emptive i.m. administration of an NSAID did not result in a significant reduction in nociceptive motor response [28].

To investigate the efficacy of topical anaesthesia to mitigate piglet castration pain when instilled into the wound and allowed a 30 s wait time, our group recently employed a method in which piglet castration was recorded on videotape, and the nociceptive motor response was graded off-line by a blinded trained observer using an NRS (0–2, based on nil, partial, or vigorous full body response) including scoring at four specific time points during the surgical procedure (i.e., during traction of each testicle and severance of each spermatic cord). Piglets were settled at the time of commencing procedures. Nociceptive motor response scores were increased at all four time points in untreated piglets, and were also shown to be significantly reduced in animals treated with topical anaesthetic via wound instillation with 30 s dwell time [32]. Together, this literature is considered to indicate that assessment of nociceptive motor withdrawal response can provide a consistent, sensitive and repeatable method for documenting piglet pain responses during the castration procedure, and the efficacy of pain management strategies.

4. Vocal Responses during Piglet Castration

A review of the literature indicates that some changes in piglet vocalisation (i) can be detected during surgical castration, (ii) can be moderated with the use of anaesthesia and; (iii) are considered to be indicative of pain (Table 3). Although piglets commonly vocalise when they are handled, and particularly when restrained, the literature indicates that during castration piglets may squeal more often, more loudly and/or at a higher frequency than piglets that undergo sham-handling [1,25,29,47,50,60–65,68]. Castration is reported to produce changes in piglet vocalisation sound parameters that are comprehensively different to those detected from handling alone [67]. A wide range of parameters have been employed to measure piglet vocal response including measurement of; duration, energy or loudness (dB), peak frequency or pitch (Hz), or highest energy (Hz), vocalisation rate, and/or the percent of piglets that vocalised. Parameters that describe a single event in a call, such as peak level or peak frequency are considered to provide more consistent results than parameters that describe an average, such as weighted frequency and main frequency [67]. Most recently, specifically designed software (STREMODO; Stress Monitor and Documentation System, Forschungsinstitut für die Biologie Landwirtschaftlicher Nutztiere, Dummerstorf, Germany) [64,87] has been developed to detect stress vocalisations in piglets. This uses linear prediction analysis [88] to differentiate stress calls, non-stress calls, or background noise.

Table 3. Summary of studies measuring piglet vocal responses during castration.

Authors	Piglets Age, Number	Castration Experimental Groups	Measurement Method	Significant Findings ($p < 0.05$)
Wemelsfelder and van Putten [61]	4 weeks	Castration without anaesthesia (CAST); female litter mates	Calls highest in amplitude	Incising the scrotum did not result in a change in vocalisation, however pulling and cutting spermatic resulted in a marked ↑ in vocalisation
Puppe et al. [64]	19; 14 days	CAST	Rate of stress calls; STREMODO automated call monitoring system	↑ Stress calls (>1000 Hz) during surgical parts of castration procedure
Weary et al. [62]	102; 8–12 days	CAST; sham-castrated (SHAM)	Mean high (>1000 Hz) and low (<1000 Hz) calls	Significantly > high frequency calls in castrated vs. sham-handled piglets Greatest differences occurred during the severing of the spermatic cords and lesser differences when the scrotum was incised and the testicles extruded
Taylor and Weary. [68]	139; 7–10 days	CAST; SHAM	Mean high (>1000 Hz) and low (<1000 Hz) calls	Significantly > high frequency calls in castrated vs. sham-castrated piglets; pulling and severing produced highest call rate
Taylor et al. [63]	84; 3, 10, 17 days	CAST; SHAM	Mean high (>1000 Hz) and low (<1000 Hz) calls	Significantly > high frequency calls in castrated vs. sham-castrated piglets; No significant age effect observed on frequency of calls
Marchant-Forde et al. [50]	32; 2–8 days	CAST; (cutting or tearing spermatic cord); SHAM	Duration, mean frequency, and frequency of peak amplitude	Significantly > peak frequency of call in castrated piglets vs. sham-handled controls
White et al. [66]	172; 1–28 days	CAST; injectable lignocaine (L)	Frequency with highest decibel level (HEF)	Ligating cord produced ↑ HEF during castration; Significantly ↓ HEF in pigs treated with L
Marx et al. [67]	70; 7, 13, 19 days	CAST; L	12 variables	Calls classified into three types (screams, grunts squeals); 2 × number of screams in untreated castrates vs. treated
Leidig et al. [60]	61; 3–4 days	CAST; SHAM; L; Procaine (P)	STREMODO	CAST pain vocalisations significantly different from other treatment groups; no significant difference (NSD) between other groups
Kluivers-Poodt et al. [25]	160; 3–5 days	CAST; L; Meloxicam (M); L + M; SHAM	Temporal, waveform and spectral parameters	CAST piglets squealed longer and louder than piglets treated with L ± M; M-treated piglets similar to CAST
Keita et al. [28]	150; mean 5 days	CAST; M	Occurrence of vocalisation during castration recorded as 'cry', 'growl', or 'silence'.	Vocalisation (crying) during castration occurred in 149 of the 150 piglets in the study. NSE of M treatment
Hansson et al. [29]	564; 1–7 days	CAST; L; M; L and M	Calls highest in amplitude	L and (L and M) piglets produced calls with significantly lower intensity than CAST, and M-treated piglets
Sutherland et al. [47]	36; 3 days	CAST; SHAM; topical anesthetic (TA); NSAID	STREMODO	Significant difference between SHAM piglets and castrated piglets (with or without treatment)
Sheil et al. [32]	40; 3–7 days	CAST; TA (with 30 s wait)	Peak dB and area under the dB/time (waveform) curve (AUC)	Significant reduction in vocal responses in TA (with 30 s wait) vs. CAST piglets during traction ad severance of first spermatic cord.
Sutherland et al. [22]	70; 3 days	CAST; SHAM; NSAID; GA (CO2); NSAID and GA (CO2)	STREMODO frequency of stress vocalisations	Percentage of stress vocalisations was greater ($p < 0.05$) in CAST and NSAID piglets than all other treatments.
Viscardi and Turner [65]	60; 5 days	CAST; SHAM; Buprenorphine (BUP); SHAM with BUP	Spectrograms from video recordings. Maximum; frequency (Hz), amplitude (μ), power (dB); and energy (dB) of each call was determined comparing skin marking, i.m. injection, skin incision, and castration	i.m. injection and castration (pulling and severing the spermatic cord) induced vocalisations of ↑ frequency (Hz), power (dB), and amplitude (u) and/or energy, than skin incision, and/or spray marking/sham-handling—all groups. NSE of Buprenorphine treatment.

Studies have reported that piglets during castration produced more high-frequency calls (>1000 Hz), (referred to as screams [67]), than non-castrated controls. Pulling and severing of the spermatic cords lead to the greatest vocalisation response, greater than those normally emitted during handling and restraint as well as during the initial incision [67,68]. Vocalisation responses were also used to compare the castration procedure itself with cutting or tearing of the spermatic cord found to have little difference on the duration of responses [50]. Interestingly, intra-muscular injection of analgesics induces vocalisations of similar power (dB), frequency (Hz), and energy as that induced by pulling and tearing the spermatic cords during castration, and of significantly greater power (dB), frequency, (Hz) and energy than skin incision [65].

The majority of studies identify that local and general anaesthesia are effective in mitigating piglet vocal response to castration. Piglets castrated without local anaesthesia produce a higher number of screams with higher frequencies compared to piglets castrated with anaesthesia [29,60,66,67]. Hansson et al. [29] used a decibel meter during castration to record the highest vocal intensity level (dB) of piglets castrated with and without a local anaesthetic (lignocaine). Piglets castrated without the local anaesthetic produced calls of a significantly higher intensity than those administered lignocaine. Leidig et al. [60] summed the total duration of stress calls relative to the total time of the procedure, finding that duration of vocalisations of piglets receiving intra-testicular anaesthesia with injectable procaine was half of that emitted by piglets without anaesthesia. Animals that have received local anaesthetic injection to the testicle on one side vocalise less when the anesthetised testicle is removed than the non-anaesthetised testicle, although there was wide variability from animal to animal [89]. Trials examining the impact of NSAID administration at or prior to castration, however, have uniformly reported little to no impact on piglet vocal responses during castration [22,25,26,36,74] compared to piglets castrated without NSAID treatment.

Despite the overall consistency of reported outcomes, the actual metrics reported by authors are very diverse and reporting of measures of variation is poor, such that it is difficult to combine these data or quantify the effect of anaesthetic interventions on vocalisation [26,74]. A confounder to studies that rely on the quality of vocalisation responses to assess pain in piglets is that, in most cases, these findings have been recorded in rooms acoustically isolated from farrowing pens where piglet castration usually takes place. Since regulatory safety and efficacy trials require demonstration in the field situation, the sensitivity of pig vocalisation measurements and the consistency of results needs to be considered against the normal background noise levels, and confounding factors of a farrowing pen in a commercial farm setting. The presence of the sow and littermates can have confounding effects on piglet vocal responses. In view of these factors, it may be anticipated that analysis of vocal responses may not be as sensitive an indicator of pain in regulatory field trial settings as in acoustically separated research environments.

We recently developed a modified method for quantifying piglet vocal responses in the on-farm setting [32]. Piglet vocal response was recorded using a decibel meter as well as time-stamped videotape recording. Off-line analysis by a blinded technician allowed generation of standardised decibel/time waveform recordings for each piglet, on which the time of various specific procedural events were able to be marked. This allowed comparison of the peak (dB) and total auditory response (area under the dB/time waveform curve (AUC)) of each piglet, during specific procedural event-time periods (e.g., piglet vocal response during traction and severing of each cord). This provided consistency and specificity to the measurement period. Using this technique, we identified that both the peak dB and AUC recording were significantly reduced in piglets (n = 20) treated with topical anaesthesia instilled to the wound followed by a 30 s wait time, as compared with untreated piglets (n = 20) during traction and severing of the first cord. A trend effect was evident for traction and severing the second cord however statistical power was affected by increased variability. This finding was in contrast to a previous report [47] in which vocal responses in castrated piglets treated with topical anaesthetics or an NSAID were compared with untreated controls (n = 10 per group) using the STREMODO system. No measurable difference had been recorded between treatment and non-treated castrated groups in this trial. This may have been due to lack of sufficient dwell time allowed for efficacy of the topical anaesthetic agents employed, and/or insufficient power. More recently, we commissioned a further trial examining vocal response to castration following wound instillation of a topical anaesthetic formulation (with 30 s dwell time) (n = 44 per group) using peak dB and area under the dB/time waveform (as above) to compare vocal response to castration between treated and untreated piglets. With increased power, a significant reduction in vocal response (peak dB and AUC) to traction and severing of both the first and second spermatic cords was recorded. (Sheil, M; unpublished observations, manuscript in preparation).

In summary, it is considered that with careful application to ensure targeting of the measurement period to coincide with the time points of pain generation, and avoidance of confounding factors (particularly duration of restraint or recordings), measures of piglet vocalisation in response to castration including; the peak dB, total vocal response (such as area under the dB/time waveform), the frequency (Hz) of call with the highest intensity (dB (A)), rate of high frequency calls (>1000 Hz) or stress vocalisations using the STREMODO system, appear to provide a relatively consistent and sensitive method of assessing procedural pain associated with castration, and pain mitigation in neonatal piglets.

5. Post-Operative Pain-Related Behaviours

In general, measures of behaviour have proven to be more reliable indicators of pain than physiological measures in animals following castration [1,51]. In other animal species, behaviours such as decreased or abnormal locomotion, turning the head towards the rump, abnormal postures including prostration (standing or sitting with head below the shoulders), "hunching" (standing with kyphosis), "stiffness" (lying with legs tense and extended or walking with a stiff gait), increased or reduced movements of the tail are considered indicators of pain resulting from castration [30,31,90–92]. More diffuse and variable responses may occur in neonatal animals, however, due to immaturity of neuronal pathways involved with pain processing [93].

Behavioural disturbances have also been examined in neonatal piglets following castration. A review of the literature however reveals that in piglets, these behavioural changes may be subtle, transient and/or variably expressed, such that findings are not always reproducible. In some cases, contradictory results have also been reported (Tables 4 and 5). Behavioural assessments usually involve either direct quiet observation and scoring of piglet behaviours by trained blinded observers, or continuous time-lapse video-recording with off-line scoring either using event monitoring software or trained blinded observers. Assessments typically include observations of piglet; (i) posture (lying, standing, sitting, etc.), (ii) location (under heat, in contact with the sow or pen mates versus in isolation), and (iii) activities, including "non-specific" behaviours (sucking, sleeping, walking, playing, exploratory or aggressive behaviour, etc., which may be divided into "active" and "inactive" behaviours) and "pain-specific" behaviours. This latter category, first detailed by Hay et al. [51] based on pain-specific behaviours reported in other species, includes; "prostration" (standing or sitting with head down below shoulder height), "huddled up" (ventral lying with at least three legs tucked up), "tremors or trembling", "spasms" (localised muscle spasm), "stiffness" (lying with legs tense and extended), "tail wagging" and "scratching" (rubbing the rump along the floor or walls, also called "scooting"). Authors have additionally included standing in "hunched" posture (i.e., with kyphosis) or walking with a stiff or abnormal gait [23,45,52]. Observations may be made by "scan sampling" (i.e., recording the general posture, position, and behavioural activity of the piglet, with frequent repetition (e.g., every 1–10 min), over a predetermined time period (generally 2–3 h in the morning and afternoon of each assessment day), and/or by "focal assessment" (scoring the presence or absence of "pain-specific" behaviours at a number of predetermined time points). As incidences of individual pain-specific behaviours are low, aggregation of "pain-specific" behaviours is commonly employed to derive a "total" or "global" pain score for each piglet over specific time periods [28,45,51,54].

Table 4. Summary of behavioral studies in neonatal piglets following castration.

Authors	Piglets Number, Age,	Castration Experimental Groups	Measurement Method	Significant Findings (NSE = no Significant Effect)
McGlone and Hellman [72]	20; 14 days	CAST; sham-handled (SHAM); Lignocaine (L)	Time lapse video recording; 3 h pre- and 3 h post-castration. Event recorder monitored general postures, position and feeding behaviour	3 h post-op ↓ standing; ↑ lying (away from heat); ↓ nursing in CAST piglets (low magnitude, no effect on weight gain)
McGlone et al. [66]	100; 1, 5, 10, 15 and 20 days	CAST; SHAM	Time lapse video recording. 24 h post-op. A digital timing and data summary program [106] was used to measure the duration of each behavior Ethogram based on [72]	↓ standing and ↑ lying and ↓ nursing 6 h post-castration in CAST piglets (low magnitude), no effect on weight gain.
Carroll et al. [58]	90, 3–12 days	CAST, SHAM	Time-lapse video recording (WJ-HD500A, 3-min scan sample immediately after castration for 2 h. Observed for "active" (running walking), lying, lying under the heat, sitting, sitting under the heat, standing, standing under the heat, and nursing (mutually exclusive).	NSE on the time that pigs spent nursing, lying, standing, or sitting, trend ($p = 0.08$) for CAST to be less active than SHAM. Overall age effect ($p = 0.01$) on the time that pigs spent standing, such that 3-day-old pigs stood more than 6-, 9-, or 12-day-old pigs. No effect on weight gain.
Taylor et al. [63]	84; 3, 10, 17 days	CAST; SHAM	Time-lapse video recording; Scan sampling. Proportion of total behaviours scored at 10 min intervalsMonitored general postures, location nursing and active/inactive behaviours.	↑ standing or sitting and ↓ lying 0–2 h post-castration in CAST piglets; ↓ lying and ↑ nursing in next 22 h. No significant effect (NSE) position (all effects low magnitude no effect on weight gain)
Hay et al. [51]	84; 5 days	CAST; SHAM (Previously tail-docked)	Detailed ethogram: posture, location, non-specific and pain-specific activity/behaviours and social isolation/desynchronization. Direct observation. Scan sampling every 10 min immediately post-op and 2 h each morning and evening for 5 days	First 2.5 h; ↑ "pain-specific" behaviours (prostration, huddled up, stiffness and trembling), ↑tail wagging ↑isolation and desynchronization ↓ suckling/udder massage, ↑ awake inactive in CAST piglets. 2–4 days; ↑scratching, tail wagging. Through-out; ↑walking and huddled up. Low magnitude ↑ kneeling otherwise NSE on postures or weight gain
Moya et al. [54] Exp 1	20; 5–8 h post-op	CAST; SHAM	Direct observation, Scan sampling every 3 min for 3 h (5–8 h post op); ethogram based on [51]	↑ total "pain-specific" behaviours (↑huddling up); ↓ walking; ↑ udder massage/exploratory activity and scratching (NSE posture or position)
Moya et al. [54] Exp 2	20; 4 days	CAST; SHAM	Direct observation, Scan sampling every 3 min for 2 h each morning and evening for 4 days; ethogram based on [51]	↑ total "pain-specific" behaviours (↑huddled up; tremors; spasms) first 0–2.5 h; Later time points ↓ sitting and ↑ trend for isolation. (Tail-wagging not recorded)
Keita et al. [28]	150; mean 5 days	CAST; Meloxicam (M);	Direct observation, Focal assessment (presence/absence) of "pain-specific" behaviours" based on [51] (prostration, tremors (trembling), tail movements and isolation) at 30 min, 1, 2, 4, and 24 h post-castration;	Greater proportion showed total global pain score '0' in M vs. CAST at 2 and 4 h (NSE 30 min, 1 or 24 h)
Kluivers-Poodt et al. [27]	160; 3–5 days	CAST; SHAM; unhandled; Lignocaine (L); M; L and M. (Piglets not tail-docked)	Direct observation, Scan sampling; 12 min intervals for 3.5 h each morning and afternoon for 4.5 days; Ethogram based on [51], tail-wagging scored separately from other pain-specific behaviours	↑"pain-specific" behaviours (2–6 h), ↑ tail-wagging in L group (3 days). ↑ sleeping and inactive behaviours in all groups in first 2–6 h post-castration. NSE suckling behaviour

Table 4. *Cont.*

Authors	Piglets Number, Age,	Castration Experimental Groups	Measurement Method	Significant Findings (NSE = no Significant Effect)
Hansson et al. [29]	398; 1–7 days	CAST; L; M; (L and M)	Direct observation scan sampling; each 10 min for 70 min. Ethogram based on [51,54,61].	(L and M) group; ↓ total "pain-specific" behaviours (huddled up, stiffness, prostration, tremors/trembling, spasms, scratching) day 1 post castration.
Gottardo et al. [45]	196; 4 days	CAST; SHAM; 2% topical tetracaine hydrochloride (THCL); 6% THCL; M; ketoprofen (K); tolfenamic acid	Direct observation, scan sampling 1 min intervals for 0–30 min and 60–90 min post-castration; Ethogram based on [54]	↑ total "pain-specific" behaviour (tremors, scratching, hunching, tail-wagging) CAST group, ↑isolation CAST and THCL groups; ↑ standing inactive all groups except K and SHAM in first 30 min. NSE 60–90 min period
Sutherland et al. [47]	36; 3 days	CAST; SHAM; topical wound anesthetic (L,B and adr administered to wound during procedure)	Direct observation, 1 min scan sampling for 180 min post-castration; ethogram based on [54,106,107] incl. Lying with or without contact, suckling behavior, general postures and "pain-specific" behaviours (huddled up or scratching).	↑ lying without contact in the CAST group
Sutherland et al. [22]	70; 3 days	CAST; SHAM; General anaesthesia (GA)-(CO$_2$/O$_2$); NSAID	1 min scan sampling 0–30, 60–90 and 120–150 min post castration; ethogram as per [47]	↑ lying without contact; CAST first 30 min thereafter CAS and CO2 piglets spent more time lying without contact than other treatments. ↑ total "pain-specific" behavior (scratching, huddling, hunched), CAS + CO2, 0–30 min.
Viscardi et al. [71]	19; 5 days	CAST; (M and EMLA® cream), (M and Placebo cream), (saline and EMLA®), (saline and placebo), prior to surgical castration, tail docking and i.m. iron injection.	Video recording 1 h pre-; 0–8 h and 24 h post-castration; analysed 15 min per hour, ethogram based on [51], behaviours analyzed separately, and grouped into "active" and "inactive" categories	↑ inactive behaviors and ↑tail-wagging all groups first 6 h post castration and docking as compared with pre-castration and docking. ↑isolation in piglets castrated without treatment as compared with treatment groups. (NSE individual "pain-specific" behaviours other than tail wagging, however small sample size).
Viscardi and Turner [69]	120; 5 days	CAST; SHAM; M; K. (Piglets not tail-docked)	Video recording 1 h pre-; 0–8 h and 24 h post-castration; analysed 15 min per hour; ethogram adapted from [51] as above. Behaviours analyzed separately, and grouped into "active" "inactive" and "pain" categories. "Pain" included; trembling, stiffness, spasms, tail wagging, and rump scratching	All groups; At 0 h, ↑active behaviours; At 5 h,↑ suckling; At 7 h ↑sleep compared with pre-op. CAST, M, and K groups; At 2, 7 and 24 h post-castration ↑tail-wagging and "pain" behaviour,. (Note "pain" category included tail-wagging). (NSE scratching or other individual pain-specific behaviours)
Yun et al. [23]	143; 5 days	CAST; No castration (left in trolley) (NoC); M; L; GA (isoflurane and M)	Video recording, analysed 10 min/h, pre- (−1 h), 0, 1, 2 h and 24 and 36 h post-castration; ethogram based off [51,54,70] and others; behaviors analysed separately. (Scans were delayed if piglets were sleeping or feeding).	Comparing pre and post castration, behaviours different only during the first 10 min observation in both CAST and NoC piglets, but not different after 1 h. Comparing CAST versus NoC- at 0 h, ↑ prostration and ↓aggression and tail wagging in CAST. At 1 h ↑prostration and abnormal walking, otherwise NSE at any time points. M, L and GA piglets; 0–2 h ↑leg crossing vs. NoC, ↓abnormal walking and prostration M v CAST. At 2 h ↑ tail wagging GA vs. NoC. Otherwise NSE

Table 4. *Cont.*

Authors	Piglets Number, Age,	Castration Experimental Groups	Measurement Method	Significant Findings (NSE = no Significant Effect)
Viscardi and Turner [65]	60; 5 days	CAST; SHAM (+saline); CAST and buprenorphine; SHAM and buprenorphine (Piglets not tail docked)	Video recording 1 h pre-; 0–8 h and 24 h post-castration; ethogram based on [51] behaviors analyzed separately, and grouped into "active" and "inactive" and "pain" categories. "Pain" included: trembling, stiffness, spasms, tail wagging, and rump scratching	All groups; ↑sleeping and ↓ walking, standing and active behaviours 4–7 h as compared with 0 h. ↑ active behaviours Buprenorphine versus other groups 0–7 h. ↑ tail-wagging and "pain" behaviours 24 h post-castration, CAST versus SHAM group. NB: "pain" category included tail-wagging.
Burkemper et al. [24]	235; 3–7 days	CAST; Lignocaine spray (LS); oral M; LS and oral M	direct observation, scan sampling each 5 min for 5 h period for 3 days post op; total pain and 5 "pain-specific" behaviours based on [51] (tail wagging, trembling, huddled up, prostrated, scratching)	↑ total pain-specific behaviours max 0–1 h post castration. No significant difference observed in behaviour between treatment groups. (Trend for ↑pain-specific behaviour in LS group)
Langhoff et al. [35]	245; 4–6 days	CAST; M, flunixin (F), metamizole or carprofen, respectively, administered 15 to 30 min prior.	Post-surgical behaviour (0–60 min and 180–240 min after castration/handling)	Tail wagging, drooping the tail, and changing the position were reduced in M and F piglets

Using these methods, abnormalities of behaviour have been documented in the early minutes and hours after piglet castration, principally consisting of a low magnitude increase in "pain-specific" behaviours and/or isolation. Although the majority of these behaviours are short-lived (i.e., observed with the greatest frequency in the first 30 min to 1 h following castration), some particular behaviours such as increased tail wagging and/or scratching tend to develop later in the post-operative period and have been observed to be increased for up to 2–5 days post-procedure in some studies [51,65,67], although not in others [23,25]. Overall in review, when comparing castrated piglets with sham-handled controls, variation in general postures and non-specific behaviours have been marginal and/or conflicting, and are generally not considered reliable indicators of piglet pain [27,51,94].

Early studies identified a number of behaviours thought to be indicative of pain in piglets, including changes in posture, position, and nursing behaviour, with reduced standing and increased lying away from heat, and reduced nursing in the early hours (3–6 h) following the procedure as compared with uncastrated controls, effects that were ameliorated by use of lignocaine local anaesthesia prior to castration [70,72]. A subsequent study [63], however, reported differently, documenting decreased lying, increased sitting, and increased nursing in piglets post-castration as compared with uncastrated controls. In all cases, however, the authors reported that effects, although statistically significant, were marginal and/or of low magnitude. Hay et al. [51] introduced a detailed ethogram for behavioural assessment of piglets post-castration. This included recording a range of indices of piglet posture and position, as well as 'non-specific' behaviours (such as suckling, walking, running, sleeping, playing, exploring, aggression), "pain-specific" behaviours (detailed above) as well as "social cohesion" (isolation and desynchronization). Using this ethogram and scan sampling over 5 days, in a study of piglets 5 days of age (n = 84) following castration, increased "pain-specific" behaviours were documented involving greater incidences of prostration, stiffness, trembling, huddled-up posture and tail wagging as well as increased social isolation and de-synchronisation, during the first 2.5 h following castration in castrated versus sham-handled piglets. Scratching and tail wagging were increased at later time points and remained elevated for 2–4 days. There were no significant changes in other variables, and it was concluded that general postures changes and non-specific activities were not reliable indicators of pain in piglets post-castration [51]. A number of studies have used similar ethograms and/or assessment of "pain-specific" behaviours to investigate post-operative piglet pain since this time (Tables 4 and 5). These have reported changes in "pain-specific" behaviours and social isolation, generally detectable only during the earliest assessment periods up to 180 min following castration. A recent study examining shorter time intervals identified significant changes in "pain-specific" behaviours were only present over the first 30 min post-castration [45]. Most studies have reported minimal [51,70] or no significant effect on suckling, and all studies have reported no effects of castration on piglet weight gain when performed on neonatal piglets > 3 days of age (Table 4). Longer-term behavioural effects have been variably reported. Hay et al. [51] reported scratching was increased with maximum frequency from 24–48 h post-operatively, and tail wagging was increased for 4 days. Wemelsfelder and van Putten [61] also documented increased tail wagging in the days following castration in 4-week old piglets. However, piglets in both these trials had also undergone prior tail docking, and it was hypothesised that prolonged tail wagging could be related to exacerbation of tail stump hyperalgesia. Viscardi et al. [65,69] recorded a significant increase in tail wagging, peaking at 24 h in non-tail-docked piglets, with no significant difference in scratching behaviour. Others have reported no significant differences in scratching or tail wagging in castrated piglets as compared with non-castrated controls up to four days post-castration [25,54].

Pre-treatment with local anaesthetic or NSAID analgesic has been shown to result in significant differences in certain pain-related behaviour in treated piglets less than 2 weeks of age in some trials [45,47,72], but not others [24,27,69]. McGlone et al. [72] reported that although the changes in behaviour were only minor, piglets castrated without local anaesthetic were observed to display significantly reduced standing, increased lying, and reduced nursing behaviours compared to piglets administered lignocaine via injection prior to castration. Hansson et al. [29] documented reductions in total "pain-specific" behaviours in piglets administered both lignocaine and meloxicam (but not alone) prior to castration as compared with untreated piglets. Sutherland et al. [47] examined the behavioural responses of piglets after castration and found that untreated animals spent significantly more time lying without contact (isolation) compared with piglets given topical anaesthetic via wound instillation during the procedure. In contrast, an alternative study [27] reported that lignocaine injection prior to castration resulted in increased "pain-specific" behaviour in the first hours after castration as compared with sham or unhandled controls, or NSAID-treated piglets. This was predominantly due to a significant increase in tail-wagging and huddling up in the early hours after the procedure, with increased tail-wagging remaining evident over the first 3 days. It was hypothesised that either the effect of the lignocaine wore off so quickly that it had no post-operative analgesic effects or the sensation of the lignocaine wearing off may have resulted in increased tail-wagging in piglets. Yun et al. [23] also reported increased tail-wagging in the first 10 min post castration piglets in piglets castrated under lignocaine or general anaesthesia. In this case, however, tail-wagging was also similarly increased in non-castrated piglets but not in piglets castrated without anaesthesia or analgesia, as compared with pre-operative values. Increased tail wagging in the early hours following the procedure was also reported following post-operative use of lignocaine hydrochloride spray to the wound [24] as compared with untreated castrated piglets. It was hypothesised that the high proportion of alcohol in the product and/or its acidic pH may have contributed to afferent nerve sensitisation. Such effects if present, may be preventable using buffering agents [95,96] and/or formulations that do not include high alcohol concentrations. Increased tail wagging was not evident in the early hours following castrated in piglets treated via wound instillation with a combination topical wound anaesthetic and antiseptic formulation (containing lignocaine along with bupivacaine), as compared with untreated piglets or pre-operative values [47] (M. Sheil, unpublished observations, manuscript in preparation). In this situation, it could be hypothesised that bupivacaine provides longer-acting sensory nerve blockade that may mitigate any sensation as the shorter-acting lignocaine wears off. Using focal assessment and an amalgamated global "pain-specific" behaviour score, Keita et al. [28] documented reduced scores at 2 and 4 h post-castration between Meloxicam-treated piglets versus those without treatment, however, there were no significant effects at 30 min, 1 h, or 24 h. Little or no difference in pain-related behaviour was seen after castration performed with or without general anaesthesia [23,47]. This is not unexpected, as general anaesthetics act primarily to prevent pain perception at the cortical level, however, they have little impact on the local cytokine response to tissue trauma that induces afferent nerve sensitisation and the development of post-operative pain, as detailed above. Hence, post-surgical inflammatory pain develops as general anaesthetic effects wear off.

It is notable that the majority of studies that have identified changes in "pain-specific" behaviours in the early hours following castration have been performed using direct observation with scan sampling and/focal assessment as opposed to continuous video recording techniques. From a scientific perspective, continuous behavioural observation is generally considered the gold standard for pain evaluation in animals, as it allows detection of deviation in normal behaviour and is considered to have the sensitivity to detect subtle or short duration behaviours [97]. Performed using video recording and off-line analysis, it also avoids the potential for confounding by observer effects on animal behaviour, and other limitations of live observations, such as reduced number of duration and frequency behaviours observed. However, video-recording may be impaired by 2-dimensionality, parallax error, and shadowing. Furthermore, behaviours may be missed when animals are grouped, hidden, or off-screen, such as may occur frequently in a farrowing pen. Such factors may all contribute to reduce sensitivity of video-recording methods to the detection of subtle behavioural changes such as are seen in neonatal piglets in the early post-operative period. It is notable that no significant differences in "pain-specific" behaviours between castrated and sham-handled neonatal piglets were evident in the first 2 h following castration in trials using video-recording techniques [65,69] as opposed to those using direct observation [29,45,51,54]. Data from these trials suggest that video-recording techniques may have high sensitivity to detect tail-wagging, however, lower sensitivity to detect other "pain-specific" behaviours such as tremors, spasms, huddling up, prostration or stiffness in neonatal piglets. Although two trials [22,47] using direct observation methods also failed to detect significant differences in "pain-specific" behaviour in piglets post-castration as compared with sham-handled piglets these trials only examined a narrow range of "pain-specific" behaviours (scooting and huddling up) as compared with the full range detailed by Hay et al. [51] and involved relatively low piglet numbers per group. This suggests that the studies may have been underpowered, and/or that important pain-specific behaviours such as tremors/trembling, prostration, spasms, stiffness and tail-wagging may have been missed. There are limited validation studies on behavioural methodologies to detect piglet pain associated with castration, however, Hay et al. [51] compared 10 min scan samples to continuous sampling on pain behaviours associated with castration and reported no difference in results when utilizing a scan or continuous methodology. Additionally, Burkemper [24] has reported low inter-observer error following observer training for direct observation of pain-associated behaviours. New studies are underway [98], using video recording techniques with event monitoring software, and comparing continuous versus scan sampling at various intervals, to better understand the sensitivity and repeatability of this method. New or alternative methods of behavioural assessment such as examining gait, locomotor performance, and latency to move are also being explored [23,52].

On this basis, our group recently examined pre- and post-operative pain-related behaviour in castrated piglets 3–5 days of age with and without wound instillation of topical anaesthesia during the procedure, across two separate trial sites (M. Sheil, unpublished observations). Direct observation using trained blinded observers was used, with scan assessments of posture and position (including pain-specific postures and positions, such as prostration, huddled-up, hunched standing, stiffness and isolation) as well as behaviours (including "non-specific" and "pain-specific" behaviours) which were recorded every 10 min for 3 h in the morning and 2 h in the afternoon; pre-castration and over the first 36 h post-castration. In addition, focal assessments of "pain-specific" behaviours were separately made pre-castration and at 1, 15, 30, 60, 90 min and, 2, 4, 6, 8, 24, and 30 h post-castration. Our results accord with those of Gottardo et al. [45], who, using similar methods, reported increased total "pain-specific" behaviour evident predominantly in the first 30 min after castration, which was mitigated by pre-administration of analgesic medication or post-surgical topical anaesthetic medication. Additionally, using similar methods, Hansson et al. [29] reported reduced total "pain-specific" behaviours in the first 70 min period following castration in neonatal piglets administered both NSAID and local anaesthetic prior to castration. These results suggest that this method currently provides the most consistent, repeatable method of identifying acute post-operative pain, and documenting pain-mitigation in the early minutes and hours following castration in neonatal piglets. We did not find a difference in "pain-specific" behaviours between groups at later times, based on focal sampling, however, scores at later times were similar to pre-operative values. This is consistent with findings reported by Yun [23] and associates, who, reported increased pain-related behaviours in the first 10 min and to a lesser extent at 60–70 min following castration, but not at other time points measured over 24 h in castrated piglets as compared with non-castrated piglets and/or pre-operative values. Using scan and/or focal assessment methods, Keita et al. [28], Hansson et al. [29] and Burkemper et al. [24]. have previously reported relatively increased "pain-specific" behaviour at later time periods following castration in untreated as compared with analgesia/anaesthesia-treated piglets, however pre-operative baseline values were not reported in the piglets under study, nor were sham-handled groups included.

Interestingly, we observed that most piglets were sleeping (~55%) or suckling (~20%) during baseline (pre-operative) scan observations. A prominent increase in piglet sleeping was evident the afternoon following castration. A similar finding has been reported by Viscardi et al. [65,69] who similarly compared piglet behaviour pre- and post-castration. An increase in piglet sleeping has otherwise been infrequently reported as a post-operative behavioural disturbance in piglets although it is, however, a well-documented response to aversive stimulation in neonates [99,100] and neuroactive steroids such as allopregnanolone, and endogenous neuropeptides such as β-endorphin, released in response to stress, are known to have potent sedative properties [101–105]. The majority of previous trials have examined piglet behaviour comparing castrated with sham-handled animals, rather than using a piglet's pre-castration behaviour as its own control. As handling and restraint are aversive to piglets (resulting in a neuro-endocrine and opiate-neuropeptide stress response), increased sleeping following handling and restraint may be common to both castrated and sham-handled animals. This could explain a lack of difference in sleep between sham-handled and treatment groups in previous trials. Kluivers-Poodt et al. [27], for example, reported a large proportion (70–75%) of piglets sleeping during scan assessments the afternoon following castration or sham-handling, however, there were not significant differences between castrated and sham-handled piglets. Trends for increased lying, with reduced standing, walking, exploring etc., and/or reduced active behaviours following castration, where reported, (Tables 4 and 5) could all be consequent upon an increase in piglets sleeping following handling, rather than being indicative of post-castration pain. It is interesting to note that buprenorphine administration prior to handling or castration resulted in a significant reduction in inactive behaviours (including sleep) and increased active behaviours in the 8 h following castration or sham-handling in neonatal piglets [67]. Buprenorphine is reported to disrupt sleep and decrease adenosine concentrations in sleep-regulating brain regions of the Sprague Dawley rat, [101] such that it could be hypothesised to have similarly disrupted sleep following aversive stimulation in piglets. A sedative response to aversive stimulation in piglets, if present, could explain the relatively low proportion of piglets exhibiting "pain-specific" behaviours over the same period, and contribute to the challenges detecting pain (and determining the efficacy of pain mitigation strategies) using behavioural observation methods at these later time points. Increased tail-wagging and scratching are the most consistently reported behavioural disturbances evident during later time periods, particularly in docked piglets, however, scratching may not be seen to a significant extent for 24 h.

Table 5. Summary of statistically significant results ($p < 0.05$) from behavioral studies examining posture, position, activity and pain-related behaviours in neonatal piglets post-castration as compared with sham-handled piglets. (Arrows indicate statistical significance. NSE = No significant effect of treatment).

Authors		Posture				Position		Activity		Pain-Specific Behaviours	Weight Gain
Compared to Sham-Castrated Piglets < 2 Weeks of Age	Time of Post-Operative Assessment	Lying	Standing	Sitting	Isolation	Heat-Lamp/(Position in Crate)	Suckling/Nursing	Active/Inactive Behaviours			
McGlone and Hellman [72]	0–3 h	Minor ↑	Minor ↓	-	-	Minor ↓	Minor ↓	-		-	NSE
McGlone et al. [70]	0–6 h	Minor ↑	Minor ↓	-	-	Minor ↓	Minor ↓	-		-	NSE
Taylor et al. [63]	0–2 h	Minor ↓	Minor ↑	Minor ↑	NSE	NSE	NSE	-		-	-
	2–22 h	Minor ↓	NSE	NSE	NSE	NSE	Minor ↑				
Carroll et al. [58]	0–2 h	NSE	NSE	-	-	NSE	NSE	NSE		-	NSE
Hay et al. [51]	5 days	NSE	NSE	NSE	↑ 0–2.5 h	NSE	↓ 0–2.5 h	↑ awake inactive 0–2.5 h ↑ walking through-out		↑ total, ↑prostration, stiffness, trembling tail-wagging 0–2.5 h; ↑ huddled up ↑ scratching tail-wagging for 2–4 days	NSE
Moya et al. [54] Exp 1	5–8 h	NSE	NSE	NSE	↓	NSE	↑ Trend	↓ walking and ↑ exploratory behaviour		↑ total, ↑ huddled up ↑scratching	NSE
Moya et al. [54] Exp 2	4 days	NSE	NSE	↓ (through-out)	↑ Trend	NSE	NSE	Trend ↓ active behaviours		↑ total, ↑ huddled up, spasms, trembling (0–2.5 h)	NSE
Keita et al. [28]		-	-	-	-	-	-	-		↑ total (prostration, tremors/trembling, tail movements and isolation) 2 and 4 h post castration.	NSE
Kluivers-Poodt et al. [27]		NSE	NSE	NSE	NSE	NSE	NSE	↑sleeping and inactive behaviours (all groups) Day 1 p.m.		↑ total Day 1 p.m. (2–6 h) (huddled, stiffness, spasms, prostrated, tremors/trembling) ↑tail wagging Day 5 a.m only (Lidocaine group ↑tail wagging day 1–3, and 5 a.m).	NSE

Table 5. *Cont.*

Authors		Posture			Position		Activity		Pain-Specific Behaviours	Weight Gain
Compared to Sham-Castrated Piglets < 2 Weeks of Age	Time of Post-Operative Assessment	Lying	Standing	Sitting	Isolation	Heat-Lamp/(Position in Crate)	Suckling/Nursing	Active/Inactive Behaviours		
Gottardo et al. [45]		NSE	NSE	NSE	↑ 30 min post-op	NSE	NSE	↑ standing inactive (30 min post-op)	↑ total (tremors, hunching, scratching, tail-wagging) for 30 min post-castration	NSE
Sutherland et al. [47]		-	-	-	↑ 180 min post-op	-	NSE	NSE	NSE (limited range = huddled up or scratching)	NSE
Sutherland et al. [22]		-	NSE	NSE	↑ 30 min post-op	-	NSE	NSE	NSE (limited range = huddled up or scratching)	NSE
Viscardi et al. [71]	Day-1–24 h	NSE	NSE	NSE	↑0–7 h (versus Day 1)	-	NSE	↑ 0–7 h (versus Day-1)	↑ tail-wagging otherwise NSE (individual)	-
Viscardi and Turner [65]	Day-1–24 h	NSE (various time effects)	NSE (various time effects)	NSE	NSE	-	NSE	↑sleeping and lying ↓walking, standing and active behaviours 4–7 h as compared with 0 h all groups.	↑ total 24 h (↑tail-wagging)	NSE
Viscardi and Turner [69]	Day-1–24 h	Various time effects	Various time effects	NSE	↑ various	-	NSE	↑active behaviours 0 and 24 h as compared with various other times both groups	↑ total and tail-wagging 2, 7 and 24 h (Note total "pain" score predominantly increased due to increased tail-wagging)	-

It is concluded that the expression of pain in neonatal piglets is subtle and confounded by behavioural responses to handling stress. Pain assessment is confounded by the lack of a validated assessment method, which has resulted in variability in the methodological approach taken in trials to date, and in the reported results. This is concerning because of the potential to underestimate both the degree of pain experienced by neonatal piglets and the ameliorating effects of analgesic medicines. In review, direct observation of piglet behaviour, pre- and post-castration using frequent scan and/or focal assessment and an ethogram that includes and is targeted to the observation of known "pain-specific" postures, positions and behaviours, including; tremors/trembling, spasms, prostration, huddled up or hunched posture, stiffness, tail-wagging, scratching, and isolation currently appears to provide the optimal method to most consistently identify a difference in acute pain-induced behaviour between castrated and non-castrated piglets, and investigate the potential efficacy of analgesics or anaesthetic medicines in the acute post-operative period. Tail wagging and scratching are the most consistently reported behavioural anomalies at later time points and appear to be equally well documented via continuous recording with off-line analysis or direct observational methods. These variables may however indicate irritation or itch rather than pain, particularly if present in the absence of other pain indicators (such as hyperalgesia) and appear to be exacerbated in piglets that are tail-docked.

6. Mechanical Sensory Response Testing

Quantitative sensory testing is a long-established and validated method of assessing the efficacy of local anaesthesia and wound analgesia in laboratory research and clinical settings [108]. The flexion reflex, or nociceptive withdrawal reflex, is a reflex response to a nociceptive stimulus resulting in the withdrawal of a limb or body part from a painful stimulus, which may be abolished by effective local anaesthesia or analgesia. In the setting of tissue injury, the release of chemical mediators such as SP, prostaglandins, and bradykinin involved in the inflammatory response, increase sensitisation of neurons to nociceptive signals resulting in the development of hyperalgesia and a reduction in the threshold for the nociceptive reflex response [109]. Afferent nerve sensitisation resulting in hyperalgesia is considered the primary pathological mechanism underlying the development of post-operative inflammatory pain [10]. The threshold for eliciting the flexion reflex may be clearly measured, including in rats [110], and pigs [111] and used to assess the development of hyperalgesia and the efficacy of anaesthetic or analgesic interventions. The reflex is evoked by stimulation of small calibre Aδ or C fibre primary afferents which transmit noxious information. The absence of the reflex response and/or a measurable change in the reflex threshold may be detected using a variety of stimuli including needlestick, heat pads, calibrated or electronic Von Frey Filaments and/or pressure algometry.

Von Frey filaments or 'hairs' are a set of calibrated filaments that bend when a certain pressure is reached, allowing a reproducible mechanical stimulus to be delivered, graduating from that inducing a light-touch sensation through to a pain-weighted stimulation of skin or tissues. Electronic von Frey anaesthesiometers are also available. Using an electronic von Frey anaesthesiometer, Herskin and Rasmussen [112] have described thresholds of mechanical nociception in the pelvic limb of pigs, using four categories of behavioural response (from slight leg movements to kicking) to detect and grade the threshold response. In addition to laboratory studies in humans, pigs, and experimental animals, modified techniques have been developed for use "in the field" for assessment of pain and pain-alleviation in association with surgical husbandry wounds in livestock species. Applied to skin in proximity to a wound at time points before and after surgery, an animals response to a fixed light touch and pain-weighted Von Frey filament stimuli can be graded (via NRS) from a nil response (0) through to; a local twitch (1), or partial (2) or full body (3) nociceptive withdrawal response. The development of hyperalgesia lowers the threshold for a response, resulting in a greater response score to application of the same stimulus. This method has provided a sensitive, consistent and repeatable method of documenting the development of post-operative wound hyperalgesia and assessing the efficacy of topical or local anaesthetic-induced wound anaesthesia/analgesia in a range of livestock species following surgical husbandry procedures, including mulesing, tail docking and/or castration in lambs [31,113,114], castration and dehorning in calves [30,114]. Using this technique, a heightened nociceptive motor response to stimulation of a surgical husbandry wound has been documented in the minutes and hours following the procedure, in lambs, calves, and piglets, as compared with sham-handled animals, and/or with pre-operative assessments, indicative of the development of post-operative hyperalgesia. Pre-operative use of injected local anaesthetic (lignocaine) and/or immediate post-operative use of topical local anaesthetic (Tri-Solfen®) applied to the wound has resulted in a significant reduction in nociceptive withdrawal responses evident within 1–3 min of application, and continuing in the minutes and hours following the procedure, indicative of significant wound anaesthesia or hypoaesthesia [31–33,113–116]. Where present, this has been associated with evidence of reduced post-operative pain-related behaviour in treated animals over the same period.

In pigs, this method has been shown to elicit similar and measurable responses to those reported in human studies, and is sensitive to the effects of local anaesthetic agents [111] (Table 6). Von Frey filaments have been employed in studies to assess the efficacy of pain mitigation in piglets following surgical castration [32,33]. Wound sensitivity testing involved the use of von Frey monofilaments of weights 4 g and 300 g and an 18-gauge needle to stimulate the wound and surrounding skin at predetermined sites prior to treatment and then at defined periods of time afterward. Involuntary nociceptive motor responses were scored using an NRS as above. Topical anaesthesia using a lignocaine, bupivacaine adrenalin combination formulation was found to provide rapid wound anaesthesia and subsequent effective wound analgesia, with treated pigs displaying significantly reduced responses compared to untreated animals [32,33] within one minute and continuing 2–4 h post operatively, and showing similar responses to wound stimulation as sham-treated piglets [33]. Pre-operative lidocaine injection (scrotal and intra-testicular), also induced early wound hypoaesthesia, with reduced responses as compared with untreated piglets for up to 1 h following castration.

Table 6. Summary of studies assessing wound sensitivity after castration in neonatal piglets. (Arrows indicate statistical significance (p<0.05). NSD = No significant difference of treatment).

Authors	Piglets Age, Number	Castration Experimental Groups	Measurement Method	Significant Findings
Lomax et al. [33]	40; 3–5 days	Castration without anaesthesia (CAST); sham-castrated (SHAM); topical wound anesthetic (TA)(TA = 5% Lignocaine (L), 0.5% Bupivacaine (B) and adrenalin 1:2000 (adr) administered via wound instillation)	von Frey filaments (4 g and 300 g) and 18 G needle; testing immediately after, 1 min, and every 30 min up to 4 h; grading on NRS for involuntary motor response	Vs SHAM: ↑ sensitivity scores post castration CAST, 30 min – 4h; L ↑ 1 – 4 h. (TA NSD). Vs CAST ↓ sensitivity post-castration; L 30 – 90 min, TA 30 min- 4h.
Gottardo et al. [45]	196; 4 days	CAST; SHAM; NSAID (NSAID = meloxicam or ketoprofen or tolfenamic acid); TA (TA = 2% topical tetracaine (THCL) hydrochloride or 6% THCL administered pre-and post-operatively.	Pressure algometry with pressure ranging between 0.1–20 kg/cm^2; testing 300 min post-castration	Vs SHAM: ↑ sensitivity to pressure post castration a) immediately; CAST and TA, b) after 15 min TA; from 60–300mins all groups. Vs CAST ↓ sensitivity to pressure in NSAID-treated piglets immediately post castration, otherwise NSD.
Sheil et al. [32]	40; 3–7 days	CAST; TA (TA = L,B and adr, administered via wound instillation);	von Frey filament (300 g) and pin-prick; testing 1 min and 1, 2, 4, 8, 12, and 24 h post-castration; grading on NRS for involuntary nociceptive response based on [32]	Statistically significant difference between treated and CAST groups at 1 min and up to 2 h post-castration with ↓ sensitivity in TA vs CAST group.

As an alternative to von Frey filaments and needlestick stimulation, pressure algometry involves applying a force to a point and measuring the pressure at which a withdrawal response is elicited using a pressure algometer. Both A and C fibers mediate pain induced by pressure stimulation [108]. Acute pain in piglets following castration and the impact of local and topical anaesthesia (tetracaine) has also been assessed by pressure algometry [45]. Efficacy of pain relief was assessed prior to and during a 300 min period after castration by scrotal skin pressure sensitivity, amongst other methods. Increasing pressure was applied to a designated point on the skin of the scrotum adjacent to the incision site and the pressure point by which a physical or vocalisation response was elicited was recorded. Results were consistent with behavioural results in which reduced pain-related behaviours documented in the first 30 min following the procedure were more prominent in NSAID than topical tetracaine-treated piglets. While one study investigating wound sensitivity in calves found a good agreement between both Von Frey filament stimulation and pressure algometry [30], other comparative studies in piglets (M. Sheil, unpublished observations) found pressure algometers were relatively insensitive due to the soft nature of the scrotal tissues. The pressure device induced discernible indents or trauma to the soft tissues at the site without consistently eliciting a response. Janczak et al. [117] examined factors affecting mechanical (nociceptive) thresholds in piglets and the stability and repeatability of measures of mechanical (nociceptive) thresholds in piglets when using a hand-held algometer to examine potentially confounding factors. These investigators reported that mechanical (nociceptive) thresholds can be used both for testing the efficacy of anaesthetics and analgesics, and for assessing hyperalgesia in chronic pain states in research and clinical settings; however, identified that in piglets age and weight affected responses to pressure algometry, particularly in the first week of life.

Whilst the number of reports of quantitative nociceptive response testing in neonatal piglets post castration are limited, direct sensory testing using needlestick and von Frey stimulation with NRS grading of the nociceptive withdrawal reflex response, has thus to date proven consistent, repeatable, sensitive and specific to the pathophysiological process generating pain, and is concluded to provide the optimal method currently available for assessing post-operative hyperalgesia secondary to peripheral afferent nerve sensitisation following castration in neonatal piglets.

Quantitative sensory testing allows assessment of an animal's response to noxious stimuli, (nociception) as an indicator of the peripheral afferent nerve sensitisation that underlies the development of post-operative pain but does not necessarily indicate the more complex cortical perception of pain, i.e., the experience of pain in the absence of a direct stimulus. Combining the use of QST with the assessment of spontaneous pain-related behaviour is recommended when assessing pain mitigation strategies, such as to provide evidence of reduced experience of pain, as well as reduction in its primary underlying pathophysiological mechanism.

7. Other Measures of Pain

Several alternative methods to assess perioperative pain in piglets have also been described.

A piglet grimace scale (PGS) was recently proposed as an alternative method to assess castration and tail docking pain in piglets [71]. Similar methodologies have previously been developed and validated for a variety of livestock species, including sheep [118] and horses [119]. The piglet PGS was developed following analysis and comparison between still images of piglet faces captured at various stages after surgical castration and the concurrent presence/absence of behaviours indicative of piglet pain. Facial actions indicative of pain were considered to be (i) drawing back of the ears from a forward position; (ii) the presence of a bulge of skin on the snout in response to cheek tightening; and (iii) orbital tightening [71]. This initial study reported a strong correlation between PGS score and behavioural activity in animals in the first several hours after castration [71]. Some doubts about the robustness of this method to consistently detect pain in neonatal piglets currently exist though. In a follow-up study applying the PGS, there were not significant differences between sham-handled and castrated piglets, and a potential cofounder in the form of piglet body weight was identified, suggesting that facial grimacing may also indicate weakness or stress related to lower body weight rather than pain [65]. It was also documented that administration of buprenorphine significantly reduced facial grimace scores as compared with both sham-handled and untreated castrated piglets. As buprenorphine also reduced sleep and increased the activity state of both sham-handled and castrated piglets, this suggests the possibility that piglet activity state (as opposed to pain) may also impact facial grimace scores. The second issue relates to inter-user operability with one study [45] revealing that the PGS method was too unreliable for use in comparative evaluation of piglet pain. It failed to show consistent inter-observer reliability in scoring in 2 of the measures while the 3rd measure, orbital tightening, did not differentiate the positive and negative control. This is therefore considered to be a promising new development however further experience and validation are needed for use in in-field trials of piglet castration pain and analgesic efficacy.

Infra-red thermography (IRT) measurement of skin temperature has also been used as a non-invasive method to assess pain responses in piglets with conflicting results reported [29,40,42,46,52]. Animals in pain lose heat from the body's periphery, measurable by IRT, due to activation of the SNS causing vasoconstriction and redirection of blood flow to the internal organs [120]. Thus, piglets experiencing significant pain via surgical castration should display quantifiably lower skin temperatures than sham-castrated piglets or piglets treated with effective pain mitigation strategies. Consistent with this hypothesis, skin temperature dropped to a greater extent immediately following castration in untreated piglets as compared with sham-handled animals and those administered both lidocaine and meloxicam prior to castration [42]. In addition,, cranial temperatures in piglets castrated and tail-docked following nursing from meloxicam-treated sows were found to be significantly higher than temperatures recorded in piglets which had nursed from placebo-treated sows up to 60 h after castration [40]. However, there were no significant differences between groups in IRT values at other sites (ear or snout-tip). Furthermore, these results conflict with an earlier study that found ear temperatures were increased in untreated piglets compared to piglets treated with meloxicam (i.m.) and/or intra-testicular lidocaine prior to castration [29]. Skin temperature measured using IRT at the wound site did not differ significantly between groups. Similarly, a separate report examining the effect of NSAID treatment administered to the sow prior to husbandry procedures in piglets, found decreased skin temperatures in piglets of sows treated with NSAID compared with piglets from placebo-treated sows at 2 and 4 h post-procedure, with no difference between groups at 1 h, or from 7–24 h following the procedure [46]. These results conflicted with eye temperature recordings in the same cohort which were increased at 1 h in the NSAID versus the placebo group, but not significantly different between groups from 2 and 4 h or up to 30 h following procedures. These investigators also identified significant temperature differences between male and female piglets and a seasonal variation in skin and eye temperature recordings.

A confounder to IRT measurements in this setting is that body temperature is also affected by the post-surgical inflammatory response (i.e., not only the SNS response to pain). Lonardi et al. [52], examined rectal temperature and eye temperature in castrated versus sham-handled piglets and documented that there was an increase in both rectal and eye temperature over time following castration or sham-handling and, although some values were numerically higher in castrated animals, there were no significant differences between the two groups. The increase in eye temperature correlated with the increase in rectal temperature. It was noted that body temperature is reported to increase in response to anxiogenic or stress-inducing stimuli or injury (surgery and trauma) secondary to endogenous inflammatory activation [121–123]. Inflammatory mediators such as TNF-α and IL-1β are considered the main endogenous pyrogens [121]. These endogenous pyrogens are increased in piglets 3 h after castration or sham-handling [54]. It was considered that this may explain the tardive hyperthermia observed in the study in both castrated and handled piglets, although other external factors interfering with body temperature, such as exposure to heat lamps or time from milk intake could not be excluded.

NSAIDs have anti-inflammatory and associated direct anti-pyretic effects and thus may have a lowering effect on temperature that may confound the assessment of any effect due to mitigation of the SNS response to pain. This is further complicated by differences in doses and methods of administration employed in trials, as well as pharmacokinetic parameters of different NSAIDs [124] and a relative lack of detail regarding effective therapeutic range for anti-inflammatory effects in neonatal piglets. General anaesthetics may also have direct effects on body temperature and peripheral vasodilation. Local anaesthetics generally do not have significant direct anti-pyretic effects, however, are commonly administered with adrenalin, which may cause peripheral vasoconstriction and similarly confound skin temperature assessment. Yet another confounder is the relationship between the body's temperature and circadian rhythms with day/night cycles influencing body temperature results in meloxicam-treated and untreated castrated piglets [40].

In view of the lack of consistency in results to date, and multiple confounders, thermography does not currently appear to provide a reliable indicator of pain in neonatal piglets' post-castration, particularly following administration of local anaesthesia with adrenalin. Thermography may be more reliable for assessment of pain or pain mitigation in non-surgical settings.

8. Conclusions

Sensitive, specific, and well-validated methods of assessing pain provide the cornerstone for developing effective analgesic medications. Unfortunately, there are few such methods available for assessing pain associated with castration in neonatal piglets. This is confounded by the neonatal piglet's physiological response to restraint, handling, and surgical stress due to tissue trauma, and the seemingly subtle, and short-lived expression of pain in the post-operative period. An understanding of the strengths and weaknesses of currently available methods for pain assessment is critical to identifying and developing effective pain mitigation strategies in neonatal piglets. Employing methodologies that lack specificity or reliability risks underestimating both piglet pain, and the efficacy of pain-relieving medications, and creates welfare concerns associated with unproductive or counter-productive research. In the absence of a validated "gold standard" method of assessment, use of a number of different methods are required and, indeed, this is a foundational requirement for any treatment method seeking regulatory approval. This review has discussed the potential strengths and weaknesses of a range of currently available methods of pain assessment in the context of examining the efficacy of different anaesthetic and/or analgesic treatment options in field trial settings.

Based on the detailed review of different methods for assessing perioperative pain associated with surgical castration of piglets, this review concludes that:

- there is a relatively short-lived (0–3 h) physiological response to castration in neonatal piglets; however, physiological parameters lack specificity for pain, and may be significantly confounded by the surgical stress response as well as response to restraint and handling. They do not provide a reliable method for assessment of pain-alleviating efficacy of general or local anaesthetic interventions. Due to differences in mechanisms of action, these parameters may however provide a more reliable method to assess efficacy of NSAIDs where confounding variables are adequately controlled.
- pain control during piglet castration may be evidenced most consistently and reliably by a reduction in spontaneous nociceptive motor response during the procedure such as by NRS or VAS scoring of intensity of motor response.
- measurement of piglet vocal response to castration provides a second method for assessing pain control in piglets during the procedure. Variables including; peak dB, total vocal (dB/time) response, the frequency (Hz) of call with the highest intensity (dB (A)), and the rate of high-frequency calls (>1000 Hz), or stress calls as documented by STREMODO, appear to provide the most consistent or reliable parameters for detection of a significant reduction in vocal response.
- for both nociceptive motor and vocal response assessments care should be taken to ensure piglets are settled prior to commencing procedures and recordings to provide a consistent baseline. It is also suggested that measures be adopted to minimise confounding factors (such as piglet responses to restraint and/or extraneous environmental stimulation) by targeting/limiting the

assessment period as closely as possible to the time of acute pain generation. This is considered particularly important if studies are required in the field situation as opposed to acoustically separated environments.

- post-operative pain control is most effectively evidenced by documenting a combination of reduced peripheral afferent nerve sensitisation with an associated reduction in pain-related behaviour.
- peripheral nerve sensitization (hyperalgesia) is currently most reliably and consistently documented in neonatal piglets using nociceptive threshold testing with Von Frey and needlestick as opposed to pressure algometry.
- post-operative pain-related behaviour may be variable, subtle, and short-lived. Careful planning of variables and time points to be measured as well as power is required. The most consistently reported behavioural changes indicative of acute pain in piglets post castration include; "huddling up", "prostration", "hunching", "stiffness" (lying or of gait), "spasms", "tremors/trembling", "isolation", "tail-wagging" and "scratching"(as defined above), which are most evident in the first 30 min to 1 h following castration. The most consistently reported abnormalities of "pain-specific" behaviour at later timer points are tail-wagging and "scratching". It is noted however that both tail-wagging and scratching may indicate itch or irritation as opposed to pain, particularly if present in the absence of other indicators of pain (such as presence of hyperalgesia) at these later time points. They may be exacerbated in piglets that are also tail docked.
- other methods in development such as facial grimace scores and thermography, hold promise in many situations however do not currently appear to provide a reliable or consistent method of documenting pain or pain mitigation in neonatal piglets following castration.

It is hoped that this review may assist the future development of more standardized methods of assessing pain mitigation in neonatal piglets, assist investigators to optimise (reduce and refine) future analgesic efficacy trials in this field, and support the development and evaluation of innovative effective and practical approaches to improve piglet welfare where surgical castration is still utilised in commercial pig facilities worldwide.

Author Contributions: Conceptualization, M.S.; methodology, M.S.; data collection, M.S. and A.P.; writing—original draft preparation, A.P.; writing—review and editing, M.S. and A.P.; supervision, M.S.; project administration M.S.; funding acquisition, M.S. All authors have read and agreed to the published version of the manuscript.

Funding: Funding for this review was provided by Animal Ethics Pty Ltd. Yarra Glen, Victoria, NSW Australia.

Conflicts of Interest: Meredith Sheil is founding Director of Animal Ethics Pty Ltd., and inventor of Tri-Solfen®, topical anaesthetic formulation. Adam Polkinghorne was funded by Animal Ethics Pty Ltd. for this project.

References

1. Rault, J.-L.; Lay, D.C.; Marchant-Forde, J.N. Castration Induced Pain in Pigs and Other Livestock. *Appl. Anim. Behav. Sci.* **2011**, *135*, 214–225. [CrossRef]
2. Andresen, O. Boar tain related compounds: Androstenone/skatole/other substances. *Acta Vet. Scand.* **2006**, *48*, S5. [CrossRef]
3. Castrum Consortium. CASTRUM—Pig Castration for Traditional and Conventional Products: A Report on Methods and Their Impacts on Animal Welfare, Meat Quality and Sustainability of European Pork Production Systems. Final Report. 2016. Available online: https://publications.europa.eu/en/publication-detail/-/publication/5fe8db00-dbb8-11e6-ad7c-01aa75ed71a1 (accessed on 20 July 2020).
4. Prunier, A.; Bonneau, M.; Von Borell, E.H.; Cinotti, S.; Gunn, M.; Fredriksen, B.; Giersing, M.; Morton, D.B.; Tuyttens, F.A.M.; Velarde, A. A review of the welfare consequences of surgical castration in piglets and the evaluation of non-surgical methods. *Anim. Welf.* **2006**, *15*, 277–289.
5. Fredriksen, B.; Font, I.F.M.; Lundstrom, K.; Migdal, W.; Prunier, A.; Tuyttens, F.A.; Bonneau, M. Practice on castration of piglets in Europe. *Animal* **2009**, *3*, 1480–1487. [CrossRef] [PubMed]
6. von Borell, E.; Baumgartner, J.; Giersing, M.; Jaggin, N.; Prunier, A.; Tuyttens, F.A.; Edwards, S.A. Animal welfare implications of surgical castration and its alternatives in pigs. *Animal* **2009**, *3*, 1488–1496. [CrossRef]

7. Dunshea, F.R.; Colantoni, C.; Howard, K.; McCauley, I.; Jackson, P.; Long, K.A.; Lopaticki, S.; Nugent, E.A.; Simons, J.A.; Walker, J.; et al. Vaccination of boars with a GnRH vaccine (Improvac) eliminates boar taint and increases growth performance. *J. Anim. Sci.* **2001**, *79*, 2524–2535. [CrossRef] [PubMed]

8. Morgan, L.; Itin-Shwartz, B.; Koren, L.; Meyer, J.S.; Matas, D.; Younis, A.; Novak, S.; Weizmann, N.; Rapaic, O.; Abu Ahmad, W.; et al. Physiological and economic benefits of abandoning invasive surgical procedures and enhancing animal welfare in swine production. *Sci. Rep.* **2019**, *9*, 16093. [CrossRef] [PubMed]

9. Bonneau, M.; Weiler, U. Pros and Cons of Alternatives to Piglet Castration: Welfare, Boar Taint, and Other Meat Quality Traits. *Animals* **2019**, *9*, 884. [CrossRef]

10. Amaya, F.; Izumi, Y.; Matsuda, M.; Sasaki, M. Tissue injury and related mediators of pain exacerbation. *Curr. Neuropharmacol.* **2013**, *11*, 592–597. [CrossRef]

11. National Research Council (US) Committee on Recognition and Alleviation of Pain in Laboratory Animals. Chapter 2, Mechanisms of Pain. In *Recognition and Alleviation of Pain in Laboratory Animals*; National Academies Press: Washington, DC, USA, 2009.

12. Brennan, T.J.; Vandermeulen, E.P.; Gebhart, G.F. Characterization of a rat model of incisional pain. *Pain* **1996**, *64*, 493–501. [CrossRef]

13. Cook, S.P.; McCleskey, E.W. Cell damage excites nociceptors through release of cytosolic ATP. *Pain* **2002**, *95*, 41–47. [CrossRef]

14. Tsuda, M.; Koizumi, S.; Inoue, K. Role of endogenous ATP at the incision area in a rat model of postoperative pain. *Neuroreport* **2001**, *12*, 1701–1704. [CrossRef] [PubMed]

15. Coutaux, A.; Adam, F.; Willer, J.C.; Le Bars, D. Hyperalgesia and allodynia: Peripheral mechanisms. *Jt. Bone Spine* **2005**, *72*, 359–371. [CrossRef] [PubMed]

16. Cunha, T.M.; Verri, W.A., Jr.; Silva, J.S.; Poole, S.; Cunha, F.Q.; Ferreira, S.H. A cascade of cytokines mediates mechanical inflammatory hypernociception in mice. *Proc. Natl. Acad. Sci. USA* **2005**, *102*, 1755–1760. [CrossRef]

17. Jankowski, M.P.; Koerber, H.R. *Neurotrophic Factors and Nociceptor Sensitization in Translational Pain Research: From Mouse to Man*; Kruger, L., Light, A.R., Eds.; CRC Press/Taylor and Francis: Boca Raton, FL, USA, 2010; Chapter 2.

18. Mizumura, K.; Sugiura, T.; Katanosaka, K.; Banik, R.K.; Kozaki, Y. Excitation and sensitization of nociceptors by bradykinin: What do we know? *Exp. Brain Res.* **2009**, *196*, 53–65. [CrossRef]

19. Raja, S.N.; Meyer, R.A.; Campbell, J.N. Peripheral mechanisms of somatic pain. *Anesthesiology* **1988**, *68*, 571–590. [CrossRef]

20. American Veterinary Medical Association. Literature Review on the Welfare Implications of Swine Castration. Available online: https://www.avma.org/resources-tools/literature-reviews/welfare-implications-swine-castration (accessed on 8 June 2020).

21. De Briyne, N.; Berg, C.; Blaha, T.; Temple, D. Pig castration: Will the EU manage to ban pig castration by 2018? *Porc. Health Manag.* **2016**, *2*, 29. [CrossRef]

22. Sutherland, M.A.; Davis, B.L.; Brooks, T.A.; Coetzee, J.F. The physiological and behavioral response of pigs castrated with and without anesthesia or analgesia. *J. Anim. Sci.* **2012**, *90*, 2211–2221. [CrossRef]

23. Yun, J.; Ollila, A.; Valros, A.; Larenza-Menzies, P.; Heinonen, M.; Oliviero, C.; Peltoniemi, O. Behavioural alterations in piglets after surgical castration: Effects of analgesia and anaesthesia. *Res. Vet. Sci.* **2019**, *125*, 36–42. [CrossRef]

24. Burkemper, M.C.; Pairis-Garcia, M.D.; Moraes, L.E.; Park, R.M.; Moeller, S.J. Effects of Oral Meloxicam and Topical Lidocaine on Pain associated Behaviors of Piglets Undergoing Surgical Castration. *J. Appl. Anim. Welf. Sci.* **2019**, *23*, 209–218. [CrossRef]

25. Kluivers-Poodt, M.; Houx, B.B.; Robben, S.R.; Koop, G.; Lambooij, E.; Hellebrekers, L.J. Effects of a local anaesthetic and NSAID in castration of piglets, on the acute pain responses, growth and mortality. *Animal* **2012**, *6*, 1469–1475. [CrossRef] [PubMed]

26. Dzikamunhenga, R.S.; Anthony, R.; Coetzee, J.; Gould, S.; Johnson, A.; Karriker, L.; McKean, J.; Millman, S.T.; Niekamp, S.R.; O'Connor, A.M. Pain management in the neonatal piglet during routine management procedures. Part 1: A systematic review of randomized and non-randomized intervention studies. *Anim. Health Res. Rev.* **2014**, *15*, 14–38. [CrossRef] [PubMed]

27. Kluivers-Poodt, M.; Zonderland, J.J.; Verbraak, J.; Lambooij, E.; Hellebrekers, L.J. Pain behaviour after castration of piglets; effect of pain relief with lidocaine and/or meloxicam. *Animal* **2013**, *7*, 1158–1162. [CrossRef] [PubMed]

28. Keita, A.; Pagot, E.; Prunier, A.; Guidarini, C. Pre-emptive meloxicam for postoperative analgesia in piglets undergoing surgical castration. *Vet. Anaesth. Analg.* **2010**, *37*, 367–374. [CrossRef] [PubMed]

29. Hansson, M.; Lundeheim, N.; Nyman, G.; Johansson, G. Effect of local anaesthesia and/or analgesia on pain responses induced by piglet castration. *Acta Vet. Scand.* **2011**, *53*, 34. [CrossRef] [PubMed]

30. Lomax, S.; Windsor, P.A. Topical anesthesia mitigates the pain of castration in beef calves. *J. Anim. Sci.* **2013**, *91*, 4945–4952. [CrossRef]

31. Lomax, S.; Dickson, H.; Sheil, M.; Windsor, P.A. Topical anaesthesia alleviates short-term pain of castration and tail docking in lambs. *Aust. Vet. J.* **2010**, *88*, 67–74. [CrossRef]

32. Sheil, M.L.; Chambers, M.; Sharpe, B. Topical wound anaesthesia: Efficacy to mitigate piglet castration pain. *Aust. Vet. J.* **2020**, *98*, 256–263. [CrossRef]

33. Lomax, S.; Harris, C.; Windsor, P.A.; White, P.J. Topical anaesthesia reduces sensitivity of castration wounds in neonatal piglets. *PLoS ONE* **2017**, *12*, e0187988. [CrossRef]

34. Marsalek, P.; Svoboda, M.; Smutna, M.; Blahova, J.; Vecerek, V. Neopterin and biopterin as biomarkers of immune system activation associated with castration in piglets. *J. Anim. Sci.* **2011**, *89*, 1758–1762. [CrossRef]

35. Langhoff, R.; Zols, S.; Barz, A.; Palzer, A.; Ritzmann, M.; Heinritzi, K. Untersuchungen uber den Einsatz von Schmerzmitteln zur Reduktion kastrationsbedingter Schmerzen beim Saugferkel [Investigation about the use of analgesics for the reduction of castration-induced pain in suckling piglets]. *Berl. Munch. Tierarztl. Wochenschr.* **2009**, *122*, 325–332.

36. Reiner, G.; Schollasch, F.; Hillen, S.; Willems, H.; Piechotta, M.; Failing, K. Effects of meloxicam and flunixin on pain, stress and discomfort in male piglets during and after surgical castration. *Berl. Munch. Tierarztl Wochenschr.* **2012**, *125*, 305–314. [PubMed]

37. Zöls, S.; Ritzmann, M.; Heinritzi, K. Einfluss von Schmerzmitteln bei der Kastration männlicher Ferkel [Effect of analgesics on the castration of male piglets]. *Berl. Munch. Tierarztl. Wochenschr.* **2006**, *119*, 193–196. [PubMed]

38. Schwab, S.; Follrich, B.; Kurtev, V.; Keita, A. Ketoprofen-practical use and efficacy for post-surgical analgesia in piglet castration. *Tierartzliche Umsch.* **2012**, *67*, 207–213.

39. Wavreille, J.; Danard, M.; Servais, V.; Art, T.; Nicks, B.; Laitat, M. Effect of preoperative meloxicam or tolfenamic acid administration on stress and pain induced by surgical castration in piglets. *Journ. Rech. Porc.* **2012**, *44*, 275–276.

40. Bates, J.L.; Karriker, L.A.; Stock, M.L.; Pertzborn, K.M.; Baldwin, L.G.; Wulf, L.W.; Lee, C.J.; Wang, C.; Coetzee, J.F. Impact of transmammary-delivered meloxicam on biomarkers of pain and distress in piglets after castration and tail docking. *PLoS ONE* **2014**, *9*, e113678. [CrossRef]

41. Courboulay, V.; Hemonic, A.; Gadonna, M.; Prunier, A. Effect of local anesthesia or anti-inflammatory treatment on pain associated with piglet castration and on labour demand. *Journ. Rech. Porc. Fr.* **2010**, *42*, 27–34.

42. Bonastre, C.; Mitjana, O.; Tejedor, M.T.; Calavia, M.; Yuste, A.G.; Ubeda, J.L.; Falceto, M.V. Acute physiological responses to castration-related pain in piglets: The effect of two local anesthetics with or without meloxicam. *Animal* **2016**, *10*, 1474–1481. [CrossRef]

43. Nyborg, P.Y.; Sørig, A.; Lykkegaard, K.; Svendsen, O. Nociception after castration of juvenile pigs determined by quantitative estimation of c-Fos expressing neurons in the spinal cord dorsal horn. *Dan. Veterinærtidsskrift* **2000**, *83*, 16–17.

44. Svendsen, O. Castration of piglets under carbon dioxide (CO_2) anaesthesia. *J. Vet. Pharmacol. Ther.* **2006**, *29*, 54–55. [CrossRef]

45. Gottardo, F.; Scollo, A.; Contiero, B.; Ravagnani, A.; Tavella, G.; Bernardini, D.; De Benedictis, G.M.; Edwards, S.A. Pain alleviation during castration of piglets: A comparative study of different farm options. *J. Anim. Sci.* **2016**, *94*, 5077–5088. [CrossRef] [PubMed]

46. Coetzee, J.F. *Evaluation of the Transmammary Delivery of Firocoxib in Sows to Alleviate Pain Assocated with Piglet Castration, Teeth Clipping and Tail Docking; Report NPB #16-118*; National Pork Board, Des Moines, Iowa: Clive, IA, USA, 2019; Available online: https://www.pork.org/research/evaluation-transmammary-delivery-firocoxib-sows-alleviate-pain-associated-piglet-castration-teeth-clipping-tail-docking/ (accessed on 20 July 2020).

47. Sutherland, M.A.; Davis, B.L.; Brooks, T.A.; McGlone, J.J. Physiology and behavior of pigs before and after castration: Effects of two topical anesthetics. *Animal* **2010**, *4*, 2071–2079. [CrossRef] [PubMed]

48. Walker, B.; Jaggin, N.; Doherr, M.; Schatzmann, U. Inhalation anaesthesia for castration of newborn piglets: Experiences with isoflurane and isoflurane/NO. *J. Vet. Med. A Physiol. Pathol. Clin. Med.* **2004**, *51*, 150–154. [CrossRef] [PubMed]

49. Kohler, I.; Moens, Y.; Busato, A.; Blum, J.; Schatzmann, U. Inhalation anaesthesia for the castration of piglets: CO2 compared to halothane. *Zent. Vet. A* **1998**, *45*, 625–633. [CrossRef]

50. Marchant-Forde, J.N.; Lay, D.C., Jr.; McMunn, K.A.; Cheng, H.W.; Pajor, E.A.; Marchant-Forde, R.M. Postnatal piglet husbandry practices and well-being: The effects of alternative techniques delivered separately. *J. Anim. Sci.* **2009**, *87*, 1479–1492. [CrossRef]

51. Hay, M.; Vulin, A.; Génin, S.; Sales, P.; Prunier, A. Assessment of pain induced by castration in piglets: Behavioral and physiological responses over the subsequent 5 days. *Appl. Anim. Behav. Sci.* **2003**, *82*, 201–218. [CrossRef]

52. Lonardi, C.; Scollo, A.; Normando, S.; Brscic, M.; Gottardo, F. Can novel methods be useful for pain assessment of castrated piglets? *Animals* **2015**, *9*, 871–877. [CrossRef]

53. Haga, H.A.; Ranheim, B. Castration of piglets: The analgesic effects of intratesticular and intrafunicular lidocaine injection. *Vet Anaesth Analg.* **2005**, *32*, 1–9. [CrossRef]

54. Moya, S.L.; Boyle, L.A.; Lynch, P.B.; Arkins, S. Effect of surgical castration on the behavioral and acute phase responses of 5-day-old piglets. *Appl. Anim. Behav. Sci.* **2008**, *111*, 113–145.

55. Saller, A.M.; Werner, J.; Reiser, J.; Senf, S.; Deffner, P.; Abendschön, N.; Weiß, C.; Fischer, J.; Schorwerth, A.; Miller, R.; et al. Local anesthesia in piglets undergoing castration—A comparative study to investigate the analgesic effects of four local anesthetics on the basis of acute physiological responses and limb movements. *PLoS ONE* **2020**, *15*, e0236742. [CrossRef]

56. Zöls, S.; Ritzmann, M.; Heinritzi, K. Effect of a local anaesthesia in castration of piglets. *Tierärztliche Praxis Großtiere.* **2006**, *34*, 103–106. [CrossRef]

57. Prunier, A.; Mounier, M.; Hay, M. Effects of castration, tooth resection, or tail docking on plasma metabolites and stress hormones in young pigs. *J. Anim. Sci.* **2005**, *83*, 216–222. [CrossRef] [PubMed]

58. Carroll, J.A.; Berg, E.L.; Strauch, T.A.; Roberts, M.P.; Kattesh, H.G. Hormonal profiles, behavioral responses, and short-term growth performance after castration of pigs at three, six, nine or twelve days of age. *J. Anim. Sci.* **2006**, *84*, 1271–1278. [CrossRef] [PubMed]

59. Horn, T.; Marx, G.; von Borell, E. Behavior of piglets during castration with and without local anesthesia. *Dtsch. Tierarztl. Wochenschr.* **1999**, *106*, 271–274.

60. Leidig, M.S.; Hertrampf, B.; Failing, K.; Schumann, A.; Reiner, G. Pain and discomfort in male piglets during surgical castration with and without local anaesthesia as determined by vocalization and defence behavior. *Appl. Anim. Behav. Sci.* **2009**, *116*, 174–178. [CrossRef]

61. Wemelsfelder, F.; van Putten, G. *Behavior as a Possible Indicator for Pain in Piglets*; I.V.O.: Zeist, The Netherlands, 1985.

62. Weary, D.M.; Braithwaite, L.; Fraser, D. Vocal response to pain in piglets. *Appl. Anim. Behav. Sci.* **1998**, *56*, 161–172. [CrossRef]

63. Taylor, A.A.; Weary, D.M.; Lessard, M.; Braithwaite, L. Behavioural responses of piglets to castration: The effect of piglet age. *Appl. Anim. Behav. Sci.* **2001**, *73*, 35–43. [CrossRef]

64. Puppe, B.; Schon, P.C.; Tuchscherer, A.; Manteuffel, G. Castration-induced vocalisation in domestic piglets, *Sus scrofa*: Complex and specific alterations of the vocal quality. *Appl. Anim. Behav. Sci.* **2005**, *95*, 67–78. [CrossRef]

65. Viscardi, A.V.; Turner, P.V. Efficacy of buprenorphine for management of surgical castration pain in piglets. *BMC Vet. Res.* **2018**, *14*, 318. [CrossRef]

66. White, R.G.; DeShazer, J.A.; Tressler, C.J.; Borcher, G.M.; Davey, S.; Waninge, A.; Parkhurst, A.M.; Milanuk, M.J.; Clemens, E.T. Vocalization and physiological response of pigs during castration with or without a local anesthetic. *J. Anim. Sci.* **1995**, *73*, 381–386. [CrossRef]

67. Marx, G.; Horn, T.; Thielebein, J.; Knubel, B.; von Borell, E. Analysis of pain-related vocalization in young pigs. *J. Sound Vib.* **2003**, *266*, 687–698. [CrossRef]

68. Taylor, A.A.; Weary, D.M. Vocal responses of piglets to castration: Identifying procedural sources of pain. *Appl. Anim. Behav. Sci.* **2000**, *70*, 17–26. [CrossRef]

69. Viscardi, A.V.; Turner, P.V. Use of Meloxicam or Ketoprofen for Piglet Pain Control Following Surgical Castration. *Front. Vet. Sci.* **2018**, *5*, 299. [CrossRef] [PubMed]

70. McGlone, J.J.; Nicholson, R.I.; Hellman, J.M.; Herzog, D.N. The development of pain in young pigs associated with castration and attempts to prevent castration-induced behavioral changes. *J. Anim. Sci.* **1993**, *71*, 1441–1446. [CrossRef] [PubMed]

71. Viscardi, A.V.; Hunniford, M.; Lawlis, P.; Leach, M.; Turner, P.V. Development of a Piglet Grimace Scale to Evaluate Piglet Pain Using Facial Expressions Following Castration and Tail Docking: A Pilot Study. *Front. Vet. Sci.* **2017**, *4*, 51. [CrossRef]

72. McGlone, J.J.; Hellman, J.M. Local and general anesthetic effects on behavior and performance of two- and seven-week-old castrated and uncastrated piglets. *J. Anim. Sci.* **1988**, *66*, 3049–3058. [CrossRef]

73. Sutherland, M.A. Welfare implications of invasive piglet husbandry procedures, methods of alleviation and alternatives: A review. *N. Z. Vet. J.* **2015**, *63*, 52–57. [CrossRef]

74. O'Connor, A.; Anthony, R.; Bergamasco, L.; Coetzee, J.; Gould, S.; Johnson, A.K.; Karriker, L.A.; Marchant-Forde, J.N.; Martineau, G.S.; McKean, J.; et al. Pain management in the neonatal piglet during routine management procedures. Part 2: Grading the quality of evidence and the strength of recommendations. *Anim. Health Res. Rev.* **2014**, *15*, 39–62. [CrossRef]

75. Lykkegaard, K.; Lauritzen, B.; Tessem, L.; Weikop, P.; Svendsen, O. Local anaesthetics attenuates spinal nociception and HPA-axis activation during experimental laparotomy in pigs. *Res. Vet. Sci.* **2005**, *79*, 245–251. [CrossRef]

76. Kehlet, H. Surgical stress: The role of pain and analgesia. *Br. J. Anaesth.* **1989**, *63*, 189–195. [CrossRef]

77. Molina, P.E. Opiate modulation of hemodynamic, hormonal, and cytokine responses to hemorrhage. *Shock* **2001**, *15*, 471–478. [CrossRef] [PubMed]

78. Desborough, J.P. The stress response to trauma and surgery. *Br. J. Anaesth.* **2000**, *85*, 109–117. [CrossRef] [PubMed]

79. O'Connor, T.M.; O'Connell, J.; O'Brien, D.I.; Goode, T.; Bredin, C.P.; Shanahan, F. The role of substance P in inflammatory disease. *J. Cell. Physiol.* **2004**, *201*, 167–180. [CrossRef] [PubMed]

80. Aloisi, A.M.; Buonocore, M.; Merlo, L.; Galandra, C.; Sotgiu, A.; Bacchella, L.; Ungaretti, M.; Demartini, L.; Bonezzi, C. Chronic pain therapy and hypothalamic-pituitary-adrenal axis impairment. *Psychoneuroendocrinology* **2011**, *36*, 1032–1039. [CrossRef] [PubMed]

81. Zacharieva, S.; Borissova, A.M.; Andonova, K.; Stoeva, I.; Matrozov, P. Role of prostaglandin E2 (PGE2) on the corticotropin-releasing hormone (CRH)-induced ACTH release in healthy men. *Horm. Metab. Res.* **1992**, *24*, 336–338. [CrossRef] [PubMed]

82. Liu, J.P.; Clarke, I.J.; Funder, J.W.; Engler, D. Evidence that the central noradrenergic and adrenergic pathways activate the hypothalamic-pituitary-adrenal axis in the sheep. *Endocrinology* **1991**, *129*, 200–209. [CrossRef]

83. Labrie, F.; Giguere, V.; Proulx, L.; Lefevre, G. Interactions between CRF, epinephrine, vasopressin and glucocorticoids in the control of ACTH secretion. *J. Steroid Biochem.* **1984**, *20*, 153–160. [CrossRef]

84. Slominski, A.T.; Zmijewski, M.A.; Zbytek, B.; Tobin, D.J.; Theoharides, T.C.; Rivier, J. Key role of CRF in the Skin Stress Response System. *Endocrin. Rev.* **2013**, *34*, 827–884. [CrossRef]

85. Harris, J.A. Using c-fos as a neural marker of pain. *Brain Res. Bull.* **1998**, *45*, 1–8. [CrossRef]

86. Dobromylskyj, P.; Flecknell, B.D.; Lascelles, B.D.; Livingston, A.; Taylor, P.; Waterman-Pearson, A. Pain assessment. In *Pain Management in Animals*; Flecknell, P., Waterman-Pearson, A., Eds.; W.B. Saunders: London, UK, 2000; pp. 53–80.

87. Schon, P.C.; Puppe, B.; Manteuffel, G. Automated recording of stress vocalisations as a tool to documented impaired welfare in pigs. *Anim. Welf.* **2004**, *13*, 105–110.

88. Schon, P.C.; Puppe, B. Linear prediction coding analysis and self-organizing feature map as tools to classify stress calls of domestic pigs (*Sus scrofa*). *J. Accoust. Soc. Am.* **2001**, *110*, 1425–1431. [CrossRef] [PubMed]

89. Gutzwiller, A. Castration of piglets under local anaesthesia. *Agrarforschung* **2003**, *10*, 10–13.

90. Molony, V.; Kent, J.E. Assessment of acute pain in farm animals using behavioral and physiological measurements. *J. Anim. Sci.* **1997**, *75*, 266–272. [CrossRef] [PubMed]

91. Mellor, D.J.; Stafford, K.J. Acute castration and/or tailing distress and its alleviation in lambs. *N. Z. Vet. J.* **2000**, *48*, 33–43. [CrossRef]

92. Crowe, M.A. *AHW.120 Review of Literature on The Relief of Pain in Livestock Undergoing Husbandry Procedures*; Meat and Livestock Australia Limited: North Sydney, Australia, 2011.

93. Hatfield, L.A. Neonatal pain: What's age got to do with it? *Surg. Neurol. Int.* **2014**, *5*, S479–S489. [CrossRef]

94. Ison, S.H.; Clutton, R.E.; Di Giminiani, P.; Rutherford, K.M. A Review of Pain Assessment in Pigs. *Front. Vet. Sci.* **2016**, *3*, 108. [CrossRef]

95. Frank, S.G.; Lalonde, D.H. How acidic is the lidocaine we are injecting, and how much bicarbonate should we add? *Can. J. Plast. Surg.* **2012**, *20*, 71–73. [CrossRef]

96. Coetzee, J.F. A Review of Analgesic Compounds Used in Food Animals in the United States. *Vet. Clin. N. Am. Food Anim. Pract.* **2013**, *29*, 11–28. [CrossRef]

97. Anil, L.; Anil, S.S.; Deen, J. Pain detection and amelioration in animals on the farm: Issues and options. *J. Appl. Anim. Welf. Sci.* **2005**, *8*, 261–278. [CrossRef]

98. Wagner, B.; Cramer, C.; Park, R.; Moraes, L.; Viscardi, A.V.; Coetzee, H.; Pairis-Garcia, M.D. Evaluating behavior as a diagnostic tool to detect castration pain in piglets. *Animals* **2020**, *10*, 1202. [CrossRef]

99. Emde, R.N.; Harmon, R.J.; Metcalf, D.; Koenig, K.L.; Wagonfeld, S. Stress and neonatal sleep. *Psychosom. Med.* **1971**, *33*, 491–497. [CrossRef] [PubMed]

100. Gunnar, M.R.; Malone, S.; Vance, G.; Fisch, R.O. Coping with aversive stimulation in the neonatal period: Quiet sleep and plasma cortisol levels during recovery from circumcision. *Child Dev.* **1985**, *56*, 824–834. [CrossRef] [PubMed]

101. Gauthier, E.A.; Guzick, S.E.; Brummett, C.M.; Baghdoyan, H.A.; Lydic, R. Buprenorphine disrupts sleep and decreases adenosine concentrations in sleep-regulating brain regions of Sprague Dawley rat. *Anesthesiology* **2011**, *115*, 743–753. [CrossRef] [PubMed]

102. Tsagarakis, S.; Navarra, P.; Rees, L.H.; Besser, M.; Grossman, A.; Navara, P. Morphine directly modulates the release of stimulated corticotrophin-releasing factor-41 from rat hypothalamus in vitro. *Endocrinology* **1989**, *124*, 2330–2335. [CrossRef] [PubMed]

103. Lancel, M.; Faulhaber, J.; Schiffelholz, T.; Romeo, E.; Di Michele, F.; Holsboer, F.; Rupprecht, R. Allopregnanolone affects sleep in a benzodiazepine-like fashion. *J. Pharmacol. Exp. Ther.* **1997**, *282*, 1213–1218. [PubMed]

104. Luttinger, D.; Nemeroff, C.B.; Mason, G.A.; Frye, G.D. Enhancement of ethanol-induced sedation and hypothermia by centrally administered neurotensin, B-endorphin and bombesin. *Curr. Biol.* **2016**, *26*, 2446–2455. [CrossRef]

105. Nath, R.D.; Chow, E.S.; Wang, H.; Schwarz, E.M.; Sternberg, P.W. C. elegans stress-induced sleep emerges from the collection action of multiple neuropeptides. *Curr. Biol.* **2016**, *26*, 2446–2455. [CrossRef]

106. McGlone, J.J. A quantitative ethogram of aggressive and submissive behaviors in recently regrouped pigs. *J. Anim. Sci.* **1985**, *61*, 559–565. [CrossRef] [PubMed]

107. Sutherland, M.A.; Bryer, P.J.; Krebs, N.; McGlone, J.J. Tail docking in pigs: Acute physiological and behavioral responses. *Animals* **2008**, *2*, 292–297. [CrossRef] [PubMed]

108. Curatolo, M.; Petersen-Felix, S.; Arendt-Nielsen, L. Sensory assessment of regional analgesia in humans: A review of methods and applications. *Anesthesiology* **2000**, *93*, 1517–1530. [CrossRef] [PubMed]

109. Ren, K.; Dubner, R. Inflammatory Models of Pain and Hyperalgesia. *ILAR J.* **1999**, *40*, 111–118. [CrossRef]

110. Fitzgerald, M.; King, A.E.; Thompson, S.W.; Woolf, C.J. The postnatal development of the ventral root reflex in the rat; a comparative in vivo and in vitro study. *Neurosci. Lett.* **1987**, *78*, 41–45. [CrossRef]

111. Castel, D.; Sabbag, I.; Meilin, S. The effect of local/topical analgesics on incisional pain in a pig model. *J. Pain. Res.* **2017**, *10*, 2169–2175. [CrossRef] [PubMed]

112. Herskin, M.; Rasmussen, J. Pigs in pain-porcine behavioural responses towards mechanical nociceptive stimulation directed at the hind legs. *J. Pain* **2010**, *1*, 175–176. [CrossRef]

113. Lomax, S.; Sheil, M.; Windsor, P.A. Impact of topical anaesthesia on pain alleviation and wound healing in lambs after mulesing. *Aust. Vet. J.* **2008**, *86*, 159–168. [CrossRef]

114. Lomax, S.; Sheil, M.; Windsor, P.A. Duration of action of a topical anaesthetic formulation for pain management of mulesing in sheep. *Aust. Vet. J.* **2013**, *91*, 160–167. [CrossRef]

115. Espinoza, C.; Lomax, S.; Windsor, P. The effect of a topical anesthetic on the sensitivity of calf dehorning wounds. *J. Dairy Sci.* **2013**, *96*, 2894–2902. [CrossRef]

116. Windsor, P.A.; Lomax, S.; White, P. Progress in pain management to improve small ruminant farm welfare. *Small Ruminant Research.* **2016**, *142*, 55–57. [CrossRef]

117. Janczak, A.M.; Ranheim, B.; Fosse, T.K.; Hild, S.; Nordgreen, J.; Moe, R.O.; Zanella, A.J. Factors affecting mechanical (nociceptive) thresholds in piglets. *Vet. Anaesth. Analg.* **2012**, *39*, 628–635. [CrossRef]

118. Guesgen, M.J.; Beausoleil, N.J.; Leach, M.; Minot, E.O.; Stewart, M.; Stafford, K.J. Coding and quantification of a facial expression for pain in lambs. *Behav. Process.* **2016**, *132*, 49–56. [CrossRef]

119. Dalla Costa, E.; Minero, M.; Lebelt, D.; Stucke, D.; Canali, E.; Leach, M.C. Development of the Horse Grimace Scale (HGS) as a pain assessment tool in horses undergoing routine castration. *PLoS ONE* **2014**, *9*, e92281. [CrossRef]

120. McCafferty, D.J. The value of infrared thermography for research on mammals: Previos applications and future directions. *Mammal Rev.* **2007**, *37*, 207–233. [CrossRef]

121. Roth, J.; Rummel, C.; Barth, S.W.; Gerstberger, R.; Hubschle, T. Molecular aspects of fever and hyperthermia. *Immunol. Allergy Clin. N. Am.* **2009**, *29*, 229–245. [CrossRef] [PubMed]

122. Olivier, B.; Zethof, T.; Pattij, T.; van Boogaert, M.; van Oorschot, R.; Leahy, C.; Oosting, R.; Bouwknecht, A.; Veening, J.; van der Gugten, J.; et al. Stress-induced hyperthermia and anxiety: Pharmacological validation. *Eur. J. Pharmacol.* **2003**, *463*, 117–132. [CrossRef]

123. Takakazu, O.; Kae, O.; Testsuro, H. Mechanisms and mediators of pschological stress-indcued rise in core temperature. *Psychosom. Med.* **2001**, *64*, 476–486.

124. Nixon, E.; Almond, G.W.; Baynes, R.E.; Messenger, K.M. Comparative Plasma and Interstitial Fluid Pharmacokinetics of Meloxicam, Flunixin, and Ketoprofen in Neonatal Piglets. *Front. Vet. Sci.* **2020**, *7*, 82. [CrossRef] [PubMed]

MDPI

St. Alban-Anlage 66

4052 Basel

Switzerland

Tel. +41 61 683 77 34

Fax +41 61 302 89 18

www.mdpi.com

Animals Editorial Office

E-mail: animals@mdpi.com

www.mdpi.com/journal/animals